Praise for

I Want a Better Catastrophe

Urgent, sobering reading.

— KIRKUS REVIEWS, starred review

The most realistic yet least depressing
end-of-the-world-as-we-know-it guide out there.

— FOREWORD REVIEWS, starred review

The book is stunning. By delivering its devastating news in imaginative,
engaging, and sometimes even hilarious ways, it marks the emergence
of a new and genuinely exciting kind of realism.

— BRIAN ENO, musician and environmentalist

A profound meditation on how to live in a world on the brink
of collapse. Boyd moves gracefully beyond the usual talk of hope and
despair to provide a startling vision of a future shaped not only
by chaos, but also by compassionate care.

— JENNY OFFILL, author, *Weather* and *Dept. of Speculation*

A heartfelt and humorous take on how to show up at
"the end of the world as we know it."

— BRITT WRAY, PhD, Human and Planetary Health Fellow,
Stanford University and author, *Generation Dread*

I Want a Better Catastrophe is unlike anything else I've ever read
about climate change, and how to keep living through it. For a start it's
extremely funny. It is also angry, passionate, curious, honest, surprising,
and very well-researched. Beyond its signature gallows humor, it brings
a kind of deeply felt "gallows love" for the beauty and wonder of the
world, and how we must fight to defend it.

— NICK HUNT, co-director, Dark Mountain Project,
and author, *Outlandish*

Time is clearly short—but *I Want a Better Catastrophe* proves it's never too late for a good laugh, a good cry, and a good call to action!
— BILL MCKIBBEN, author, *The Flag, the Cross, and the Station Wagon*

Through expert interviews, compassionate analysis, and deliciously dark wit, Boyd beats a path through the messy emotional and psychological terrain we must travel in order to face the future.
— ONNESHA ROYCHOUDHURI, author, *The Marginalized Majority*

A rowdy, taboo-busting get-together of climate emergency thinkers.
— JOSEPHINE FERORELLI, co-founder, Conceivable Future

A must read for its wit, and for the insights it offers.
— PAUL D. MILLER, aka DJ Spooky

As a personal report from the frontlines of the climate struggle, Andrew Boyd's *I Want a Better Catastrophe* is heartfelt, incisive, and highly illuminating. As an anthology of interviews, it's a superb resource that gathers divergent positions into a coherent mosaic. If you want a no-bullshit guide to the best current thinking about climate change, this is the book to read. Courage, passion, hope, despair—it's all here.
— JAMEY HECHT, Psy.D., Ph.D., LMFT

Exactly the kind of paradoxical truth-telling we need
now more than ever!
— FRANCES MOORE LAPPE, author, *Diet for a Small Planet* and *Hope's Edge*

Boyd nails our predicament with enough bald truth and gallows humor to maybe, finally, penetrate our thick skulls.
— VICKI ROBIN, host, *What Could Possibly Go Right?*
and author, *Your Money or Your Life*

One of the most heart-breaking, heart-fun, and thought-full books I've read in a long time, filled with joy, grief, despair, and, yes, hope.
— MANDA SCOTT, host, *Accidental Gods*,
and author, *Boudica* dreaming series

The best contribution to the climate-anxiety discourse
that I've read so far!
— Jamie Henn, co-founder, 350.org and director, Fossil Free Media

Rather than simply despairing, the author finds the insights,
the perspectives, even the jokes, that make the journey endurable
and might lead us to a different path altogether.
— Richard Heinberg, author,
Power: Limits and Prospects for Human Survival

Andrew Boyd shows just how important humor can be
as a tool for grappling with deadly serious matters.
—Jonathan Matthew Smucker, author, *Hegemony How-To*

Read *I Want a Better Catastrophe* for insight, pathos, dark humor,
and an honest inquiry as to where we are on the journey.
— Chuck Collins, program director, Institute for Policy Studies,
and author, *Altar to an Erupting Sun*

Anyone can prophesize the future. Andrew Boyd's *I Want a
Better Catastrophe* can help you live in it.
— Nostradamus

I WANT A BETTER CATASTROPHE

NAVIGATING THE CLIMATE CRISIS
with Grief, Hope, and Gallows Humor

*An existential manual for tragic optimists,
can-do pessimists, and compassionate doomers*

ANDREW BOYD

new society
PUBLISHERS

Cover design by Diane McIntosh.
Cover images © iStock.
All figures © Andrew Boyd except where noted.

Photo credits: p. 31 Pauline P. Schneider, p. 41 Tim DeChristopher,
p. 85 Margaret Wheatley, p. 133 Gopal Dayaneni, p. 203 Jamey Hecht,
p. 253 Anjali Pinto, p. 285 Adrianna Ault.

Printed in Canada. First printing December 2022. Second printing April 2023.

Inquiries regarding requests to reprint all or part of *I Want a Better Catastrophe*
should be addressed to New Society Publishers at the address below.

To order directly from the publishers, order online at www.newsociety.com

Any other inquiries can be directed by mail to:
New Society Publishers
P.O. Box 189, Gabriola Island, BC V0R 1X0, Canada (250) 247-9737

LIBRARY AND ARCHIVES CANADA CATALOGUING IN PUBLICATION

Title: I want a better catastrophe : navigating the climate crisis with grief, hope,
and gallows humor : an existential manual for tragic optimists, can-do pessimists,
and compassionate doomers! / Andrew Boyd.

Names: Boyd, Andrew, 1962– author.

Description: Includes bibliographical references and index.

Identifiers: Canadiana (print) 20220424950 ⌐Canadiana (ebook) 20220425655 |
ISBN 9780865719835 (softcover) | ISBN 9781550927764 (PDF) | ISBN 9781771423724 (EPUB)

Subjects: LCSH: Climatic changes. | LCSH: Climate change mitigation. |
LCSH: Sustainable living. | LCSH: Human ecology.

Classification: LCC QC903 .B69 2023 | DDC 363.738/74b

Funded by the   Financé par le
Government   gouvernement
of Canada   du Canada

Canadä

New Society Publishers' mission is to publish books that contribute in fundamental
ways to building an ecologically sustainable and just society, and to do so with the least
possible impact on the environment, in a manner that models this vision.

I can't be a pessimist because I am alive.
To be a pessimist means that you have agreed
that human life is an academic matter.
So, I am forced to be an optimist. I am forced to believe
that we can survive, whatever we must survive.

— James Baldwin

Contents

Acknowledgments

They say "it takes a village." This book took a medium-sized city.

Without the many friends and colleagues who listened to me and encouraged me; who read and edited the book; who suffered alongside me (and suffered me), this book would simply not exist. It is a co-creation of us all. Huge thanks to every one of you.

Let me especially thank the eight remarkable people—Guy McPherson, Tim DeChristopher, Meg Wheatley, Gopal Dayaneni, Joanna Macy, Jamey Hecht, adrienne maree brown, and Robin Wall Kimmerer—who were willing to sit down with me and my voice recorder, share their ideas, and trust me to tell their stories. Beyond "interview subjects," they are the spirit guides of this work.

Mad respect to my patient and ruthless editors-for-hire, Nick Hunt and Virginia Vitzthum, who each came in at a critical moment in the book's development and got me to kill off some darlings, and see it all with fresh eyes. They made the book leaner and wiser. If you run across a passage that seems extraneous or tone deaf, definitely blame me, not them.

To the many, many reader-editors—Laura Dresser, Janice Fine, Dave Cash, Rae Abileah, Duncan Meisel, Britt Wray, Dave Mitchell, Josh Bolotsky, Logan Price, Onnesha Roychoudhuri, James Levy, Eden James, Leah Marie Fairbank, and others who spent a portion of their short lives reading (and in some cases, re-reading) the manuscript at various stages, I am forever grateful. I owe you all big time.

Every Don Quixote needs a Sancho Panza, every Bertie Wooster needs his Jeeves. Mine was Harry Cash, who not only resampled our jpegs and dotted our "i"'s in fine Chicago Manual of Style style, but like his literary forebears, was maybe the wiser of the pair.

Big thanks to the team at New Society, including Murray Reiss, Sue Custance, Diane McIntosh, Greg Green, John McKercher, and especially Rob West for choosing Marmite and taking a risk on this unusual book.

Deep appreciation to all the movers and thinkers who contributed ideas and stories to me directly, including: Richard Heinberg, John Jordan, Rachel Schragis, Gan Golan, Paul Kingsnorth, Joshua Kahn Russel, Josephine Ferorelli, Jeremy Sherman, Meg McIntyre, Carson and Benjamin Donnelly-Fine, Brett Fleishman, Bob Rivera, Charlotte Du Cann, Michael Barrish, Dan Kinch, Greg Schwedock, Alejandro Frid, and Paul Kiefer.

A shout out to the many beautiful places who offered me a place to write, including: the Blue Mountain Center, the Mesa Refuge, Lacawac Sanctuary, Photon Farm, the New York Writers Room, the Inn at Richmond, Mud Cafe on East 1st, the Suffolk Street Community Garden, and the good people of Gloversville, New York.

Big appreciation to Sarah Mason, Adrian Carpenter, Will Etundi, Andy Menconi, Alex Kelly, Matthew Hinders-Anderson, Josiah Werning, Joel Pett, Gaia Kile, Jason Stewart, Simone O'Donovan, Jake Ratner, Movement Generation, the Hemispheric Institute, Robert van Waarden, Raul de Lima, Twyla Frid Lotenberg, and others for their multifarious assistance.

A very special thank you to Katie Peyton Hofstadter for having my back, expanding my vision, and putting up with me and my dark musings.

Deep bow to my comrades at Beautiful Trouble and the Climate Clock for holding down their respective forts during the times I had to bury myself in the manuscript.

A signal thank you to Lois Canright and Chuck Collins for keeping vigil with me all along the way—including going on a "virtual hunger strike" (yes, they did that, and the pics looked real)—until I finished the book.

And finally, a supreme thanks to everyone, friend and stranger, striving against the odds for a better world.

IT'S THE END OF THE WORLD.
NOW WHAT?

New York City, Summer 2014

"There's no PLANet B," said the sign, roughly lettered in green magic marker on white poster board. The young woman holding it pumped her other fist in the air. And in unison with the throng around us, we all chanted "Change the system, not the climate! Change the system, not the climate!" our voices echoing off Manhattan's steel canyon walls.

We could barely move, the crush of bodies was so dense, stretching for dozens of blocks behind us and in front of us. The crowd was youth of color from the frontlines of hurricanes and pipeline blockades; scientists in lab coats wheeling along a huge "The evidence is in!" blackboard; workers from the building trades keen to build a green jobs future; children in polar bear costumes; clergy from a host of faiths rolling along a "We're all in the same boat" Noah's Ark.

It was September 2014, the largest climate march in history. We were 400,000-strong along Central Park West; from 50 countries and every walk of life; a living, pulsing testament to all that was at stake.

Could this finally be the game-changing moment? Could we galvanize public opinion and put world leaders on notice? Could this build an unstoppable momentum for a global treaty in Paris the following year that would prevent climate catastrophe and give us a shot at a more just and sustainable world?

Yes! sang out the great hope that filled the streets that day, lifting all of us up and onwards.

And yet, even in that jubilant throng, even at that moment of Peak Hope, part of me felt hopeless.

1

Pick up any book on climate change and you'll invariably see a statement that goes "if we don't do X by Y, we're Z." If the book was published five or ten years ago, you can actually check it against the historical record. When you do, you'll invariably see that we haven't done X, Y has long since past, and Z is some version of "fucked." And yet we keep writing these books. Hey look, I'm writing another one.

I'm no stranger to hopelessness. I'm no stranger to that burned-out, frustrated, the-odds-are-stacked-against-us-and-our-side-is-so-disorganized-and-self-defeating-that-well-fuck-it kind of hopelessness. I'd felt it many times before in my long career of causes. But then I'd eventually rally myself or the situation would shift. And my sails would fill again with hope and we'd (miraculously) win. "It always seems impossible until it's done," Nelson Mandela tells us. And social movements that I'd played my tiny part in had done many impossible things, including overthrowing apartheid, calming the nuclear arms race, electing a black president, and winning marriage equality.

But the hopelessness I feel in the face of our climate circumstances is different. "Hope," says theologian Jim Wallis, "is believing in spite of the evidence, then watching the evidence change." [1] But this time the evidence was already in. Yes, I could bring all the defy-the-odds exuberance I'd brought to previous fights, but how was that going to change the fundamental science of our climate situation? The earth chemistry was clear:

Protestors swarm the streets of New York in 2014 demanding Climate Justice.

Credit: Robert van Waarden, Survival Media Agency

decades of greed and negligence had already hard-wired a catastrophe into our climate future. If our task was to prevent catastrophe, it wasn't just impossible to the cynical and faint of heart, it was *in fact* impossible. We'd simply run out of time.

This kind of realization is hard. It doesn't just hit you once, and then OK you get it, and then you're done. It has to hit you again, and yet again, before you get it in your bones. And that summer, as we geared up for the big march, it was more than my little heart could handle. It also put me in a moral dilemma.

Activists are hope-mongers. It's our special power. Against all the odds we hope. You say we can't fight City Hall? You say we can't save the world? *Yes. We. Can.* What's less known is that when we fire up a crowd, we're not only getting our community and constituents to believe victory is possible, we're getting ourselves to believe it as well. We're both hope pushers, and self-dealing junkies, too.

But I couldn't get myself to believe this victory was possible.

The public mantra of the climate activist: *There's still time.* It's what we tell the crowds at our rallies and what we write in email after email to our lists. It's the basic assumption needed to motivate the public (not to mention ourselves) to action. "If enough of us choose to act, we can accomplish the impossible!" But did I believe it? *Really* believe it? No, not really.

So what to do?

I didn't want to bullshit myself, and I didn't want to bullshit anyone else. But I also didn't want to break ranks. A social movement is a fragile thing. If we lose hope, we lose everything. I didn't want to be the one who told everyone else that they—I mean, we—were living a lie. I had no trouble telling folks who believed that, say, Area 51 was full of little green men, that they were living a lie, but I didn't want to tell some of the most noble, selfless, and compassionate people in the world that we were all on a fool's errand.

You see, *I* needed other folks to have hope. I may have lost hope, but that didn't mean everyone else should. It's a bit twisted and unfair, I know, but I figured: if everyone else still had hope, there'd still be a movement, and with a movement, I could still hope.

Of course, it wasn't a simple case of hopeless me and hopeful everyone else. Many of even my most hard-core activist friends had come to their own doom-filled conclusions. That summer we would have made for an interesting psychological study—in, hypocrisy maybe, or schizo-phrenia. Extremely well-meaning schizophrenia.

In preparation for the big march, we were pulling together auditoriums full of people, and convincing them—as well as ourselves—that this was finally going to be the tipping-point moment that would win the day. And yet, after the coalition meeting, or the community-outreach session, or the long day painting banners, a few of us would invariably fall back to some bar and the great social lubricant would loosen our hearts and tongues.

Even if we pull this off, isn't it already too late?
Aren't we already doomed?
What's the point?

We'd fall into silence, some sad country singer wailing away on the juke-box. Until one of us might say: "Giving up on the future would be a dick move." And another: "I'm owning my doom." And a third: "Every morning, I claw my way out of despair and get back to work. Even if the chance of winning this fight is infinitesimal, I've got to try."

Here we were, about to bring nearly half a million people into the streets of New York. And yet there we were, in our cups, riven with grief, afraid all hope was already lost.

And finally the big day arrived, and the streets of New York filled with wave upon human wave of hope and purpose. In that great sea of humanity, alongside the Amazonian elders in blue and yellow-feathered headdresses, folks from the building trades in green hard hats, clergy in their blacks and whites, there, too, was the man who clawed his way out of despair every morning. And the woman who was owning her own doom. And the man who hadn't given up on the future yet because it would be a dick move to do so. And there was me.

And I realized that day: a social movement is just a crush of people carrying each other forward, each of us fighting our inner demons, the temporarily hopeless tag-teaming the temporarily hopeful, and trading back again, in a constant existential solidarity pact.

Up until now that has worked for me. I'm optimistic enough to have been an activist for 40 years. From the abolition of nuclear weapons in the 1980s to battling sweatshops in 1990s to the fight for affordable healthcare in the 2000s to a Green New Deal today, I've kept the faith in cause after cause, decade after decade. And yet, given what I've come to know—already—about our climate future (basically, that a catastrophe of unimaginable proportions is already baked into the atmosphere), it's become difficult for me to hope in the way I used to hope. As an activist, my M.O. has always been, keep your eyes on the prize, and eventually things will come around. For the first time, that has felt untenable. So what do I do?

I don't want to encourage premature despair with talk of doom, but I also don't want to bullshit myself (or anyone else) with false hopes. "Of course, we have to do everything we can!" says my natural enthusiasm and fighting spirit. "But why bother doing anything?" says my rational tabulation of the data. The more I've come to feel boxed in by these paradoxes and ironies, the less sure I am how to act in good faith. I need something more. Not so much a reason to hope, it seems, as a *way* to hope. Or, maybe, a way to not need to hope. Or at the very least, a soulful approach to our fate.

> *We're in for a catastrophe,* I said to myself that summer.
> *I can't give up hope,* I said back to myself that summer.
> *But it's too late.*
> *But it still matters.*
> *No, nothing matters.*
> *Yes, all the particulars matter.*
> *What particulars?*
> *How soon it comes, how bad it is, how we treat each other when it*
>   *arrives…*
> *Dude, it's a catastrophe!*
> *I know, I know, but there are better catastrophes and worse*
>   *catastrophes.*

Mass extinction and social collapse were a hard set of lemons to be making lemonade out of, but I was trying. If we were locked in for catastrophe, I reasoned, we needed to set our sights on the best catastrophe possible. Not an easy thing to "hope" for. Or devise policies around. What story do you tell? What strategy do you plot? How do you rally a social movement around a "better catastrophe"? And how do any of us—at the soul level—reconcile ourselves to such a dark aspiration? How do I get my heart to want, to actually *want*, a "better catastrophe"?

This book, which started to take shape that summer, is a search for answers to these questions.

Glasgow, Scotland, Winter 2021

Fast forward seven years to December 2021. It was in these darkest days of the calendar that I got the news my book about facing the climate catastrophe—the one you're now reading—would be published. The Omicron Covid wave was peaking, the anniversary of the January 6 insurrection was approaching, and the utter fiasco of the COP26 UN climate conference in Glasgow the month before (more on that below) was still fresh and bitter in my mind.

Across those intervening years, many of the "if we don't do X by Y, we're Z" formulations had grown even more dire. Yet humanity had also responded to our existential challenge with growing intensity and focus. The big march in New York had in fact helped trigger a wave of people-powered government action that led to a historic climate treaty in Paris. Water Protectors across the Great Plains united to block the Dakota Access Pipeline at Standing Rock. Greta Thunberg went on school strike in Sweden, sparking millions of youth to follow her in Fridays For Future (FFF) strikes for climate around the world. Extinction Rebellion (XR) took over the streets of London and then exploded across the globe. Solar and wind became the cheapest source of new energy, undercutting coal, oil, and even natural gas.[2] The Green New Deal burst on the scene, uniting labor and environmentalists around the promise of quickly transitioning the US economy off of fossil fuels and creating millions of green jobs in the process.

I put my shoulder to the wheel, joining XR and FFF in street demonstrations and direct actions; helping people turn their climate grief into action with the Climate Ribbon global story-sharing project (see page 61); swarming Congress with the Sunrise Movement for a day of mass sit-ins and lobbying for the Green New Deal; occupying the New York State Governor's mansion in Albany with NY Renews (which eventually led in 2019 to the passage of one of the nation's most aggressive clean energy laws); trying—and failing—to save the 991 trees in my beloved local East River Park from the chainsaws of developers;[3] and launching the Climate Clock, first in New York, then across the world.[4]

The Climate Clock counts down the time remaining to prevent global warming rising above 1.5°C (currently six and a half years and closing), while simultaneously tracking our progress on key solution pathways (renewable energy, Indigenous land sovereignty, and others). The Clock became a household name when we installed a monument-sized version

Credit: Raul de Lima

Glasgow during COP26, November 2021.

of it in New York's Union Square in 2020, and we spent most of that year and the next growing the idea into a global network of teams in 30 countries, putting clocks in the hands of climate champions from Greta Thunberg to Washington state Governor Jay Inslee to President Addo of Ghana, all synchronized around our critical timeline for action. Maybe it was already "too late," but here I was still doing my damnedest to get the world to #ActInTime.

But in spite of all these fierce, beautiful efforts by millions around the world, we were failing. The Clock makes it very clear "what we need to do by when," but the powers that be weren't doing it. The Standing Rock Sioux resisted courageously, yet crude was now flowing through the Dakota Access pipeline. The Green New Deal captured America's (and Joe Biden's) imagination, yet we were barely able to get a watered-down version of it past Joe Manchin and the Fossil Fuel Lobby in Congress.[5]

The signing of the Paris Accords in 2015 after 20 years of failure, was a truly historic accomplishment. Still, it was such a weak compromise that Bill McKibben likened it to Chamberlain's infamous appeasement of Hitler.[6] It's non-binding; aviation and shipping were not included; and nowhere in the document's 27 pages do the Accords even mention "fossil fuels"—the literal poison at the heart of the crisis.[7] Moreover, even if all signatories were to voluntarily follow through on their commitments (many are not), it would still leave us on track for a catastrophic 2.7°C of warming.[8]

Glasgow was supposed to correct this. Already delayed a year by Covid, it was heralded as humanity's "last chance" to save the world from runaway climate change.[9] Alongside thousands of activists, our team showed up in Glasgow hoping that the Climate Clock's unique combination of cultural cachet and scientific legitimacy could be one of the missing ingredients that would finally get the world on the timeline that science and justice demand.

What the world needed out of Glasgow was a hard commitment to basically end the burning of fossil fuels by 2030.[10] Instead, we got another mushy compromise between humanity's survival and the interests of the fossil fuel industry. What we needed was an agreement to cut global greenhouse gas emissions by 7.6% every year, or face the likelihood

Credit: Raul de Lima

Glasgow during COP26, November 2021.

of cascading environmental collapse as Earth's systems pass critical thresholds.[11] Instead, we got business as usual and a whole lotta "blah blah blah." Analyst George Monbiot called the resulting agreement a "suicide pact." Greta Thunberg had pronounced the conference a failure even before it began, and given that the largest single delegation—larger than that of any other country—came from the fossil fuel industry, you could understand why.[12] (And COP27 in Egypt the following year, in spite of a breakthrough on Loss and Damage financing for the most vulnerable nations, was just more of the same.)

Alongside this colossus of failure, we activists tried everything to land a better result—street demos, impassioned speeches, event-crashing, walkouts, and alliances with the few heads of state (Denmark, Palau, Costa Rica) who seemed to "get it." All the while, we piled up a mountain of single-use plastic waste from the endless Covid lateral-flow antigen tests we had to take every morning just to get into the necessary venues.

Did we accomplish anything? Yes, we probably caused the final agreement to be "less worse" than it otherwise would have been, including brief nods to "phasing down" coal and fossil fuel subsidies, and a little bit of money for nations of the Global South to "adapt" to the climate crisis.[13]

Yes, climate activists from all over the world came together, bonded, shared strategies, and drew strength and solidarity from each other. Yes, the term "fossil fuels" actually got mentioned in the text of the final agreement this time. But by the end of the two-week conference, there'd been no honest reckoning with the reality of our situation, and the Climate Clock was now two weeks closer to the end of the world.

It felt like humanity's last stand, a Battle of Stalingrad waged against ourselves, and we'd just lost. Again. Our leaders (Biden, Macron, and a few others) had barely put up a fight. Traitors (Australia, Saudi Arabia, and arguably everyone else) had broken ranks. Our MC-in-chief (Prime Minister Johnson) had fled the field of battle early, flying home in a private jet. Meanwhile, the global economy was still on track to cause catastrophic levels of warming, and false solutions like "net-zero by 2050" and other escape hatches for bad climate actors were the dogma of the land. Seven years ago, my hope was already hanging by a thread. Now what? At what point do you simply call it "game over"?

Over these seven years, I've been turning over my hope and hopelessness, my grief, my persistent desire to make a difference, and holding the contradictions up to the light. What began with that gut-check moment in the streets of New York became a full-scale quest. I dove into the literature, spoke with Americans from many walks of life, and convened a series of "hopelessness workshops"—loose forums where people could voice their climate anxieties and try on life philosophies that don't depend on good outcomes.

I also tracked down and interviewed eight "Remarkable Hopers and Doomers"—doomer scientist Guy McPherson, climate activist Tim DeChristopher, wisdom teacher Meg Wheatley, grassroots strategist Gopal Dayaneni, eco-philosopher Joanna Macy, collapse psychologist Jamey Hecht, "organizational healer" adrienne maree brown, and Indigenous botanist Robin Wall Kimmerer. What ways had they found, whether through ancient wisdom or science-informed belief or fierce necessity, to live on the cusp of catastrophe? And could the rest of us live there, too?

With their help, I try to unpack the thorny paradoxes at the heart of our predicament: how it's already too late, and also never too late; how, "we're all in this together—not!"; how we must "learn to be good ancestors." Along the way, I uncover a few home-grown philosophies—from tragic optimism to can-do pessimism to compassionate nihilism[14]—to help us on our journey.

The book can be read straight through, or approached like an existential toolbox you rummage about in for the idea or story you need at that moment. Or it can be traversed via the sprawling "Navigating Our Climate Predicament" flowchart included in the book (see chart following page 98). Follow the paths, and read the associated pieces. Or treat the flowchart as a huge apocalyptic advent calendar, with each logic-box opening to a gallows-humor bonbon. There's lots of ways to approach the book. And maybe something in it for everyone.

It is no longer a secret that human civilization as we know it is doomed. Deadly droughts, heat domes, once-in-a-millenium floods every few years, and climate refugees from Syria to Central America to California, have tipped off the world that climate breakdown is not just our future, but already upon us. What I offer here is a small head start on the grieving process—and some help answering the question, What is still worth doing?—as we face the consequences of the destruction we humans have leveled on our beautiful planet.

The voices collected here are dispatches from a world grappling with an impossible new reality, drawn from interviews, public conversations, and my own dawning realization that climate catastrophe is no longer preventable.

Unlike quite a few of the people profiled herein, my life is not a daily struggle to survive at the sharp end of climate chaos. As a city-dwelling, well-off, white guy with a questionable sense of humor and the spoils of empire at his fingertips, the lottery of birth has privileged me in ways I'm still uncovering. My story—like everyone's—both blinds me and allows me to see. Nonetheless, I strike out gamely in search of some big truths.

I leave to others the hard policy prescriptions and the nitty-gritty of the science. This book is an existential map of our predicament, a matrix of our ethical and emotional options, an inventory of how the climate crisis is making and remaking our inner lives. It's a guide to why—and why the hell not!—to act in the face of climate catastrophe. May it help you find your bearings—both spiritual and strategic—as you seek out a "better catastrophe."

IMPOSSIBLE NEWS

We've got to live, no matter how many skies have fallen.

— D. H. Lawrence

We are where?!

Let's begin by getting ourselves situated:

We are here

Credit: Ref. 1

...a sentient denizen of a blue-green dot orbiting a humdrum star in the outer suburbs of a galaxy of 200 billion other stars; a galaxy that is itself but one of 100 billion galaxies in a universe expanding at 68 kilometers per second per megaparsec (and if you thought megaparsecs were just a term *Star Trek* screenwriters came up with to sound scientific, you're not alone). But we are alone! God is dead, or so Nietzsche has told us, and as far as we know we're the only creature in the Universe able to fathom its unfathomable vastness.[1]

It's confusing. It's an absurdity. But hardly the only one: We fall in love. We know we are going to die. We do strange things like make art, and dream, and put each other in prison, and cut ourselves when we're depressed. And try to be kind when we can.

And maybe even do something to make our little home a better place for the next wave of existentially challenged humans that follow. Because we know how hard it can be. And because we believe things can get better. Because we've been told, and many of us still believe, in Progress. We (note: in this book, when I say "we" I usually mean the dominant culture on the planet) believe that History, in the very broadest sense, works like this:[2]

OK, it's not quite as smooth or linear as that; maybe more of a one-step-back-two-steps-forward kind of deal:

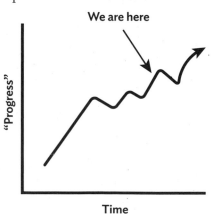

Or maybe—in the spirit of Martin Luther King's celebrated (and surprisingly line-graph-friendly) claim that, "the arc of the moral universe is long but it bends towards justice"—it's more like this:

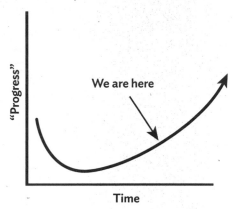

In any case, weirdly graced with an opposable thumb and the gift of gab, we rose from the primordial muck to burn fire, legislate laws, and paint paintings. With a fervid mix of violence, care, farming, and metallurgy, we upswung through Time, through Ages of Stone, Bronze, Iron, Steel, and Plastic; until we were bending rivers, splitting atoms, and replacing hearts. Along the way we drove the woolly mammoth and the passenger pigeon to extinction and committed unspeakable acts against one another; yet we also eradicated small pox, defeated Hitler, codified universal human rights, wrote symphonies, and went to the Moon. The past keeps handing us gifts and responsibilities, which we keep both honoring and squandering, and then passing on to the future. Our track record is decidedly mixed, but this continuity across Time, this Great Chain of Being, gives our smallish lives an extraordinary sense of meaning.

So, what if it turns out that, actually, we are here:

What if Progress is a lie and we're on the cusp of a historic-level catastrophe? This would be very unwelcome news indeed. Profoundly disorienting. Almost impossible to hear. How could this be? Well, here's how Richard Heinberg, scientist, author, Senior Fellow at the Post-Carbon Institute, and one of the foremost analysts of our energy future, explained it to me:

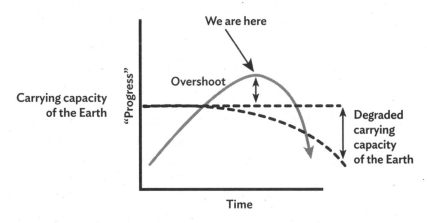

In short, our rate of consumption is overshooting our planet's sustainable sources of production. According to the Global Footprint Network, humanity is currently using the equivalent of 1.75 Earths to provide the resources we consume and the waste we generate.[3] Even worse, if everyone else in the world had US rates of consumption (and most countries are trying their damnedest to), we would need five Earths.[4]

Well, we don't have 1.75 Earths let alone five. (Newsflash: we only have one.) So we compensate by drawing down the future supply of non-renewable resources, which degrades the Earth's ecosystems and impoverishes future generations.

To avoid—or at least lessen—the catastrophe we're setting ourselves up for, we will have to make a double adjustment:

1. to the simple fact that the Earth has a limited carrying capacity; and
2. to the slightly less simple fact that, because of how irresponsibly we are currently managing it, that limited carrying capacity is itself being degraded.

We not only have a five-Earth appetite, we're not replenishing the one Earth we do have. Heinberg's conclusion, and that of many others who've analyzed the same trends: We need a planned "degrowth" of the world's richest economies. We need to partly "power down" our civilization. (Along with a fairly radical redistribution of wealth to soften the blow for us non-billionaires.)

Wait, God is dead, we're alone in the Universe, we know we're going to die, and now you're telling me that after finally clawing our way up the Ladder of Progress to some kind of half-decent (for some of us) civilization, it's all going to fall apart again?

We-e-ell, the exact mix of chaotic falling apart vs. thoughtful restructuring is partly up to us, but, in a word, yes.

Well, fuck you. Fuck you, and the data you rode in on.

Actually, it's even uglier and more complicated than all that because we are also here:

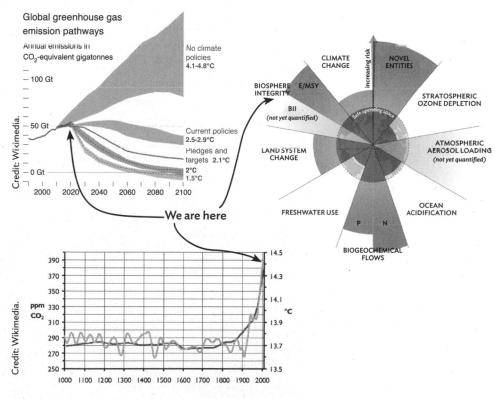

Andrew Boyd, Ref. 2

That's what climate change looks like all at once in cereal box recipe-sized type. There's a bunch of math in there. A lot of numbers and data and science and hockey-stick-shaped graphs and possibilities and probabilities and trend lines and scenarios and it can be hard to sift through it all. But here's the short version:

We're fucked.[5]

And here's a slightly less-short version, cribbed from the opening line of David Wallace-Wells' 2019 blockbuster, *Uninhabitable Earth*:

It's worse, way worse, than you think.

Indeed it is. And to explain how much worse, here's the slightly-longer-but-still-fairly-short version, from Nathaniel Rich's 2018 *New York Times Magazine* special feature, "Losing Earth":

> The world has warmed more than one degree Celsius since the Industrial Revolution. The Paris climate agreement—the non-binding, unenforceable and already unheeded treaty signed on Earth Day in 2016—hoped to restrict warming to two degrees. The odds of succeeding, according to a recent study based on current emissions trends, are one in 20. If by some miracle we are able to limit warming to two degrees, we will only have to negotiate the extinction of the world's tropical reefs, sea-level rise of several meters and the abandonment of the Persian Gulf. The climate scientist James Hansen has called two-degree warming "a prescription for long-term disaster." Long-term disaster is now the best-case scenario. Three-degree warming is a prescription for short-term disaster: forests in the Arctic and the loss of most coastal cities. Robert Watson, a former director of the United Nations Intergovernmental Panel on Climate Change, has argued that three-degree warming is the realistic minimum. Four degrees: Europe in permanent drought; vast areas of China, India and Bangladesh claimed by desert; Polynesia swallowed by the sea; the Colorado River thinned to a trickle; the American Southwest largely uninhabitable. The prospect of a five-degree warming has prompted some of the world's leading climate scientists to warn of the end of human civilization.[6]

Oh, and climate change is just one of several ecologically destructive macro-trends. Combine it with deforestation, overfishing, habitat encroachment, ocean acidification, biodiversity loss, and plastic pollution, and the very web of life is under assault. The UN estimates that over a million species, including key pollinators, are now at risk of extinction in the next few decades, with dire consequences for humanity.

Of course, ecological catastrophe is not the only catastrophe we're facing. If you're the apocalypse-inclined type, there's many to choose from. A COVID-19-like pandemic, but with a far more virulent pathogen, could rage beyond efforts to contain it and rip through the human population. Some perfect storm of surveillance ubiquity, terrorism, and collapse of democratic norms could usher in a 21st-century fascism that might be next-to-impossible to turn back the clock on. AI and automation could develop past some mysterious "singularity" threshold and flip the switch over to a society where humans are literally slaves to their robot-superiors and can never regain control. I'm not a conspiracy theorist, nor a doomer by inclination, I'm just obsessed with climate because it has a hard, relentless trajectory that I can't wish myself around.

Whether these other apocalypses happen or not remains to be seen, but the climate apocalypse is already happening, and happening fast. The facts are brutal.

To stay under 1.5°C, the global economy can burn only a fixed amount more carbon, which at time of publication, the Mercator Research Institute on Global Commons and Climate Change estimated at approximately 275 gigatons and dropping fast.[7] This is our carbon budget. The maximum amount of coal, oil, and gas the world can burn and still stay under the 1.5°C limit.

However, the world's already known carbon reserves are many times more than this, which means the lion's share of these reserves must stay in the ground. The rub, of course, is that, while the carbon may be in the ground, its monetary value is already above ground, circulating in the economy—and most significantly, already counted on the balance sheets of the world's fossil fuel companies and petrostates. If you can't burn it, it's worth nothing, which means Exxon, Shell, Kuwait, et al., are likely sitting on a $14 trillion "carbon bubble" write-down.[8] And we have little evidence of those players forgoing profit for the greater good.

And so the battle lines are drawn: A fossil fuel industry whose survival depends on burning every last drop, even if it means killing the planet versus an increasingly mobilized citizenry trying everything it can to "keep it in the ground"—everything from carbon taxes to fossil-fuel divestment to pipeline blockades to rapid deployment of renewables to a

Green New Deal. Meanwhile, the vast majority of people are still sitting on the sidelines.

If industry prevails and we stick with a business-as-usual scenario— and pretty much all data indicates this is what we are doing—we'll blow through our carbon budget in less than seven years.[9] Global temperature rise will continue on its trajectory of 3°C[10] increase or worse by the end of the century, and, in environmentalist Bill McKibben's words, "we won't be able to have civilizations like the ones we're used to."[11]

But—and if you follow the climate story at all, you know there's always a "but"—that's not the only scenario. If a perfect storm of technical progress, enlightened government and victorious social movements can force a tipping point of action and a WWII-level mobilizatin to move the global economy off of carbon, we could keep global temperature rise under, if not 1.5°C, then maybe 2°C. That would prevent the very worst outcomes.

But at this point even the very best outcomes are not very good. Which means in the decades ahead my hometown of New York City will be inundated by rising sea levels. It means humanity's one and only home will be plunged further into a sixth mass extinction; and droughts, fires, flooding and irreversible damage to critical habitats will give rise to hundreds of millions of climate refugees, widespread social chaos and, very possibly, some degree of civilizational collapse.

And so it seems we are here:

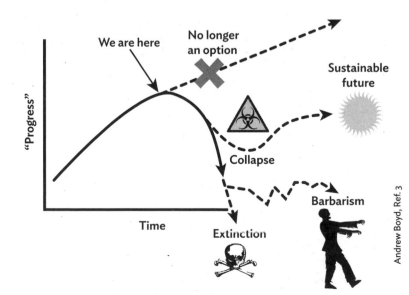

Andrew Boyd, Ref. 3

Catastrophic climate change is going to happen. Period. Whether we're locked-in for the fatal worst-case scenario or still have a shot at a better-but-still-pretty-terrible-case scenario depends on who you talk to. Much of the uncertainty lies in the difficulty of accurately modeling a complex system like our climate as it undergoes unprecedented conditions.[12] Feedback loops, tipping points, melting permafrost, and the possible release of ocean-floor methane as the oceans warm, are especially volatile unknowns. The other great unknown: How will humanity respond to the crisis?

In all scenarios, however, we are in for some kind of catastrophe, whether it is fatal to the human species, or *merely* to civilization as we've known it. It is no longer hyperbole to speak of "the end of the world"—it is already in motion.

I do not refute the logic of the science, but deep in my heart I'm still in denial. I know this because every time I stop for long enough to let myself squarely face these truths my body jolts anew into a sickening awareness. While hiking in Iccland in 2005, our guide paused to show us how far the glacier had receded in just the last decade. Then tells us to listen to the nearby ice-melt.

We all go quiet. For the longest time, all we can hear is the wind and the drip-drip-drip of the glacier. "That's the sound of global warming," he finally says with a frown. And then it hits me:

This is really happening. To the planet my feet are standing on right now. Panic, shame, and rage clutch at me, each in its own turn, until eventually giving way to a quiet grief. I wonder: *Is there any hope?* I find myself reaching—almost instinctively, almost in spite of myself— for hope. But is there really any hope? And what would it mean if there weren't?

Why it's so hard to hope these days.

It's always hard to hope. But there are times when it's particularly hard. Our time is one of those times. We know what we need to do to prevent climate catastrophe, and not only are we not doing it, but even if we were doing it, it still wouldn't be enough because we needed to start doing it 30 years ago. It's hard to hang your hope on preventing climate catastrophe—as I and so many of us have—when there is next-to-no basis for actually preventing it.

We can still hope in an arbitrary, disconnected-from-reality way (God will intervene; science will invent some magic process; people power will win the day), but not in a way that is consonant with an objective understanding of the situation. And by "objective" I don't mean a cynical, realpolitik, business-as-usual understanding of the situation; I mean the cold scientific facts of even the most optimistic scenarios. So, if you're hanging your hope on preventing catastrophe, you're hanging your hope on an illusion. This is a brutally sobering realization. It's taken me years to come to terms with it, and at some level I'm still failing to do so.

So, what then are our options? Well, we can hope against hope. Just hope anyway. Why not? It might be a better way to live. Or we can give up hope, and find a way to live without hope while still remaining true to ourselves—and decent to our neighbors—as things fall apart. Or we can hang our hope on something beyond our own time. Not on preventing catastrophe, but simply surviving it; on keeping a tenuous thread of civilization going across the next many generations; on some of us getting through the horror and wreckage of it, to some other mode of living profoundly different from anything we know. This is the "hope beyond hope" that British writer and ecologist Paul Kingsnorth (see page 228) glimpses through his grief. Hope is always hard, but this does seem a particularly hard kind of hope.

In our everyday personal lives, when we hope, we're generally hoping that things will get better in some recognizable way. That our children will be economically better off than us; that our second marriage will be happier than that tumultuous first one; that after knee surgery we will be able to play basketball again. In our collective lives, it's not so different. Think of the lifeline that at-risk queer youth grab hold of: "It gets better!" Or the rallying cry of the Latinx worker rights movement:

"¡Si, se puede!" (Yes, we can!) Historically, it's been these kinds of hopeful, positive attitudes that have powered social movements to resilience and victory. The only problem here, is if we're talking about the next 100+ years, then no, it won't be getting better; it'll definitely be getting worse. And if we're talking about preventing climate catastrophe, unfortunately, "No, we can't."

It's not about being optimistic or pessimistic; it's not about having a pro-social or an anti-social attitude. Because we already know we're in for a catastrophe and, while there's much we can do to slow it down or make it less terrible, there's nothing we can do to prevent it. So, hope—at least the kind of hope we're familiar with from our daily lives and our historic social struggles—is not going to get us through the next 100+ years. We need something different. Maybe we need the kind of hope you call upon in an emergency. The hope you call upon when you're going in for cancer surgery, or battling addiction, or taken hostage, or lost at sea. The hope we associate with grace, redemption, deliverance; the hope, not that things will get better, but that we will simply, doggedly make it through to the other side.

How do you build a social movement on this kind of hope? What vision do you offer people? What promises can you make—that you can actually keep?

But maybe even this kind of hope is not enough. Because most of us are not yet caught in a storm; and if we're addicted, it's to carbon; and if we've been taken hostage, it's by the fossil fuel industry, and our little individual Stockholm Syndromes are, you know, complicated. It's hard to know how to hope when your own everyday life helps inch the apocalypse forward. Because we're not just the carbon victims, we're also the carbon perps. We've heard how essential it was for those in the Nazi camps to hold on to hope. But their stories can only inspire us so far because we're also moonlighting as guards. We're half-prisoners half-guards—what strange hybrid hopes do we harbor? That History will never catch up with us? That, if it does, some tribunal will absolve us? That we will rise up against ourselves before we lead ourselves to the ovens?

The cognitive scientists and behavioral psychologists and social-marketing experts and professional climate communicators will share their findings: more dismaying facts don't help; individual eco-morality is a turn-off; personal legacy matters. They'll suggest small, concrete, positive steps communities can take; that we tell a story not of predicament and grief and collapse, but of hope and possibilities and a grand opportunity to redesign society. Yes, I see their logic and appreciate their realism (and I've been part of sit-ins in Congress and knocked on neighbors' doors in the snow to try to make it happen), and I half-believe it might just be the best we can do, yet I'm still left wondering what good any of it is, if, when you add it all up it's a whole lot of too little too late.

In 2014's *This Changes Everything*, Naomi Klein makes the case that "tackling climate chaos is not only necessary to protect our critical eco systems, but can lead to a better life for all."[13] And she's right. Our task is not only necessary but full of promise.

But. We're. Not. Doing. It. Not even close.

And we needed to do it many yesterdays ago.

How do you fix a predicament?

Once you've looked squarely at the climate science, it's hard not to feel like Sarah Connor in the dream sequence from *Terminator 2*. It's a beautiful day, kids and parents (as well as the younger Sarah) are in a park playing on the swings. With her fingers curled around the wire mesh backstop, she's shouting, trying to warn everyone (including herself) of imminent doom. But everything is just so lovely and normal, and who is this madwoman shouting her crazy thoughts? Then the flash and the fireball, and all is lost.

Many of us are living a quiet, less dramatic, slower-motion version of that dream. Today, for example, it's glorious in New York. I'm biking around the edge of Manhattan. I pause to lie on the grass by the water. Hudson River Park, built on landfill reclaimed from the river, is the picture of tranquility: runners, toddlers, kids playing Frisbee, all framed by blue sky and handsome buildings. I try to imagine the sea wall the city elders will try to build here, before they abandon lower Manhattan—and maybe much of the city—to the chaos of Frankenstorms and rising tides.

Like Cassandra, Sarah Connor had some secret, time-travelly knowledge from the future. Of course, people thought she was mad; of course, no one listened to her. But I'm no such Cassandra. Any news I might bring has already been brought. Thousands of scientific papers. Millions of newspaper column inches. Anyone who cares to pay attention already knows that we've broken Nature, and the world we know will soon end. This park, for one, is done for. This city I love, home to almost nine million, and one of humankind's most extraordinary creations, will, under pressure from extreme weather and system-wide collapse, be wrecked and made uninhabitable by the end of the century.[14] Possibly within my lifetime.[15]

Either way, I'm living in a ghost town. The ghosts aren't from the past, though, they're from the future. Why do I see them, and no one else? Actually, I think everyone sees them. We're all inverse-Cassandras: We can secretly see that the world is going to end, but no one wants to say it out loud because then it will really end, and we'll have to take responsibility for killing it, or at least failing to save it. Instead, we nod to our ghosts and carry on.

No one is happy about this. No one thinks this is the right way to live. But we don't know what to do. We don't know how to feel. And so, a part of us falls silent. We play tricks on our soul. We slide into a strange double life, "caught," says eco-philosopher Joanna Macy, (see page 187) "between a sense of impending apocalypse and the fear of acknowledging it." She elaborates:

> In this "caught" place our responses are blocked and confused. On one level we maintain a more or less up-beat capacity to carry on as usual…and all the while, underneath, there is this inchoate knowledge that our world could go at any moment. Unless we find ways of acknowledging and integrating that anguished awareness, we repress it; and with that repression we are drained of the energy we need for action.[16]

Millions of us are caught in this place. Who wants to dwell on such terrible news? Who wants to be the radioactive person at the party? Who wants to open themselves to all the grief waiting for us? So we don't. And this elaborate act of self-misdirection has many an accomplice:

- Governments that can't bring themselves to announce the news like the existential emergency it actually is.
- The niceties of everyday life which shun the grief-struck herald.
- The paid agents of Big Oil and Gas who have spent millions to cast shade on news we already wish weren't true, and are hellbent on convincing us that we consumers are the main problem.[17]
- The mystifications of late capitalism, that train us to act as if we weren't aware of our own contradictions even though we acutely are.
- And maybe, most of all, the structure of the Climate Crisis itself:
  - Its relentless trajectory: To bring it on, all we have to do is, um, nothing.
  - Its overwhelming complexity: To fix it, not only do we have to *do something*, but, as Naomi Klein has said, we pretty much have to "change everything" about how our economy and society operates.
  - Its asymmetries of power: Those of us most historically responsible for causing the problem—wealthy, mostly white, high-carbon-footprint folks in the Global North who burned dirty coal for two centuries to build up our economies—are, for now, the least impacted, while those who did the least to cause it—poor communities, people of color, and those who live in the Global South—are suffering the most.

- Its pernicious decoupling of causes and effects: Millions of years of evolutionary programming have hard-wired us to react to immediate threats with a fight or flight response, but here we are, stuck in a slow-motion catastrophe whose worst effects many of us alive now won't feel for decades, if ever.

As I lie in the grass along the Hudson River, the sky is blue, the sun is shining, the kids are playing. In spite of a rise in extreme weather events, catastrophe still seems far off and abstract. Our backs are objectively up against the wall, but it rarely feels that way. We sense the doom, but only vaguely; at some essential level, what's happening remains unbelievable. Our scientists and our most prescient leaders and even our own consciences are telling us that we must act, but our bodies don't feel the urgency. We're not even listening to what we ourselves have to say.

I came to political adulthood during the wave of protests against nuclear power and nuclear weapons that swarmed the Western world through much of the 80s. Back then it also felt like we were facing a doomsday scenario. The gravest threat was a catastrophic meltdown at a nuclear plant, or an escalating arms race triggering all-out nuclear Armageddon. Apocalypse loomed, but it was far from inevitable—in fact, just the opposite: it would require an accident or a war. There were things we could do—and did do—like phasing out nuclear power plants, de-escalating the arms race—to help prevent the worst outcomes. But it was always possible that nothing too apocalyptic would happen. Not so with the climate crisis. Given the carbon emissions path we are currently on, all we have to do is carry on just as we are, and climate apocalypse will be upon us. It doesn't require an extraordinary accident, just a slow business-as-usual march into the future.

So what can we do? At one level, there's a quite a number of things we actually *can* do, both individually: bike more, fly less, recycle, compost, go vegan, put solar on your roof; and collectively: divest your self/workplace/city/church/school from fossil fuels, make a community resiliency plan, block a pipeline, sue an oil company, pass a Green New Deal, sign an international agreement limiting carbon emissions and vote people into office who'll uphold it. Just to name a few. But at another level, we sense that even if all this were happening, it still wouldn't be enough to prevent catastrophe, as it's simply too late now. So, again, what can we do?

Well, we must keep doing all of that, but also take a deep breath, step back, and try to get honest with ourselves. Here's Paul Kingsnorth trying to get honest with himself:

> Is it possible to observe the unfolding human attack on nature with horror, be determined to do whatever you can to stop it, and at the same time know that much of it cannot be stopped, whatever you do? Is it possible to see the future as dark and darkening further; to reject false hope and desperate pseudo-optimism without collapsing into despair?[18]

He then gamely answers his own question: "It's going to have to be because that's where I'm at right now."[19] And that seems to be where more and more of us are at right now, myself included. And it's not an easy place to be. It's a heartbreaking mess of a place to be, actually. A whipsaw of competing emotions and commitments that are hard to hold all at once.

The hardest part for Type-A, can-do, eyes-on-the-prize activist me is letting go of the expectation that we *can* make this right. Because I really, really want to make this right. I want to fix this problem, and make things better. I want, as they say, to "save the world." And to let go of that possibility—to even think of letting go of it—is a blasphemy, a kind of death. But the thing is, climate catastrophe, and the broader civilizational crisis of which it is a part, is not a *problem* we can fix; it is rather something quite different: it's a *predicament* we must face.

This distinction was brought home to me by collapse theorist, sci-fi novelist, and, former Archdruid of North America, John Michael Greer.[20] "A controlled, creative transition to sustainability might have been possible," Greer argues, "if the promising beginnings of the 1970s had been followed up in the '80s and '90s." But our politicians and CEOs failed us mightily in those decades and since, and so, "our predicament in the early 21st century includes the very high likelihood of an uncontrolled transition to sustainability through...collapse." In other words, there was a window when there might have been a "solution" to the "problem" of climate change and the general unsustainability of our civilization but that window was squandered, leaving us in a predicament. Here's how Greer explains the difference:

> A problem calls for a solution. A predicament, by contrast, has no solution. Faced with a predicament, people come up with

responses. Those responses may succeed, they may fail, or they may fall somewhere in between, but none of them "solves" the predicament, in the sense that none of them makes it go away.[21]

Greer notes the striking irony of a civilization that believed it could turn every human predicament—poverty, sickness, even death—into a problem to be solved by technology, that is now "confronted with problems that, ignored too long, have turned into predicaments." But in this irony he finds a silver lining. Unsolvable predicaments—particularly the inevitability of our own deaths—are the stuff of the human condition, and our reckoning with them has arguably given rise to what is most beautiful and profound in human culture. Could the predicament of industrial civilization, Greer wonders, "push us in the same direction—toward a maturity of spirit our culture has shown little signs of displaying lately, toward a wiser and more creative response to the human condition?"[22] Could it? Theoretically, yes. Will it? Who knows. But, here, at least, was something worth hoping for.

And so, our story comes into focus: Decades ago, our politicians and engineers and other problem-solvers failed to build us a bridge to the future when they had the chance. Now, stranded here in the early 21st century, a chasm opening up in front of us, we must find a different path between the worlds. Caught in the teeth of an unsolvable predicament, facing a future "dark and darkening further," we must still walk forward. But how? Neither pessimism nor simple optimism is going to cut it for us. Something more robust is needed.

We live in a strange time that asks difficult things of us. On the cusp of a long descent, in the face of radical uncertainty, each of us must find an ethos for the path ahead. Do we just say "fuck it"? Can we find a way to hope in spite of it all? Must we settle for the stoic satisfaction of helping things get worse as slowly and humanely as possible? Facing a catastrophe we can mitigate but not prevent, and unable to know—ultimately— whether we are hospice workers or midwives, what is still worth doing?

The Hudson River laps at the edge of Manhattan. A seagull cries. I gather this last paradox and all the others into my bag. Behind me the early-afternoon sun splashes across the water. Beneath me the Earth turns imperceptibly on its axis towards twilight. Our unborn ghosts keep vigil here. They know things we cannot yet see. Not just about the future, but about me. I'm trying to listen.

MEETINGS WITH
REMARKABLE
HOPERS AND
DOOMERS

DR. GUY
MCPHERSON

*"If we're the last of our species,
let's act like the best of our species."*

I was on an odd quest. I had no ring to drop into the fires of Mt. Doom, but there was doom aplenty. There was no Minotaur to slay, but there was still a labyrinth of the self to descend down into. And maybe yet a world to save. In search of ways to face our collective fate, I went to speak with people whom, in a nod to Gurdjieff, I called, "Remarkable Hopers and Doomers."

My first meeting was with climate scientist Dr. Guy McPherson. As the foremost spokesperson for the view that abrupt climate change will result in near-term human extinction (NTHE), he was the Dark Prince of Climate.[1]

Many strongly dispute his claims, but as a scientist—and specifically, as a conservation biologist, and one of the first to identify and elaborate the dynamics of climate change—he had professional standing to bring such a dire prognosis to the debate.[2] His views are considered quite the outlier (climate journalist David Wallace-Wells describes him as on the "fringe," while climate scientist Michael E. Mann calls him a "doomist cult hero"), but that doesn't mean his views are necessarily wrong, and to write him off as a contrarian or crackpot seemed against the spirit of my two-part mission: (1) to face the truth of our climate predicament, no matter how dark or disorienting; and (2) to meet deeply informed people who've found a way to live with that truth. So, why not jump in at the very deep end?[3]

31

As with all these interviews, I'd wanted to do this one in person, and was hoping to catch Guy at his sustainable homestead outside of Tucson, but he's almost never in Arizona and so we spoke on Skype.[4] I asked for an overview of his work.

Guy: Okay, I'm Guy McPherson. I'm a Professor Emeritus of Conservation Biology at the University of Arizona. After leaving active service at the University of Arizona, I continued to conduct research on what I call the twin sides of the fossil fuel coin: energy decline, sometimes known as peak oil; and climate change. My ongoing research on climate change, which I've conducted since I was in graduate school in the mid-1980s, has led me to the conclusion that we're in the midst of abrupt climate change.

This phenomenon has precedence in planetary history, although not in the history of our species. This abrupt event, which is clearly underway now, is taking us to a point where it's difficult for me to see how the planet will be able to continue to sustain human life.

As a conservation biologist, I tend to look at things through the eyes of speciation, extinction, and habitat. Habitat is rapidly disappearing for humans already. We're already about 1°C above the baseline of where we were in 1750. When we get another few degrees above that, it's difficult for me to imagine we'll have habitat for our species in the future.[5]

Andrew: I know you've said all this to audiences hundreds of times before, but, right or wrong, hearing you say it now, hits me hard. I'm interested in how such a conclusion plays out at the psychological and philosophical level for you. When you were a teenager you didn't think that humanity would be extinct in a few short decades, but you think that now. That's a profound shift. Were there specific moments or stages of heartbreak and awakening? Why do you think you've landed where you have?

Guy: That's a great question, a serious question. There were many steps along the way. I began with an Al Gore-style understanding of climate change as a linear process—where, yes, there's still time to do something about it. That gradually shifted, until, ultimately, 14 years ago, I concluded we were in the midst of abrupt climate change.

As I indicated, I started studying climate change as a graduate student because it tied in with my research in field biology. Wherever I looked, I kept coming across a similar pattern: events, well-documented in the historic record, that showed how rising and falling precipitation levels contributed to the increase, decrease, and sometimes the complete removal of populations—and ultimately species—from the planet. This was some of the earliest work of its kind.

In 2002, I found myself, co-editing a book[6] that specifically focused on the role of precipitation in climate change, an aspect that had, up until that point, been largely ignored. There'd been a lot of research on temperature because that's the most obvious factor, but there hadn't been much research on precipitation. It was while co-editing this book, that I came to the conclusion that, given the degree of habitat we're expected to lose, we're headed for human extinction in the not so distant future.

Andrew: Not the kind of "Eureka!" moment most scientists dream of.

Guy: No.

Andrew: What was it like to come to this realization? Did you and your co-editor see things the same way, and how did your larger community react?

Guy: At the time, I was pretty quiet about it. It was a way-outside-the-box notion, and not one that others shared. Though, since then, it appears the science has pretty steadily and emphatically caught up with me. My co-editor, a former graduate student of mine with three children, looked at the same data, edited the same manuscripts right alongside me, but we reached different conclusions. That's fine. Scientists are noted for reaching different conclusions based on evaluating the same evidence. At some point, though, the evidence becomes simply overwhelming. Then any legitimate scientist will capitulate to the evidence. At the time, 2002, the evidence was hardly what one would call overwhelming. It was mostly through intuition that I reached the conclusions I did, but the data has since caught up with me.

The impact over the course of the next several years was devastating for me. Ultimately, it led to the loss of all of the relationships I held dear. I no longer interact with any of my former colleagues or co-workers at the University of Arizona, neither personally nor professionally. I lost essentially all of the personal relationships I had. It was a shattering experience. That was one of the reasons that I didn't talk about it much in the early days because when I talked about it, it didn't go well. Later, when the evidence just flat overwhelmed me, I couldn't ignore it anymore. I sort of had to talk about it. I'm at that point now.

Andrew: Wow. That's quite intense. I think a lot of people are going through *now* some version of what you went through *then*. A similar, and very human, process of being shattered, and feeling exiled. And whether or not they're coming to the same factual and scientific conclusions as you have, it's all very heavy, and there's a lot of grieving going on. People are really struggling to find their footing. Thank you for being so upfront about your story.

You said that for so long you held your tongue and then you felt like you *had* to talk about it. This is a corny word, but would you say that has become your *mission*? You speak and blog and write books about this.

Guy: I'm a teacher at heart and I can't help myself. When I was six years old I brought the Dick and Jane Reader home from school and I plopped in front of my four-and-a-half-year-old sister and I pointed at a picture and I said, "What's that?" She said, "It's a dog." I don't know this from memory. I just have been told this story dozens of times by my parents. Apparently, I said, "No, that's Sp, Sp, Sp, Spot." I was already frustrated and angry with her because she couldn't distinguish between any animal dog and the dog Spot, the hero of the whole series, right? So, even back then, I felt this obligation to present information to people. Whether they wanted it or not, I was going to shove it down their throats. I look back on that now, and it's just a horrible thing.

Now, we've got something a little different. Now, this is truly an existential moment. We're talking about human extinction. This is the most important phenomenon in the history of our species. Why do I do it now? Part of it is that I'm a teacher at heart and I can't help it. Part of it is, I'm not the captain of this ship, but it seems to me that all the other captains of the ship—the government, the media, et al.—are neglecting their duty to tell people the ship's going down.

I was a firefighter in my youth. The lookout, when spotting a fire, is supposed to call that in so that people can go put out the fire. In this case, the fire can't be put out, but I still think people have the right to know that the fire is upon them. That not only are our individual lives short, but our run as a species is about to be abruptly shortened as well. This has led me to this bizarre practice for the last several years of going around and reminding people that they too are going to die. You'd think that people would, in fact, recognize that and take it to heart and not view it as something that was utterly bizarre. But, I've been surprised by the response in many cases.

Andrew: It's more profound than we're ready to deal with. It's so close to the bone. A lot of us live in great comfort, historically speaking. We're not prepared to live—or even think—differently. Among the responses you get, what surprises you the most?

Guy: Built into our culture is the notion—and it goes without question by most people—that we're going to live a long time. That civilization will persist forever. That economic growth is just a given, and every generation will have more than the generation before. Which I guess shouldn't be too surprising. Since nobody comes on the television or the nightly news and says, "You know, we might be actually reaching limits to growth," or "There might be consequences for civilization."

The people we might have looked to for leadership failed us. Even those who are willing to go the next step and acknowledge that we're in the middle of abrupt climate change, suggest we can still survive on an outpost on Mars, or ride it out in a nuclear submarine. Then I realized that at some level, they're not radicalized enough. They're not willing to go that next series of steps and acknowledge that abrupt climate change is going to take the human species down, just like we're already taking down with us a couple hundred species every day to extinction.

Andrew: So, do you believe that human extinction is a 100% certainty? Is there a path—even an infinitesimally narrow one—by which we survive? Do we have any wiggle room?

Guy: I'm 100% certain that our species, like every other, will go extinct. I no longer put a date on when our species will go extinct. I just say, "in the short term." It will be "faster than expected" because that's the tag line for almost every newspaper article I see these days with respect to climate change: basically, fast things happening quite a bit faster than expected.

Andrew: If, as you say, we have no "expiration date," could you imagine us surviving for another 1,000 years? Or 10,000 years?

Guy: No and no.

Andrew: Oy.

Guy: Oy, indeed.

Andrew: The subtitle of your blog is "Our days are numbered. Passionately pursue a life of excellence." Completely independent of climate catastrophe, this seems like a fine life philosophy. Some of the leading thinkers out there argue that facing up to our current climate reality can also bring us closer to our truest humanity. Do you agree?

Guy: Absolutely. Even if we live what we would consider a long time in this culture, maybe 100 to 110 years, even if our species were to persist for a few thousand more years, our days are still numbered—as a species, and as individuals.

If our species persists for another couple thousand years—and, again, it's hard for me to imagine that it could, but if we do—our run will still be about one-seventh as long as a typical species of mammal on planet Earth. The Universe, as near as we can tell, has been around for about 13.8 billion years. Our species showed up about 200,000 years ago. Yet we have the hubris to think this whole thing is about us? It seems a little ridiculous to think the universe sat around for 13.8 billion years—well, 13.7 and change—just waiting for us to show up, just waiting for the good news that is humanity.

Of course, our lives are short. Of course, our days are numbered. They've been numbered from the day we were born. Of course, one should passionately pursue a life of excellence, no matter how long one has on the planet.

Andrew: So, climate change, an otherwise unmitigated disaster for humanity, has a silver lining?

Guy: I've been told by many people that my message actually reminds them that they've got to start living with urgency. We've had 60 years or so—since approximately World War II—of the system working in a certain way but there's no guarantee that it's going to work that way in the future. We tend to buy into the dominant narrative which says that we're all going to go to school, and grow up, and go to college, and work summer jobs, and then get a real job, start saving for retirement, work until we're 60 or 65, then retire, and travel the world. I'm reminding people that no matter how this turns out, it's not going to be like that anymore.

I hear from people who, after they hear me speak, after they see my writing, think: "Yes, now is the time. Now I'm going to start living with urgency. I'm not going to wait for another ten-years'-worth of tomorrows before I start doing what I actually love to do." So, yes, for me and for many other people, that's a silver lining.

Andrew: Yes, there's something freeing about facing the inevitability of your own death. But it's one thing to consider that at the level of your own mortality, of your own individual lifespan—and another thing entirely to do it at the level of our entire civilization or species. Isn't there some gray area in the science? Maybe we can adapt our way through the brutal changes to come? Maybe nature's vicious feedback loops will not accelerate as horribly as you expect? Or maybe virtuous feedback loops on the human adaptability front will give us a fighting chance?

Maybe, just maybe, there's a sliver of a sliver of a window of survival here. If so, we should do everything we can to abate emissions and build communities so that some of us can make it through, no? To focus on just checking things off my bucket list takes me away from that absolutely essential task of survival, possibly the most important task any human generation has ever had. That's the dilemma that I and many of my peers struggle with. Does that make sense to you?

Guy: Sure. Absolutely. I have never suggested inaction in response to the evidence, and the conclusion I have reached. I have remained an activist myself. I take to heart Tim Garrett's excellent work,[7] indicating that only the collapse of industrial civilization will prevent runaway climate change. And there are many, many other good reasons to terminate this set of living arrangements. According to a United Nations report from August 2010, we're driving to extinction 150 to 200 species every day. That's an outstanding reason to stop operating how we're operating right now.

I tend to take a Buddhist perspective on this issue, as I do with many issues. I think we should determine what is right and do what is right. We should take "right action"; then, inspired by another Buddhist principle, we should not be attached to the outcome. Because the outcome is ugly.

The runaway train that is industrial civilization has run away. It has not just left the station, but has gone over the trestle, off the tracks and now we're negotiating about who gets the best seat for the best view on the way down. From my perspective, that's no reason to act indecently. That's no reason to push grandma out the window so she dies first. That's no reason to gather up a bunch of overweight people so I have something soft to land on when we hit bottom. I still think that there are things we can and ought to do in light of this situation.

Even if the situation is truly hopeless, does that remove meaning from my life? Does that mean I should stop acting decently? No, of course not.

Andrew: So, losing hope doesn't mean you start acting immorally or indecently. Okay. And even if the metaphorical train has already gone over the metaphorical cliff, you're saying there's still much we can do. There are still a lot of ways we can care for the people around us: We hold each other's hands on the way down; everyone helps each other to brace for the fall, is that it?

Guy: Not only that but what better judge of our character than how we act in the face of impossible odds. Comforting the afflicted and afflicting the comfortable on the way down seem like perfectly reasonable strategies to me.

Andrew: Muckraking at the end of the world.

Guy: *Especially* at the end of the world. If it's the end, then how do we want to go? If we're the last members of our species, then shouldn't we act like the best members of our species? On the way down why should we not act with decency, with compassion, with passion towards what we view are right acts? Why not? What better time than now to demonstrate the best in all of us?

The essential question is: How do you choose to act in the face of the unspeakable? We are at the end of Time. We're all going to die. But that's not what this moment is all about. The focus must be on how we live.

Andrew: Is this approach rooted in what you call your "Buddhist perspective"?

Guy: It's interesting because I have reached my conclusions from a strictly left-brain scientific perspective. My perspective, when I speak and when I write, is still deeply rooted in rationalism, although, curiously, my message comes across as spiritual to a lot of people. In the early days, when I was doing public speaking, I found that insulting. I don't find it insulting anymore. I'm actually gratified that some people find my message a spiritual one. They think I'm a spiritual teacher of some kind. I'm definitely not a guru. I'm not interested in that sort of guru-like status, but if I am actually capable of changing people's behaviors and values, well that's what I've been shooting for for a long time. If that falls into the category of spiritual, then so be it. I can live with that.

Andrew: Thank you. This has been, I don't know, a melancholy, but very beautiful conversation. I really appreciate you being so open with me. I have one last question: Would you consider yourself hopeful?

Guy: I consider hope to be a bad thing. I believe that hope and fear, both four letter words, by the way, are the twin sides of the "I can't predict the future with certainty" coin. I will either hope for some outcome that I have no influence over or fear some outcome that I have no influence over. I view both of those notions as being not only unnecessary but unhelpful. I want to have agency. I want to take action. I don't want to hope for some good outcome that depends on a future that will never happen; nor do I want to fear that terrible future. I see hope as wishful thinking, and fear as its opposite: the nightmare of the future. I'm trying to avoid both those things in my life. Maybe that makes my message more difficult to swallow for many listeners and readers. In any event, my version of honesty is pointing out that hope is useless, or worse.

The interview[8] was not what I expected. I'd expected tons of numbers. I'd expected a somewhat rattled, insistent personality, out on a limb and defensive. Instead, far from the Dark Prince of Climate, far even from your standard-issue contrarian who likes the sound of his own voice, I found Guy to be accessible, humble, and touchingly grief-stricken; all in all, a profoundly decent person.

Here was a man bearing the worst kind of news. Nobody wanted to hear it. Even he wished it weren't true. It brought him no "I told you so" satisfaction to have been first out of the gate, and have the data slowly move in his direction. He'd suffered for it. He'd lost friends and colleagues; he'd been forced out of his beloved profession.

And yet he'd landed in a profoundly life-affirming place. He'd turned all this news of death and doom into an invitation to live more vividly, to see the preciousness of life with awakened eyes. Most curious of all: his relentless, level-headed, duty-bound pursuit of rationality had landed him in a place his readers and listeners experienced as spiritual. He was a strange prophet indeed.

Yet in spite of Guy's equanimity, I was pretty wrecked by our interview. Was it really too late? Were we all doomed—and so soon? Was it true that the best we could do was to try to live ethically while the ship sinks? In the face of extinction, where does the motivation for that kind of right living even come from?

Guy's a scientist. His habitat graphs may not lie, but they're not tracking all the data that matters. They don't include what humans are truly capable of when everything is on the line. Who could speak to that intangible quality without using it as an excuse to bullshit me into false hopes?

TIM
DECHRISTOPHER

*"It's too late—which means
there's more to fight for than ever."*

Climate activist and divinity student Tim DeChristopher seemed awake to our predicament in a way few others were. He'd been part of social movements most of his adult life, and understood their power, but he was no Pollyanna. He spoke publicly about how "it was already too late"—and yet he was still blocking pipelines and organizing in his community. I trusted him to tell me straight up whether he thought we still had a chance at avoiding or at least surviving the worst outcomes and, if so, what we could do to up those chances. We'd briefly crossed paths over the years, and now it was time for a deeper listen.

Tim came to national prominence in 2008 for dramatically foiling a contested auction of oil and gas leases in Utah's red rock country in the closing weeks of the George W. Bush administration. Tim showed up intending to physically disrupt the event, but as he walked through the door, an attendant asked if he was there to bid. "Why, yes, yes I am," he answered, and the attendant gave Tim a bidding paddle.

It took him a while to screw up his courage, but he started to bid. He saw how easily he could drive up the prices. Eventually, realizing he could be even more disruptive by winning bids, he began to outbid everyone else.[1] Tim won about a dozen lots in a row until the auctioneer realized something was wrong, suspended the proceedings, and had Tim arrested.

After Obama took office, his administration investigated the auction for "irregularities," and a federal judge cancelled the sales. Tim's inspired act of civil disobedience singlehandedly saved many precious acres of Utah wilderness from destruction.

His subsequent trial and eloquent defense of civil disobedience inspired millions. Articles were written, and a documentary film—*Bidder 70*—was made. In between his trial and his sentencing, Tim was invited to address the annual PowerShift gathering in Washington, DC, in 2011. In front of over 10,000 youth climate leaders, Tim laid down the following challenge:

> The truth that our movement has not been willing to talk about, is that it's probably too late for any amount of emissions reductions to prevent the collapse of our industrial civilization. This doesn't mean that all hope is lost. This doesn't mean there's nothing left to fight for. This means that there's more to fight for than ever. Because that collapse...could be an opportunity to build a better society on the ashes of this one; it could be an opportunity for mass reflection, and to decide that maybe greed and competition aren't the best values to be basing our civilization off of.
>
> So, we're at this key point. Where our challenge is not just to reduce emissions...but now we have this other challenge: of maintaining our humanity through whatever lies ahead. Maintaining our humanity through that period of ugliness and desperation that we are inevitably on track for.[2]

The crowd went wild. What does it mean when the apocalypse is announced at a gathering of the most hopeful people in the world and everyone cheers? The crowd wasn't cheering the news, of course. They were cheering the straight talk. They were cheering the call to step up our game to meet the moment. They were cheering Tim for having the courage to tell us what we already knew in the dread-filled chambers of our hearts.

I wanted to know how he'd come by that courage, and how, seeing our dark future so plainly, he could stay so fired-up with hope and conviction.

As I prepped for our interview,[3] I read a 2011 interview[4] he did with poet and essayist Terry Tempest Williams in *Orion Magazine*. Tim described a pivotal moment in his evolution as a climate activist when he attended a

March 2008 presentation of findings from the IPCC's Fourth Assessment Report by one of its lead authors, climate scientist Dr. Terry Root:

> I went up to her afterwards and said, "That graph that you showed, with the possible emission scenarios in the twenty-first century? It looked like the best case was that carbon peaked around 2030 and started coming back down." She said, "Yeah, that's right." And I said, "But didn't the report that you guys just put out say that if we didn't peak by 2015 and then start coming back down that we were pretty much all screwed, and we wouldn't even recognize the planet?" And she said, "Yeah, that's right." And I said: "So, what am I missing? It seems like you guys are saying there's no way we can make it." And she said, "You're not missing anything. There are things we could have done in the '80s, there are some things we could have done in the '90s—but it's probably too late to avoid any of the worst-case scenarios that we're talking about." And she literally put her hand on my shoulder and said, "I'm sorry my generation failed yours." That was shattering to me.

Tim entered a paralyzing period of despair. "I was," he said in the *Orion* interview, "grieving my own future, and grieving the futures of everyone I care about." Ultimately, however, that death sentence had a galvanizing effect on Tim. "Once I realized that there was no hope in any sort of normal future," he said, "I realized that I have absolutely nothing to lose by fighting back. Because it was all going to be lost anyway."

"But if it's true," Tempest Williams wanted to know, "that there is no hope—then what's the point?"

"Well, there's no hope in avoiding collapse," he answered, but "I have a lot of hope in my generation's ability to build a better world in the ashes of this one. And I have very little doubt that we'll have to."

Tim's story had carried him through grief and depression to an unusual mix of hope and hopelessness. I wanted details. I needed to know more precisely how this kind of hope worked for him.

From a distance, Tim had cut a larger-than-life figure. When I finally met him, he lived up to it. He was soulful, plainspoken, and quietly charismatic: one of those people you were immediately drawn to. At one point in the interview, he told me he had a very hard time lying to himself. This, I believe, was his secret power.

After serving his two years in prison, Tim had moved from Utah to Cambridge, Massachusetts, and then Providence, Rhode Island, enrolled at Harvard Divinity School, and co-founded the Climate Disobedience Center. Nearby, in Boston's West Roxbury neighborhood, Spectra Corporation was digging up the streets to put in a Liquified Natural Gas (LNG) pipeline, much to the dismay of the local community and most local elected officials. As we walked from Penn Station to the High Line park on Manhattan's west side, Tim began telling me about his involvement in the effort to stop that pipeline—his first direct action arrest since the expiration of the three-year probation that had followed his release from jail.

Tim: As this campaign of civil disobedience against the pipeline was going on, there was an article in Reuters about a heatwave expected in Pakistan. The previous year the heatwave there had been so intense that people were dying faster than they could be properly buried. So this year, they had dug anticipatory mass graves in preparation for the heatwave. One of the gravediggers was quoted in the paper saying, "Thanks to the grace of god we are digging the graves now." You know, being a climate activist is heartbreaking work, but this broke my heart in a whole new way. It seemed to me that entering the age of anticipatory mass graves was some kind of dark milestone in the progression of the climate crisis that we had to recognize and grieve.

I also saw that the picture of the long trench they were digging in Pakistan for a mass grave looked a lot like the trench they were digging in Boston for this new pipeline, and it struck me that they were linked in some way. If climate change were already so far along that we were in the age of anticipatory mass graves, then every new fossil fuel project like this pipeline would mean more mass graves somewhere, and those impacts were likely to be felt in the most vulnerable places like Pakistan. So when we laid our bodies in the Boston pipeline trench in the same way the bodies had been laid in the mass graves in Pakistan, we were both blocking that pipeline and mourning the loss of the current and future victims of fossil fuels.

Andrew: Ooof. Indeed. And knowing that this kind of thing will only get worse, how do you stay hopeful?

Tim: Part of it is how we define hope. The point is to get beyond optimism rather than beyond hope.

Andrew: How do you see the difference?

Tim: I think optimism is the expectation that things are going to be OK. That we're going to get a good outcome. Hope is much more about meaning; hope is the will to hold on to our values in the face of difficulty. Optimism is one kind of hope, a rather flimsy sort of hope. What we need now is a more resilient kind of hope, one not based on an expectation that things will be OK.

Andrew: Can you really base hope on the certainty—or near certainty—that things will not be OK?

Tim: I don't think our hope is based on that certainty. I think our hope is despite that certainty. I got some insight on this from a hospice chaplain in Boston who said that it's never too late to redefine our hope. This was the underlying motto for a lot of her work in hospice, where it's clear that the person you're caring for is going to die, that someone's loved one is going to die. You need to find a reason for continuing to care for someone when it's not based on an expectation that they're going to be OK, that they're going to get better. Of course, we all do that with the people we love. We continue to wipe a dying person's brow not expecting that they're going to get any better because of it, but because that's how we express our love for that person. That's how we show that we care.

I think there's a way for that to be translated into the broader challenge of facing the mortality of our industrial civilization. Our actions to fight the fossil fuel industry, our actions to build a more sustainable world, and work for justice, cannot necessarily be based on an expectation that things will work out. It's rather an expression of how much we love the people around us, of how much we love the world around us, of how much we love the people who will come after us.

Andrew: Even if we don't know those people, even if they're not born yet. And isn't it also how much we owe the people who came before us?

Tim: Yeah, I think it's particularly an expression of gratitude for those who came before us and for the legacy that they gave us—physical, social, and cultural. It's also related to our individual mortality and the mortality of our civilization. Which was really clarified for me by reading and studying *The Road*.

Andrew: Cormac McCarthy. Brutal.

Tim: That book—and the discussion that followed—were the most revealing moment so far at divinity school.

Andrew: How so?

Tim: It came at a time when I was very much asking myself these kinds of questions. I remember a few months earlier I'd been out in Oregon and was talking a lot about this with Sara Van Gelder, the editor of *Yes Magazine* [the "solutions journalism" quarterly founded in 1996]. We were grappling with the reality that it's too late to stop climate change. She said, "Well, if we tell people it's too late to prevent collapse, then what's going to prevent them from just abandoning their values, and letting go of the struggle for justice, and just living as hedonistically as possible with whatever time we have left?" And I didn't really have a very good answer for her. And I thought: "We need to have an answer for that or we're going to become increasingly irrelevant to people who are paying attention." And that became a driving force for a lot of my work for a long time.

Andrew: Particularly during the divinity school period?

Tim: Yep. So, we read *The Road*—an incredibly bleak story in an incredibly bleak world with really no basis for optimism whatsoever. No expectation that things are ever going to get better. One of the central questions in that book is whether or not it's ethical to keep striving, to keep walking this road, to keep holding on to their hope, and in particular, to keep raising their child in such a bleak and seemingly hopeless world. It's framed as a debate between the father and the now dead mother who took her own life because she felt like it wasn't ethical to continue.

The last page or so is a lyrical passage about brook trout and about the vermiculite, worm-eaten patterns on the back of brook trout which map out something that has always been coming. It can never be made right again. It can never be put back. In other words: Death is written on our bodies from the moment we are born. We were always going to die. In a world like *The Road*, it's really obvious that death is waiting for them. In that bleak of a world, it's clear that's the outcome.

So, it invites the question: Is it ethical to keep striving when we're just going to die? Of course the reality is that this has always been the case. It might be easier to ignore that reality in our world of comfort and excess. But if our system of meaning doesn't make sense in a world like *The Road*, then it's really never made sense. Our system of meaning must be based on the here and now; the meaning can't hang on some future outcome. Our actions must be worth the effort here and now, period.

The Road presents that meaning via gratitude, a theme woven throughout the book. In the sharing between the father and the son, who share everything they have. And in the way everything they come across is basically a gift. Because the other stark reality of their world is that they can't produce anything themselves. Nothing will grow in that world. There's really no one left that they can trade with or have any reciprocity with. All they do is find stuff that has been left behind by those who've come before them. They can't even express a thank you.

Andrew: At least a "thank you" that can be heard.

Tim: Right. All they can do is just accept the gift and receive it with gratitude. And part of living out that gratitude is continuing to walk that road, continuing to take care of one another. In our world of relative comfort and convenience, it's easy to convince ourselves that we have created everything for ourselves, that we've earned all this. We've lost sight of the fact that this life is a gift, that all that sustains us is a gift. We've lost touch with that sense of gratitude, the same way we've lost touch with our mortality. The book really connects the two, intertwines them into one.

Andrew: Part of our challenge—at least for those of us with some privilege—is to be living now in this world of comfort and convenience, as you say, but all the while knowing that this other world—harsh and full of death—is coming. This next world that's coming at us may not turn out quite as bleak as *The Road, inshallah,* but it's taking us in that direction. Now, for whatever reason, you, Tim, have chosen to straddle those worlds, live in both of them, about as fully and consciously as anyone I've met. I know it didn't come easy. Is everybody really up for that? What kind of person is willing to go through all that by choice, before calamity forces it upon us?

Tim: That's a question a lot of people are struggling to answer. I actually haven't found much of a connection between people's knowledge about climate change and their willingness to grapple with the despairing reality of it. However, there is often a recognizable trait in the people who are willing to sit with suffering. It's hard to identify, but I know it when I see it.

Andrew: In Terry Tempest Williams' interview with you in *Orion* she referenced a Syrian myth of going down into the Underworld, going through our own Hell, and we emerge with "death eyes." "Eyes turned inward," I think she called them, that allow us to look at suffering without turning away. Like that?

Tim: Yeah, it is like the death eyes.

Andrew: Hemingway talked about "being strong in the broken places." Maybe that's similar to the quality you're talking about. I lost my brother young. That was an untimely and incredibly wrenching death. Then my father, and most recently my mother. And I've had my share of romantic heartbreaks. By welcoming in that grief and sorrow, rather than shying away from it, I grew up. All that loss worked mud out of my soul, burned away some of my ego and bullshit and self-preoccupation. Made me more sinewy or "resilient." I think that kind of approach has made me a better person, and maybe also better able to face and handle the grief in the larger world.

Tim: I see that in a lot of folks that are willing to go to those despairing places. I don't think it's absolutely necessary. I actually haven't had that much personal loss in my life. The worst of it happened when I was building the motivation to take the Bidder 70 action back in 2008. In the months leading up to that, my girlfriend, who I'd been living with for a couple years, left me. Then my mentor died unexpectedly about a month before the auction. He was hiking in Zion on Thanksgiving and stumbled off the trail and died.

He'd been a big part of helping me process my despair when I was trying to come to terms with what it means to be too late to prevent climate catastrophe. The deepest, darkest period of despair about climate lasted about six or eight months, right after Dr. Terry Root told me how bad it was, backstage, after her presentation, in 2008. Since then, I feel like I've cycled back through it with a tightening spiral.

Andrew: Grief does seem to have a spiral structure, doesn't it?

Tim: Right. The next cycle through was shorter, and then shorter, until it just became inherent in me. Until I just carried that despair around with me. That time period was generative for me. It was necessary. I think it was liberating in really important ways. But I don't think everybody needs to go through it the way I did, or to that degree, or for that long.

Andrew: What strikes me here is that you're not trying to move through the despair, as if it were something you could just "get over" and leave behind. It's rather something that you have integrated, that you "carry around" as you say. Your despair is always there, but it doesn't prevent you from showing up.

Tim: Right. Just the opposite, in fact.

Andrew: You might be living proof, exhibit A, that the scientists should just come out and tell us how bad things really are.

Tim: They're very reluctant to do so. I caught up with Dr. Terry Root a couple years after she told me the truth about our climate future that kicked off that first cycle of despair. It was 2011 when we were making *Bidder 70*. I said, "Do you remember that conversation?" She said, "Yes." I said, "Look: You didn't scare me into paralysis. You scared me into action. Are you still hesitant to tell people the full truth?" She said, "Yeah. I still don't want to scare people into paralysis."

Then she and I were at the Mountainfilm Festival in Telluride together, and she was giving the presentation there: "This is the biodiversity loss that we would have with 1.5 degrees…and this is the biodiversity loss we would have if we let it get to 6 degrees." I'm sitting there in the audience thinking, *Terry, tell them. Tell them that you know we're going to 6 degrees.* She didn't, and I talked to her afterwards. "Terry, do you think there's any chance we're going to keep it under 6 degrees?" She said, "No. We're going to 6 degrees."[5] I said, "Why can't you tell people that?"

She said the same thing. "I just can't bear the responsibility of pushing people into despair." To her credit and the credit of a lot of climate scientists, who hold back on those implications, nobody got into atmospheric science 30 years ago because they want to help people through the human grieving process. That's not their skill set.

Andrew: In another era, [nuclear scientist and "father" of the H-bomb] Edward Teller said, "Morality is not my expertise."

Tim: Right, and this was part of my motivation for going to divinity school. In order to tell people the truth, I realized I had to be willing to hold them in their pastoral care needs.

Which is where I think ritual comes in—as a way to help us grapple with where we're headed. That's what the pipeline action felt like to me: a ritual reflection on mortality. I lay in the trench for an hour and a half before the firefighters extracted us. Initially, there was this wave of adrenaline as we ran and jumped into the trench. But then it shifted.

Reverend Mariana White-Hammond stood up in the trench and started preaching and I shifted into a more meditative space: reflecting on the loss in Pakistan and the mass graves there, but also reflecting on my own life and death. I felt like I was lying in a grave. All these policemen and pipeline workers were standing over my end of the grave. Then being lifted out by the firefighters. They brought in the stretcher because I refused to walk out. As I was being lifted out of there, it kept reminding me of my baptism. I was baptized with a full immersion baptism when

I was 19 years old. It was that same ritual of descending into a symbolic death and being lifted back out, letting a part of yourself die, letting go and being lifted back into the world to see it in a new way.

Andrew: The "death eyes" again.

Tim: Yeah. When I was in the trench, most of the time I was in the shade. It was a very hot day. I was kind of tucked under the pipe. As they lifted me up, I came into the full sun and it was like re-entering the world. It was like being reborn into the world. Ritual sometimes allows you to cultivate a relationship with death. We're going to need more ways to do that, ways for folks to relate to death, relate to sorrow, before they have to experience it directly. Not just wait around for the people they care about to die off.

Andrew: Folks on the frontlines, in Pakistan, the South Pacific, New Orleans, they're already in the thick of the impacts, they've already got so much to mourn. Meanwhile, those of us removed because of our geography or relative privilege know it's coming—we know that everything we love is on the line—but we're cut off from one another and afraid to go there. There's a lot of alienation; a lot of disaster porn. We need grieving rituals so we can go through our despair together, so we can face the losses we know are coming.

Tim: The value of all this is to give us some sense of how we will feel down the road. I think a lot of people's hesitation with engaging on climate change is this fear of how they will handle the despair. There's a fear of being paralyzed by despair, overwhelmed by it. These rituals and similar kinds of practices, give people the tools to grapple with their despair, and ultimately build this sense that, "hey, you can handle this."

Andrew: Ritual is one powerful container to help people move through their climate despair. What are others?

Tim: One of [LGBTQ+ poet and activist] Andrea Gibson's great lines is, "What I know about living is the pain is never just ours." She says that in her poem about suicide. Everyone is broken. So, part of our task is to create places where people can be open with each other, and express where they're at in front of other broken people who're in that same place. It helps when people see they're not alone in it, that other people are feeling it too.

We need to show people that facing our climate future is not the end of the line. People are afraid of being paralyzed by despair. But that's not what happens. Or, at least, that's not what happens for long. Also, it's critical to provide people opportunities to take action while they're still in despair. It's easy to fall into the trap of thinking, "Well, I need to work out all of this stuff myself before I can go and take action." The truth is you never really work it out completely, especially with a constantly evolving crisis, where there's always new news—like they're digging anticipatory mass graves in Pakistan—that breaks your heart in a whole new way. You've got to keep processing it.

Andrew: It's a lot to ask of ourselves.

Tim: I use the metaphor of riding a bike. There's the internal work and the external work we need to do. The internal work, dealing with despair and that sort of thing, is like getting your balance on the bike. Going out and taking action is like pedaling. If you get on your bike and you try to get your balance perfect before you start pedaling, you never go anywhere and you can't find your balance. At the same time if you just pedal and don't worry about balance, you're going to crash and get hurt.

You're always going to be a little bit off-balance, but the forward momentum helps you balance, and the balance sustains forward momentum. In the same way, if you're trying to figure out all your own internal shit before you start fighting back, you'll never get there.

Andrew: So, taking action, fighting back, steadies you?

Tim: Yes. Exactly. Edward Abbey says, "Sentiment without action is the ruin of the soul." It's one of my favorite quotes, a personal motto of sorts. It really came home for me during the Bidder 70 action. Prior to disrupting the auction, I'd sign petitions, go to a rally every once in a while, grow some of my own food—actions that felt nowhere near in line with how bad I knew things were, and what I felt needed to be done. It wasn't until the auction action, when I was suddenly risking everything, that I felt my actions were finally in line with how I really felt. I realized that not only is "sentiment without action the ruin of the soul," but the positive flip-side is also true: "Action in line with my sentiment is the healing of the soul." I felt that immediately that day.

Andrew: You also mentioned hospice as another way to approach our situation. Our world is dying. We can't "fix" it, but we still have to operate with love and care.

Tim: I think hospice has its place, but it's not the approach that carries me through my dark periods. In part because I'm not one of those who thinks we're "at the end." I don't believe we're facing near-term extinction, like, say, Guy McPherson.

Andrew: He and I spoke last month. Whether the severity of his specific predictions pan out, people seem to appreciate that someone with standing is naming our worst fears.

Tim: Well, I think he raises some good points about how the climate movement is shirking the harsh truths. But he makes assumptions about the limits of human adaptability that are unfounded. I think a lot of people land in this end-of-the-world "near-term human extinction" position because the consequences are so unthinkable that they don't have a context for understanding it in any other way besides complete annihilation.

Andrew: Agreed.

Tim: We're heading towards a global population of nine billion people by 2050. At that point, if not before, climate change will start putting real constrictions on the population. Now, that's utterly unprecedented in our historical memory. Even during the great crises of the 1930s and 40s, even including all the deaths in WWII, the Nazi Holocaust, Stalin's famine and purges, even through all that, the global population increased.

Now, we might hit nine billion in 2050, but by 2150, if you look at some of the projections of the planet's reduced carrying capacity over that next century—projections that incorporate the effect of climate change—some of those come in at two billion or less. Which means over the course of a century (from 2050 to 2150), we will lose as many people as are now on the planet. We could lose seven billion people. We'll lose what we think of as the entire world.

Andrew: It's almost inconceivable.

Tim: Yes, exactly. In so many ways it's beyond our comprehension. It pushes people into that notion of the end of the world.

Andrew: It's easier to think the world will end than to contemplate all the suffering to come if it were to actually continue. Not to mention all the work we're going to need to do, political and otherwise, in order to carry on.

Tim: Yeah. But on the other side of that unimaginable loss we would still have two billion people left, which is more than there had ever been before the 20th century. It's this paradox of losing the whole world and still having the whole world left to fight for. So, the hospice metaphor doesn't quite fit for me. [Eco-philosopher] Kathleen Dean Moore has talked a lot about the need for new myths and new stories. We've had these hero myths that appeal to men. She often says that what the new story needs to be is a love story—and that has always been the more genuine analogy for me. I realized this about myself some years back when my girlfriend and I were splitting up.

Andrew: How so?

Tim: It was impossible from the start. She split up with me 15 times the first couple months we were dating. I knew it would never work out. But I loved her anyway. And we did last a couple of years. The point is: I allowed myself to become totally committed even though I knew it was impermanent. It was one of the most important lessons I ever learned, and you could say it's in that spirit that I am a climate activist.

So, one of the most celebrated climate activists in America takes up his burden like a doomed lover. We often say, "Better to have loved and lost, than never to have loved at all." But Tim had flipped that sentimental, backwards-looking view into something heartbreakingly forward-looking—"better to choose to love *knowing* you will lose…"—and applied it to our collective predicament. In so doing, he'd found a supremely humanist attitude for how to live in the 21st century. Yes, there is so much to lose. Everything, in fact. The whole world. And yet, as Tim so eloquently says, there's a whole world to fight for, too.

In spite of the grim future Tim had outlined, I found his take both grounding and soberly empowering. Are we in for a catastrophe? Yes. Is the future going to break our hearts? Yes. But around the margins, there is much we could do—on both the resistance and resilience fronts—and it is essential that we do it. Not just for society and the future, but for ourselves. Because only by getting on that bicycle and riding it, can we find our own balance.

Toward the end of the interview it began to rain furiously. My iPhone recorder beaded up with rainwater. I sheltered it under my shirt and we calmly moved under a tree in the park. "This is the other half of the 'flash droughts' we're now seeing in the Midwest," Tim said, and proceeded to explain how summer temperatures over the Great Plains were no longer dropping down at night as they used to. Thus, critical moisture was getting sucked out of the top layers of the soil, parching the ground there, and eventually depositing the extra moisture here in the East in just this kind of downpour.

Then, checking the time, he hugged me quickly and ran off through the rain to catch his train.

Should I tell people how bad I think it is?

Everything is going to become unimaginably worse and never get better again. If I lied to you about that, you would sense that I had lied to you, and that would be another cause for gloom.

— Kurt Vonnegut

In a time of universal deceit, telling the truth is a revolutionary act.

— George Orwell

Many of us run on hope, and hope can be a fragile thing. Dr. Terry Root, the climate scientist whose 2008 presentation sent Tim into his spiral of despair and awakening, had been extremely reluctant to share her findings. To get to the truth, Tim had to corner her after the talk.

"Why aren't you telling people how bad it is?" Tim wanted to know. "I feel like if I told people the truth," she'd said, "people would just give up."

Unable to lie to himself, Tim had demanded the truth. And it crushed him. For a while. Until he found a new footing, a new way to hope. Now he carries around this "impossible news," and he must decide how, and when, and with whom, to share it. Part of what impelled him to head to divinity school was to learn how to do so with care.

What about the rest of us?

Some of us trim the truth to fit the most hopeful storyline we can get behind. Others keep our worst thoughts to ourselves, hoping that by not saying anything too depressing about the bigger picture, we'll all just keep plugging away at our little piece of the problem and things might just work out.

Is there any way to keep faith with the unpleasant facts and still offer people a viable strategy forward? And if there isn't, should we still tell people how bad we think it is? Yes, Guy McPherson and Tim DeChristopher have concluded.

What have I concluded? Inside my head it goes something like this:

I'm going to tell people exactly how bad it is.
Good for you.
I mean, doesn't everyone have a right to know what's what?
They certainly do.
Then they can decide for themselves what they want to do, or not do, about it, right?

OK, so tell them.
Because, who am I to decide what other people can and can't handle, right?
Like I said, tell them.
But won't that crush the hope out of folks?
So, are you going to tell them, or not?
Because if enough people have hope —even if that hope is based on a lie—then maybe there actually is hope (and it's no longer a lie). Because that's one of the strange ways hope works.
Alright then, don't tell people.
But another strange way hope works is it's only when enough people realize that the impossible is necessary, that the impossible becomes possible.
OK, so—and I'm pretty sure I said this before—tell them!
But what a terrible, mean, heartbreaking bummer of a thing to do to someone.
So—and I'm pretty sure I said this before, too—don't tell them!
But someone told me, and I'm still here. I know how bad it is and I'm not demobilized. I'm still in it, doing what I can, I'm still trying to win back the world. Why wouldn't they?
Fine then, tell people how bad it is. But can we decide already, and just get on with the telling or not-telling part?!

A climate activist colleague of mine has become quite wary of talking to people about how bad she thinks it is. People just don't want to hear it. Her friends get angry at her for bringing it up, and for bringing everyone down. She is seen as a "party pooper." And if she does bring it up, she feels forced to give the facts a bright, hopeful spin that she herself doesn't share.

Climate scientists have it even worse. Not only do they have the heartbreaking responsibility of being among the first to tell the world the bad news, but they must balance that against the degrees of uncertainty in the data, the careful and scrupulous conventions of scientific research, and the inevitably brutal attacks[6] of the climate-denial lobby. (If you think the process should be as straightforward as publishing your peer-reviewed research and letting it naturally inform public policy, well, you get a pony.) In the late 1990s, Michael Mann[7] published the core research undergirding the unprecedented nature of global warming and the upward-trending temperature curve whose shape has been likened to a hockey stick. Since then, he's been attacked professionally, placed under investigation, received death threats, and had every aspect of his life placed under scrutiny. There's a very well-funded opposition that picks scientists apart for any step they take "too far," whereas if they underestimate, there's little cost, except, of course, to the billions of people who will have to live through the second half of the 21st century.

In a heartbreaking 2015 *Esquire* investigation into what it's like "when the end of human civilization is your day job," John H. Richardson tracks down several of the world's leading climate scientists about the harassment they've faced and how they handle the grave responsibility of knowing the worst before the rest of us.[8] Richardson finds them wary of journalists, bruised from political fights, and heavy-hearted with knowledge of the future. He subtitles the article: "Things are worse than we think, but they can't really talk about it." He presses them. *How bad is it, really?* he wants to know. While most of the scientists are exceedingly reluctant to give him a straight answer, they are more forthcoming about their coping strategies—from goal-oriented focus on the work, to professional detachment, to deliberate compartmentalization, to "brightsiding"—and it's while exploring these various forms of half-denial that the impossible news leaks out.

"The glaciers are going to melt, they're all going to melt," NASA climatologist Gavin Schmidt finally admits, but "what is the point of saying that? It doesn't help anybody."[9] Schmidt takes a relentlessly "bright side" approach, and has won a Climate Communication Prize from the American Geophysical Union for his efforts. Michael Mann, on the other hand, while acknowledging the need to connect with the public, disagrees with Schmidt's approach. It's not really scientific to focus on the middle of the curve, Mann argues. "We're supposed to hope for the best and prepare for the worst, and a real scientific response would also give serious weight to the dark side of the curve." Sometimes, Richardson concludes, "strategy is one thing, and truth is another."

This same tension between truth and strategy came up in a very public way in the firestorm response to David Wallace-Wells' 2017 piece in *New York* magazine, "Uninhabitable Earth: When Will the Planet Be Too Hot for Humans? Much, Much Sooner Than You Imagine." It quickly became the most-read piece in the history of the magazine, and set off a fascinating conversation about whether an article that "sketches the bleakest possible scenario"[10] is ultimately helpful, and if not, what kind of story should we be telling about our climate predicament?

In his response in the *Atlantic*, Robinson Meyer wonders, Why is it "so hard to talk about the worst problem in the world"? While acknowledging the ample "research suggesting that people respond better to hopeful messages," he nonetheless concludes, "I don't think journalists should frame the truth to better inspire people."[11] Others disagree.

Climate activist and citizen journalist Nathan Thanki's take was short and to the point: "Fuck Your Apocalypse." "What good is our analysis," he asks, "if we cannot even see past our nihilism to ideas about how we might possibly fight and win."

> Yes, the situation we find ourselves in is far worse than we think and maybe worse than we can imagine.... And yes, the actions offered by our governments amount to "too little too late"...And yes, the promise of technological salvation delivered via mystical "market forces" is complete nonsense...Yes, that may all be true. And that should make you want to put your head in the oven. But what is the point of saying that—and nothing more?[12]

Thanki suggests that we focus our attention instead on "the reality of resistance around the world." Calling it "instructive" that communities on the frontlines of climate impacts "have at every stage of this apocalypse we are living in decided that their lives are worth fighting for."

And so, it seems we have four options:

We can tell people, "It's not my job."
This is the path Dr. Root has taken. She presents her findings, but doesn't explicitly draw out their dark conclusions. She doesn't want to take responsibility for sending people into despair. It's not her job.

Tell people the balls-out truth and let the chips fall where they will.
This, more or less, is the path taken by Wallace-Wells in his *New York* magazine piece. Tell everyone exactly how bad you think it is. Whether because you think it will terrify people into action (even though most research says such an approach is more likely to terrify people into *in*action), or simply because it's the truth and therefore it simply needs to be told. Tell them the whole ugly, apocalyptic truth of our situation and let the chips fall where they may.

In many quarters, the news may not be welcome. You could lose friendships, or receive death threats. You could send people (including yourself) into a spiral of depression and hopelessness. But, hey, it's the truth. I'm trying to deal with it; we all need to deal with it. And, who knows, amidst all the heartbreak and depression it causes, the balls-out truth could turn out to be the sobering, clarifying first step we need to lead us towards action and our eventual survival. In any case, anything else is bullshit, anything else is living a lie, and what's the point of that?

Tell people only the most hopeful version of the truth.
For most activists, professional climate communicators, and environmental groups like 350.org, the answer to the question *Should I tell people how bad it is?* seems to be "Yes, but." As in: "Yes, but don't be a jerk about it." Or: "Yes, but embed it in a solutions framework." Try to be as positive and pragmatic as you can be. Tell the best possible version of events. Focus on the promise and potential of the moment. And fight like hell and hope for the best. Here's some advice from the Yale Program on Climate Change Communication that typifies this approach:

- Invoke a strong sense of human agency and responsibility. We're causing it. We should fix it.
- Foster engagement and efficacy. Futility is the enemy of responsibility, and it's rampant in our political culture. But people remain hungry for solutions, and eager to participate. Pollyannish optimism? No. Can-do determination to build a better future? Definitely.
- Embed (don't bury) climate in the challenge of freeing ourselves from fossil fuel dependence. Almost everyone at least suspects that fossil fuel dependence is a dead end, and feels victimized by the forces that perpetuate it. Climate solutions can free us![13]

This approach is more than good strategy. As actress-turned-activist Ashley Judd reminds us, "It can be abusive to highlight a problem, without also underscoring a solution."[14] In other words, raw doom-saying can be a form of cruelty. If you're going to tell people how bad it is, you become ethically responsible for also offering some kind of solution or hopeful story or winning strategy. If you're unable to do that, you must at the very least learn to hold people in their despair.

Tell people the full and awful truth, but with great care.
If you can't embed the awful truth of our climate predicament in a solutions framework (because, say, you don't actually think there is a solution, per se), you need to be prepared to catch folks in their grief. You need to

offer them a way to live with the impossible news—a story or an ethos that can hold it, and maybe a community to hold it with. "If I'm going to tell folks the truth," Tim DeChristopher said to me, "I have to be able to offer them the pastoral care and counseling to see them through it."

None of us want to hear the impossible news; nor do any of us want to deliver that news. However, if we must hear it, and if we must tell it (and I believe, on both counts, we must), then it's best to first ask: how would *we* ourselves like to be told such news, because with news like this, you're always both the teller and the one who must be told.

THE FIVE STAGES
OF CLIMATE GRIEF

We don't do grief, grief does us.

— Rob Delaney

Guy McPherson said we're locked into a path unavoidably leading to extinction. Tim DeChristopher: Not so fast. It's only industrial civilization that's going to collapse. Yes, it's too late to prevent catastrophe, but if we step up our game, over the long-term we can still build a new, more decent society on the ashes of the old. When someone tells you that it's "only" civilization that's going to collapse, and the news comes as a great relief, you know you and your species are in for a world of hurt.

These first two meetings gave me future-scenario whiplash, but reminded me, whichever path we're on, it's a path of grief.

Per Espen Stoknes, Norwegian social psychologist and author of *What We Think About When We Try Not To Think About Global Warming*, argues that regardless of whether there's still time to turn things around, the only way we will be able to take adequate action (never mind honor the reality of the moment we're living through), is if we first go through a "Great Grief," about the destruction of our natural world.[1] Naomi Klein agrees: "There has to be a space in which we can grieve," she says. "And then we can change."[2]

Whether hopeless, still hopeful, or agnostic, all agree: we must break our own hearts; we must grieve the end of the world. But how?

To help people confront their climate grief, I and a handful of fellow writers, artists, and activists began a storytelling project in 2014 called The Climate Ribbon (theclimateribbon.org). We asked participants to consider, in a deep and personal way, the question "What do you love and hope to never lose to Climate Chaos?", then write their responses on ribbons and share them with each other, often by tying them to a living tree, or a tree-like sculpture. The whole process unfolded like a ritual, harkening back to the many cultures around the world who affix fabric to trees in prayer or hope.[3]

On her ribbon Lynda Montoya from South Florida simply wrote: *Miami, my city*. Jon from Milwaukee: *The hope that my children can live healthy lives*. A. J., a ten-year-old boy from the Rockaways in Queens, New York, who'd been displaced by Hurricane Sandy, wrote: *My books, my toys, my best friend, my apartment, my mom*. He'd seen the ocean rise up and try to take all that away once; he didn't want it to happen again.

When I had to write my first ribbon, I thought of that beautiful, ghostly day in Hudson River Park and wrote: *The coastline of Manhattan*. Later on I added others: *Winter as a distinct season* and *The time we need to make mistakes*. And, finally: *The kindness among strangers*. This last one—as I imagine the flood of refugees and the panic and crunch of scarcity turning people cruel—is, for me, the most precious, the most heartbreaking.

When it came time to share ribbons, people took the time to read through each other's until they found one that truly moved them, then tied it to their own wrist, pledging to be the defender of what this other person—sometimes a friend, and sometimes a total stranger—most loved and didn't want to lose. This last step sometimes moved folks to tears, and connected people in what we began to refer to as an "intimate solidarity."

People from many walks of life participated in the project, sometimes in a church basement, sharing ribbons around a small circle; other times amidst the great climate mobilizations in New York in 2014, Paris in 2015, and San Francisco in 2018, where people exchanged ribbons with thousands of others from all over the world. Everyone has something to lose, and when all those losses were assembled in one place, the overwhelming scale of the catastrophe—as well as its essential humanity—was palpable.

For some, like A. J., the ten-year-old from the Rockaways, the Climate Ribbon is in part a memorial to what he'd already lost; for others it's a kind of "prefigurative mourning"—a way to carry their open, beating hearts into the future, past the data and all the abstractions, to grieve the losses yet to come. People are rarely willing to do this alone, but alongside others making themselves vulnerable in this same way, many were able to pierce through their fear, isolation, and hopelessness—and the simple overwhelmingness of it all—to reckon with all they stood to lose.

Some groups went one step further, encouraging each other to declare one action they were committing to take, and witnessing it with a chorus of "Your commitment strengthens mine."

Whether people walked away more hopeful is hard to say, but they did walk away more awakened, more grounded, more connected—to the Earth and to each other. People came to better understand, in their bones, what was at stake, and if they were inclined to fight, they came away strengthened for the task.

You have to go through all five stages of climate grief, except they're not stages, there's more than five, and you might have to walk backwards.

From DeChristopher to Klein, our leading climate thinkers are telling us grief is a necessary reckoning with a 21st century increasingly gutted by Climate Chaos. Rituals like the Climate Ribbon show us that grief can be a path of compassion and community. Mystics and poets across the ages go further, claiming that, at some essential level, grief is what makes us human. "Before you know kindness as the deepest thing inside," says poet Naomi Shihab Nye, "you must know sorrow as the other deepest thing."

But how human do we have to get? How much sorrow can we take? We're losing as many as 200 species a day;[4] the Arctic ice will be gone in a generation; New York and Shanghai will be underwater by the end of the century; billions of humans may die. For any of us alive now, who've known the beauty of our world and the relative stability of our climate, how could our grieving ever end? Maybe it can't. Maybe it's just something we're going to need to keep doing our whole lives, and just try to get good at.

Elisabeth Kubler-Ross formulated her five stages of grief—denial, anger, bargaining, depression, acceptance—to track the stages an individual goes through as they reckon with the news of a diagnosis of terminal illness. These stages can also be a helpful guide as we reckon with all we're losing in our larger world. However, as anyone who's been through a deep grief knows, it's never a simple scripted march through any set of stages, and then you're somehow healed and back in action. It's a tangled meandering; less a linear process than it is a spiral or even a fractal one (where at every stage you have to go through sub-stages of all the stages, ad infinitum). And we never quite arrive, we're never quite done with grief.

When it comes to climate grief specifically, British eco-journalist Gregory Norminton suggests[5] we may even need to walk backwards:

Could it be that we [environmentalists] are undergoing the five stages of grief in reverse? We began with acceptance—of the science, and of what the science means for us. Then came depression, since the future is a place we would not wish to bequeath to our worst enemies let alone our children. In defense against

64

depression, we entered the bargaining stage: surely with enough treaties, tax incentives and political will we can avoid the worst.

Yet the political will is lacking and every week the evidence mounts that we are hurtling towards the abyss. This makes us angry with ourselves and our elites for failing the greatest test our species has ever faced. And where do we go from anger if not into denial—a soft-focused, open-hearted denial without which we would have to abandon all efforts to forestall the apocalypse?

As climate catastrophe rolls toward us, we're not just grieving our devastated world, we're also grieving something trickier, more hidden, inside ourselves. Because what I suspect we may be most in denial of—and most angry at, bargaining with, depressed about, and, yes, need to accept—is not so much the facts of climate change, nor even the reality of all that we stand to lose, but the *burden* it places on us.

We didn't ask to be alive at this time in history. We didn't ask to be responsible. But here we are, and we must find a way to live with what we know is coming. And, whether we find ourselves hopeful, hopeless, or something in between, we must find a way to act. In fact, maybe all our grieving has a secret purpose: to clear room in our hearts for this unwelcome burden.

Whether climate chaos has already caused you grievous loss, or whether it's just tugging on your conscience from afar; whether you're walking through the stages backwards, or following a spiral, or caught in an endless fractal, in the 21st century we're all going to have to walk a path of grief.

1. Denial: The wisdom of denial.

Denial helps us to pace our feelings of grief.
— Elisabeth Kubler-Ross

Climate denial ain't just a river in Egypt; it's a cynical multimillion-dollar fossil-fuel-industry lobby in Washington, DC. Exxon and Shell[6] might play dumb, but for more than 30 years they and their ilk have known that humans were warming the planet, and to put off their day of reckoning, they've poured over $80 million[7] into an orchestrated campaign of climate denial. For them, almost any argument to prevent meaningful climate action will do, a propaganda strategy one could neatly sum up (with help from the good folks at rationalwiki[8]) like so:

> First they say global warming isn't happening, so we don't have to do anything about it. Then: global warming is happening, but it's not caused by humanity—so we don't have to do anything about it. Next, it's on to: global warming is happening, it is caused by humanity, but China and India aren't doing anything—so we don't have to do anything about it. When that doesn't work: global warming is happening, it is caused by humanity, and maybe China and India are willing to do something, but "science will find a way"—so we don't have to do anything about it. Until finally: Sorry folks, global warming *was* happening, it *was* caused by humanity, and previous governments could and should have done something, but it's too late now!

This kind of "denial" is nothing but a strategic pose, a cynical ploy to keep the profits flowing. These "Merchants of Doubt" (as historians Naomi Oreskes and Erick Conway call them)[9] and their political allies have effectively declared war on the Earth and the future of humanity and we must be merciless in debunking their lies.

But there are other kinds of denial that might deserve more compassion, in part because they're not destroying the planet for their own greed. Consider the busy parent who's just not paying attention. The "solution-averse" libertarian who doesn't want the government to mess with his thermostat. Or the 39% of Americans, according to a 2014[10] Public Religion Research Institute poll, who believe God will intercede to stop humans from destroying the earth. These are deeper stories. And, if past history is any indication, they're deep enough to hold strong, year after year, even in the face of overwhelming evidence to the contrary.

In *The Worst Hard Times*, Tim Egan writes about the 1930s Dust Bowl, arguably the worst human-caused U.S. environmental disaster prior to the one we are creating now. Unlike the Joad family of *Grapes of Wrath* fame and other "Okies" who packed up and went west, Egan focuses on the people who stayed put and rode out the disaster. Alongside useful survivalist lessons (What do you eat after four years of crop failures? Key: rabbit stew), the book is a revelation about the persistence of denial.

Even after eight years of horrible conditions—and powerful evidence that the Dust Bowl was a human-created disaster (the result of humans plowing up arid plains and grasslands not suited to agriculture)—many persisted in their belief that it was just God's will, and what they needed to do was behave better.[11] Others continued with a ghoulish boosterism that things would get better. A full third-to-half of the regional population never fully grasped the scientific explanation.

One could imagine, a hundred years from now, even after rising tides have taken our homes, even after there's no fossil fuel industry left to defend, that some hard-core believers will persist in denying the climate science. And why not? If denial helped folks (at least psychologically) get though one of the worst hard times of the last century, why wouldn't denial help us get through the even worse hard times coming in the next century?

For those of us who accept mainstream climate science, and recognize the clear and present danger, it's easy to rail against those who don't "get it," and clump them in with the Merchants of Doubt. In doing so, however, we not only demonize a lot of decent folks, but also conveniently set ourselves up as the righteous smarty-pants in this story.

Well, not so fast.

Because if we're honest with ourselves, I think we'd have to admit that we're in our own kind of denial—not factual denial of the science, but emotional denial of the full reality of the carnage to come, the difficult changes we need to make in our own lives, and the possibility that it's already too late. Of course we're in denial. The consequences of catastrophic climate change are daunting, almost too terrifying to contemplate, and denial can help us through it.

There's a wisdom in denial. It's as much an act of psychic self-preservation as it is a refusal of reality. My mind knows the climate science is true, but my body hasn't felt the consequences yet. In spite of being in New York during Hurricane Sandy, I have no climate scars. Meanwhile, my heart, like the valve it is, slowly titrates the impossible news. *I too am a climate denier.* I say this softly to myself. It feels weird. But it's true. At some level we're all in denial. It isn't the irrational deniers vs. the rational rest of us. No, we are all on a spectrum of denial. In fact, we all *need* to be in denial.

Denial buys us the time our hearts need to gradually break open to the terribleness of it all. In this sense, denial is an odd kind of courage—a handmaiden to the will, hope's necessary accomplice. If hope is a somewhat wanton decision to believe in the impossible (or at least the highly improbable), then how could hope possibly accomplish this extraordinary non-rational task without a certain kind of denial working alongside it?

Consider the difference between these two statements:

1. "There's nothing wrong here."
2. "There's nothing wrong here we can't fix."

They're both made in denial, but with a crucial difference. One pretends there's no problem; the other pretends the problem can be solved (or at least mitigated). One leads to paralysis; the other to action. Both protect us emotionally; the second one, however, might, at the same time, also help protect us in reality. If we can shift our allegiance from the first to the second, we find that the protective wall of denial we need is still there. Only now, instead of fear and junk science, it's made out of hope, and solidarity, and a Tesla Megapack battery. You're still emotionally protecting yourself from our horrible future, but now, every day you're also helping to make that horrible future slightly less horrible.

In the end, maybe hope itself is just a beautiful, human, productive, uplifting—and, yes, necessary—kind of denial. And if we have to be in some kind of denial, let's choose wisely which kind.

2. Anger: We can! We must! We won't?!

In leadership school they tell you: "Don't lose your temper, use your temper." Yeah, well, fuck that. Anger is a tonic, a truth serum, a cleansing fire. It can electrify your senses and put the world on notice. It can unlock your bile and out your secrets—secrets that were secret even to you. "I rebel, therefore we exist," said Camus, in a provocative détournement of Descartes. Anger—and the great existential "No!" it makes possible—is an essential part of our humanity. Without it, we tend to passively accept the givens, and never find our voice.

And in our time of rising climate chaos, we have reason to be angry at nearly everything and everyone. We're angry at the messengers—at the scientists who tell us we're doomed, the cynics who tell us it's too late, even the scolds who tell us to go vegan. We're angry at ourselves—at our own guilt, shame, and helplessness; at our own contradictions as we suck at the teat of the very fossil fuel industry we rail against. We're angry at other people—at the folks who do nothing, the politicians who do next to nothing, and the doomers who say there's nothing to be done. And, hell, we're angry at reality in general for, well, handing us this sack of existential shit and telling us to man up.

In our time of rising climate chaos, there's so many wrongs to rage against. But what gets me angrier than the wrongs is all that could be right. Because, while climate chaos is happening, and it will get worse, the fact stands that if we can keep the rise in global temperature below 2°C, we can prevent total climate catastrophe. We imagine ourselves stuck on a train barreling forward towards a cliff, and there's no way to stop it from going over. But here's the thing—and it's kind of an amazing thing—we *can* stop it.

An average 1.8% drop in emissions per year would keep global temperature rise under 2°C.[12] Sure, that's 80% stricter than the Paris Agreement targets that most governments are already failing to meet, and even that only has a 50% chance, but still! That's all it will take to avoid totally ruining our one and only home, and sentencing our children and grandchildren to mass starvation. We must do this. We must. And we can. But—and this is what kills me again and again and again—we won't.

We can. We must. We won't.

We can: History says we can. We did it during WWII. Twenty million victory gardens say we can do it again. The technology is there. The price of solar has been falling dramatically for decades.[13] It's now competitive with fossil fuels. The roadmap is there: Scotland already gets nearly 100% of its electricity from renewables.[14] Germany is on track to get there by 2035.[15] Professor of civil engineering Mark Z. Jacobson and his team of researchers at Stanford University have mapped out a detailed transition plan[16] for the switch to renewables for every state in the US. No one can say we can't do this. It's more than theoretically possible: we have the technology, the historical precedent, and any number of plans to choose from. So: We can. And given that the alternative is unthinkable, we must.

You'd think: if we can, and we must, then we would. Kind of a no-brainer. And yet my gut says: we won't. Everything I know about capitalism, bureaucracy, denial, habit, how much slower our psyches move than our machines, says—and it breaks my heart, and my shoulders slump as I write this—we won't.

We can. We must. We won't.

Are you fucking kidding me? Is that going to be the epitaph of humanity? *We could have. We needed to. We didn't.* I revolt at the insanity of it. Because we are better than this. Most of us are good, decent people. If a neighbor needs a cup of sugar, or a jump-start in the winter, we are there. And we're smart: hell, we went to the moon—and we didn't even need to! We've got to be able to fix this. And in the most secret chamber of our hearts, even the most cynical and doom-filled among us, even me, even face-squarely-the-horror-before-us me, has to believe: We will.

To imagine anything else is insane. But maybe, as a species, we are insane. Or at least the economic system we've created is insane. Well, that for sure. And maybe it rules us more than we rule it. But, maybe, just maybe, Naomi Klein is right and the climate crisis "changes everything." Maybe it really is the perfect storm to prod humanity to heroically rise up and get its house in order. And for a moment I go: We can. We must. We will!

Look at the good work anger did here. Anger helped separate right from wrong. Anger put this shard of hope in my hand. But I'm still angry. Angry at the whole damn situation. Angry that simply being angry isn't enough. Angry that it's falling on me to do something about this mess we've made. Because it feels like there's so little I can do, but I also know there's so much I could do, but some days I just don't want to do any of it. I hate the burden of it all. I hate feeling like I should do something when it might already be too late to do anything. Because who wants to be played the sucker by History?

But my psychic inbox is flooded by endless texts, emails, do-gooder junk mail, and that moral foghorn in my own head called my conscience, all nagging at me, going "Americans of the future will look back..." "Always consider the next seven generations, whenever you..." "What kind of world will we be leaving our..." Till I'm just: Shut! The! Fuck! Up! Can the future just fuck off and die, already! Can it just leave me the fuck alone?!

But am I gonna play things that way? No! Of course fucking not! I'm still gonna do my stupid little part to try to dial back this absurd catastrophe we've inflicted on ourselves. And if all our efforts end in fucking failure, at least I'll know that the little I did to slow things down will have been a fucking kindness to someone or to some creature. Okay, Mr. Future, are you fucking happy now?

3. Bargaining: Is it now yet?

Supposedly, if you drop a frog in a pot of boiling water it will jump out, but if you gently lower a frog into a pot of tepid water and then very slowly raise the temperature, it will let itself be boiled to death. It turns out, however, that this story—invoked for decades by everyone from author Daniel Quinn to Al Gore to warn us of complacency in the face of slow-motion environmental threats—is a lie. Scientific observation has shown that no matter how gradually the water is heated, the frog will eventually jump out.[17]

(I hate to think how many frogs these scientists had to *try* to boil to death to find this out, but there you have it.)

The boiling frog story is a myth, but as metaphor, it's still going strong. Why? Because the metaphor is about us, and it turns out, we are dumber than frogs. The frog is paying attention and will do what it needs to do to survive. We're the ones not paying attention; we're the ones slowly cooking the planet while we convince ourselves everything is OK. What are we saying to ourselves? Why won't we jump out?

"I've built a life here in the pot."
"No one else is jumping out."
"Surely, it can't get any hotter, right?"
"If I jump out, who knows what I'll jump into…it could be worse."

Who are we bargaining with? The water? The pot? The rising temperature? We're bargaining with ourselves, of course. Irreconcilable parts of ourselves are furiously negotiating over our impossible new climate reality.

It's hard to negotiate with the impossible. We slide quickly into metaphysical self-dealing. How many of us, after yet another doom-filled scientific announcement, have said to ourselves, "Can't I just pretend I didn't hear that?"

No matter how much I rail against the government and industry to get its act together, I can feel inside me a lingering belief that Daddy is still in control: "They're gonna figure something out, right? They have to. There's no way they're going to let something like this happen, right?"—

when I know full well that there is no "they" and they're doing no such thing.

And every time I put out the recycling, I can sense a secret part of me offering up a prayer to the deaf and implacable God of Ecocide to witness my puny act of eco-do-good-ish-ness and please spare my foolish species. It's a ridiculous notion, I recognize its irrationality, but I do it anyway.

The classic book on negotiations is *Getting to Yes* by the Harvard Negotiation Project.[18] But when you're negotiating with the impossible, the best you can usually get to is "Yes, But." *Yes*, climate reality is real. *Yes*, it's really bad. *Yes*, it might be too late. And, *yes*, we need to do everything we can. And, *yes*, we need to do it now. *But*. But what? *But* is it really now yet? Can it please not be now yet?

This is where we make our last stand. It's always now, of course—and with climate change, it needed to be now decades ago—but a part of us never wants it to be now. I don't want it to be now yet. I like my life, and the relative normalcy I've been able to carve out of this crazy world, and for a little while longer I want to keep pretending I don't have to live that life in the glare of a huge, blinking-red global emergency button that changes everything. I know now is already here—but can't it just be already here a little later?

It's not unlike how we feel about death: We know we're going to die, but does it have to be now? Can it not be now, please? Can it just be, always be, a little later? Until exhausted, stalemated, and without reconciling anything, we finally surrender. And like the wise frog, we jump out of the pot into the unknown.

4. Depression: Despair is our only hope.[19]

*Despair is the only cure for illusion. Without despair
we cannot transfer our allegiance to reality — it is
a kind of mourning period for our fantasies.*

— Phillip Slater

I used to believe. As a kid I trusted that everything was more or less Okay, that progress happened, that the people in charge were trying to make things better, and the good guys would eventually win. *Hawaii 5-0* was my show.

As a young man I realized that the people in charge were not trying to make things better for everyone, just for themselves. And so—because I'm a hopeful kind of guy—I came to believe that the people *not* in charge could get together and change all that. I loved the movie *Hair*. I used to believe the revolution was just around the corner, that before I turned 30 we'd be celebrating in the streets.

Well, that didn't happen.

Into my 40s, I still had faith in humanity. Not blind faith, not even a faith in our essential goodness. But I believed that we would somehow stumble through, that the small acts of kindness among people would somehow make up for the evil and folly of the gangland of States and Capital. I could still see a future, maybe not a better one, but no worse, either.

I'm now in my 50s, and I've lost even that meager faith. Now I binge-watch *Game of Thrones* and *House of Cards*. I have no illusions about how power operates. I've done the climate math, and I see no way forward. I am filled with despair.

And then a strange thing happens: I feel fierce. I feel clear. I feel free. I don't give a fuck anymore. I've got nothing left to lose. I'm willing to take risks that I wasn't before. I say true things, things that are true but that you're not supposed to say. And people notice. Hell, *I* notice. It turns out despair is its own kind of power, its own kind of freedom. And then I think: If enough of us fall into a dark enough despair, who knows what we can do together. This is the only hope I have left.

5. Acceptance: We must awaken to our burdens.

If you have a chance to help others and fail to do so,
you are wasting your time on this earth.
—Roberto Clemente

I've seen gobs of apocalypse films. Whether it's a comet, virus, aliens, zombies, nuclear war, or climate chaos that wrecks us, I can't seem to look away. I don't want it all to end, of course. But there's something strangely satisfying about disaster movies. Our world is broken, corrupt, monotonous, and full of junk—from Congressional gridlock to YouTube comments. It almost feels like a farce to be alive sometimes. What a relief, then, what an unburdening, to go to the movies and see it all destroyed in a kind of giddy, mesmerizing horror.

Decades before CGI and Industrial Light and Magic—never mind nukes, global warming, and all the rest—Walter Benjamin had our number. "Our self-alienation," he said, "has reached such a degree that it can experience its own destruction as an aesthetic pleasure of the first order."[20]

Sitting in a movie theater, it's easy to imagine the apocalypse as a kind of liberation, a final exit, an end to all the human bullshit and useless striving. But there's nothing simple or pure or quick about how this is going to go down. It's not going to be a clean all-or-nothing anything. As Naomi Klein says, "Climate Change isn't just about things getting hotter…it's about things getting meaner."[21]

It takes but one example to bring this truth home: Congolese park ranger Rodrigue Katembo,[22] winner of the prestigious 2017 Goldman environmental prize. Tricked into being a child soldier at the age of 14 during the civil strife in the Democratic Republic of the Congo in the 1990s, he had to watch his little brother die and then carry this news to his mother. Eventually he managed to escape and pull together just enough schooling to join the park service—a line of work that has proven nearly as dangerous.

Outnumbered ten to one by poachers and hostile militia, 160 rangers have been killed in the last ten years trying to protect the extraordinary wildlife in the DRC's parks. Government corruption and illegal mining also pose a grave threat. In 2013, soon after stopping an illegal oil development, Katembo was arrested, tortured for 17 days, and put through mock executions by local militia operating in cahoots with foreign companies. Yet he remains stubbornly courageous and dedicated. "It is really worth it to make this sacrifice," he says, citing the progress he and his crews have made at the Virunga and Upemba national parks, expanding the number of hippos from 500 to 1,700, and the number of elephants from literally zero to 68. Under constant death threats, he is often separated from his family, but he persists in the work. "If I left," he says, "that would feel like a betrayal to the protections the wildlife and national parks deserve."

This is what it's like to swim at the deep end of the 21st century. And for those of us still in the shallows, there's a rip tide inexorably pulling us into deeper and deeper water. The intersecting catastrophes unspooling all around us don't offer an escape from reality, but an intensification of it.

So we have a choice: (1) Accept this reality. Accept the full toxic soup of conditions we've put ourselves in, as well as the thick, messy, profoundly human dramas playing out amidst it. And awaken to the burdens—of grief, hope, solidarity, and action—it places on us. Or: (2) Keep to our bubble, hunker down in our theater of self-alienation, and stay mum. What's at stake? Arguably, the fate of the world. Certainly, the fate of our own humanity.

But wait, you say, *how can I possibly open my heart to all this suffering? It's overwhelming. Where do I even begin?* Let's begin with something disarmingly simple: The Serenity Prayer.

> Grant me the serenity to accept the things I cannot change, the courage to change the things I can, and the wisdom to know the difference.

You might be familiar with it from AA, or from personal counseling. But why shouldn't it work for public life as well? It's an inner dialogue between our head and our heart, a dialectic of hope vs. fact. It's a way to take our moral pulse, a way to focus and steady ourselves in the face of the impossible. As we stare into the darkening waters of climate catastrophe, the Serenity Prayer may not be able to get us over our grief (what could?), but with a modicum of courage and wisdom it may help us to carry that grief into battle.

No one else can tell us what we can—and can't—change. We have to decide that for ourselves. And, then, once we've drawn that line, we not only have to get on with the hard work of changing what we can, but the possibly even harder work—especially in the era of climate chaos—of accepting what we can't. Because it's one thing to accept that you're an alcoholic, or that your marriage is over, or even that your cancer is terminal and you're on track to die. But how do we accept that the dystopian world in the latest sci-fi disaster film is actually going to be your grandchild's future? Or that ocean acidification could kill off most marine life by the end of the century.[23] Or run-away global warming will make 1.2 billion people homeless by 2050.[24]

In spite of the relentless trend lines driving these predictions, some things are simply unacceptable. We refuse them their inevitability. And where does this refusal land us? If we stay within the prayer, we always have two options. Either we accept the unacceptable, or we go back to where we previously drew that line, and we decide: actually, I *can* change more things than I thought I could. Of course, as the prayer says, we must then find the courage (as well as a game plan) for this larger task. It's a constantly iterative process, both open hearted and hard headed. It's a process of awakening to our burdens.

Try it yourself.

6. The Sixth Stage: Gallows humor.

*Our age is both comic and tragic: tragic because it
is perishing, and comic because it continues.*

— Søren Kierkegaard

It was the dead of winter in New York, but outside it was balmy, bizarrely
so. I didn't realize it until, all bundled up, I'd walked out of my apartment
door onto the street. I was sweating. And stunned. "More signs of the
Apocalypse," I mumbled reflexively, half to a neighbor standing there,
half to the sky. He chuckled and shook his head in wobbly agreement.
Humor, or at least tragic self-mockery, is the way many of us try to cope.

"Fall canceled after 3 billion seasons," goes the *Onion* headline. "A
beloved classic comes to an end." Ha! A four and a half foot rise in sea-
level "can mean only one thing," deadpans Conan. "Gary Coleman is
going to drown." Ha ha! When I heard that last one, I laughed out loud.
But climate change is no joke. In fact, in objective terms, it is the farthest
thing from a joke the human species may ever experience. It's a tragedy
of unimaginable proportions. But I have to laugh. If I can't laugh, I can't
function.

Humor's classic equation is tragedy plus time. But what do you do
when that time hasn't happened yet? When the worst of the tragedy is
still in the future? You see it in the distance, but it's yet to arrive. Do you
mourn preemptively? Do you laugh at the broken feeling inside, and the
black wall coming at you? Do you mock yourself for even caring? For the
sheer ridiculous idiocy of the situation? I do. I do all of it. It's the only way
I can keep going. "Look Ma, no ice cap!" Ha…ha?

My mom had a pretty bleak view of things. Even though she counted
optimists Elizabeth Warren and Bella Abzug among her heroes, at heart
she was a pessimist, even a misanthrope. She was down on America,
down on capitalism, down on any possibility of progress. When things
were looking bad in the news (and to her they always were), I'd try to
cheer her up with a recitation of all the little positive things she may have
missed. It never worked.

In her later years, there was yet more to get her down: Her heart
arrhythmia, her shortness of breath, the arthritis in her hips (she'd had

both replaced and then those replacements further "revised"), her ulcerative colitis (they'd removed her colon and for years she'd had to walk and sleep and eat with a shit-bag tied to her belly), her sinusitis ("Andrew, it's like having a cold for the rest of your life!"), and on top of it all, the crappy black and white news dropped every morning on her doorstep.

In those last years, I tried a different approach: "Ma, it could always be worse." "Worse?!" "Yeaaah, I mean, you could be dead." "Dead?" "You know, like, most of your friends." All 4-foot-11-inches of her glared at me. I carried on: "Or, I don't know, the next time you're squeezing shit out from that hole in your stomach, instead of eventually flushing the toilet and going back to your day, a crazed Ebola-infected alligator could lunge up through the pipes and bite off piece after piece of your fleshy ass in its slimy putrid jaws." A chuckle. "Things could be so much worse, ma, think about it: While reaching up to adjust the dial on the kitchen radio you could lose your footing, and trying to catch yourself, accidentally spin the dial to a 24-hour Rush Limbaugh Special and then lie there on the kitchen floor in a crumpled paralyzed heap with the I've-fallen-and-I-can't-get-up pager just out of reach listening to the whole thing while your colostomy bag overflowed." Finally, she would laugh, and, shaking her head in wonder at why her loins had chosen to produce this particular human to caretake her in her final years, she would be ever so slightly cheered.

If it worked for her, with all she had to bear, couldn't this same approach work for us as we face the bleak prospects of our climate future? When you hear that the Great Barrier Reef is 30% bleached to death,[25] take note that there's a whole 70% death-bleachedly worse it could be. But it isn't. Shocked and dismayed that a climate denier was elected President of the world's largest economy? Just think: He could have been reelected. But he wasn't. There's so many, many ways things could be worse. But they're not. Yet.

No matter what the universe throws at us, no matter how terrible a mess we've made of things, if we can laugh in the face of it—even if we're also crying; even if we're also furious at the injustice of it all—then there is daylight between our circumstances and our victimhood. More than just a balm, gallows humor is an act of cosmic defiance, a reframing, a choice to tilt our gaze to see the human-all-too-human muck of things in a cosmic, tragicomic light.

Gallows humor is also a kind of hope. If we can still laugh, we can still hope. Not with a "things are going to be OK" kind of hope. But with a no-matter-how-terrible-things-are-they're-somehow-still-OK kind of hope. That kind of hope requires a deep laugh, not a cheap laugh. The cheap, dismissive, don't-go-there-just-keep-it-lite laughs are actually a brittle kind of denial. To paraphrase Woody Guthrie,[26] "we don't need more of those kind of laughs; the world is already too full of them." In fact, that kind of humor—life-cheapening, product-selling, business-as-usual humor, and the whole culture of distraction that it's part of—helped get us into this mess in the first place. What we need is the deep laugh. The tragic laugh. The rueful, heartbroken, accursed laugh. The laugh that has all the sadness of the world in it, and still laughs.

This Trickster-Hangman can sometimes show up unexpectedly, as he did one day in January 2015 at one of my "hopelessness workshops." It was a few months after the People's Climate March. 400,000 people from all over the world had marched through the streets of New York for climate justice, and while the climate math hadn't changed, everyone was feeling a little more upbeat.

At one point I asked everyone to spread out across the room in a line depending on where they fell on the question "is catastrophe inevitable?" When the room had sorted itself out, my friend and 350.org staffer Duncan Meisel had stationed himself on the gloomiest side of the room, staking out the far edge of the question.

Duncan, who in his activist circles was already used to being the prophet of doom, shared his garden-variety "we're already in the midst of a rolling catastrophe which, while survivable, is not preventable" perspective. Standing next to Duncan was a middle-aged guy who introduced himself as "Dan, with bad news." A few people chuckled nervously. Then Dan, his lips pursed, his voice flat, told us that he didn't give the human species more than 15 to 20 years. The room caught its throat. There were a few beats of frozen silence. Duncan eventually turned to face the contrarian in our midst, listed a few points of disagreement, and took three large steps towards the middle of the room. Dan, seeing the gap between his perspective and everyone else's, moved in the opposite direction, pressing himself up against the far wall. Whatever the room thought of his Near Term Extinction arguments, we could see it was hard to be that guy. I thought: *this* is not funny, but I hope *he* is.

Since that workshop, I've gotten to know Dan a bit better, and it turns out he is funny. Or at least funny enough to put together a one-man clown show about the apocalypse called "Hospice Earth." Something he needed to do, I think, just for his own sanity. And I went to see it on Broadway. Not the Broadway of *Cats* and *Phantom of the Opera*, but the Broadway of a one-time-booking-in-a-tiny-black-box-theater-on-the-far-west-edge-of-42nd-street-where-half-the-audience-are-friends-and-family. Dan took the stage in a Grim Reaper death mask and a clown outfit, holding a hockey stick in place of a scythe. Sporting a polka-dot onesie and oversized bowtie, he narrated the familiar litany of planetary ecocide with balloon sculptures of extinct animals to soften the blow.

His humor began with the more psychologically observational: "There's two kinds of people in the world: Those who want to know everything, no matter how bad the news. And yeast." Yeast, he pointed out, blindly follow biological orders, reproducing and consuming their petri-dish contents until there's nothing left, and then they die. "Now, what good would knowing the worst do a yeast?" Then he moved on to full gallows humor: "If I drown myself in a bathtub that's half-full does that make me an optimist?" He ended on a sincere and loving note. "I can be dispassionate about the science, but not about you, dear friends."

We laughed. Uncomfortably. Nervously. Ruefully. But we laughed. And then we went for beers. And as he and I and his in-laws made small talk underneath the bar's 23 big screens as both the Jets and the Giants lost, it all felt horribly surreal. We were a table of inverse-Cassandras, the impossible news hovering like an eerie deja-vu that couldn't find a place to land. We spoke about who showed up and who didn't, what it was like to work with that particular production house, anything, absolutely anything, but the bright pulsing neon shout-out-loud fact: in 20 years (if you believe the Guy McPherson-sourced data Dan built his show around) there won't be a human left on the planet. For a moment, the darkest of laughter had let in the darkest of news, but then, as per the survival machines we are, we'd resealed the hatches, and slid back into our Big Lies and small joys.

Al Gore did his famous slideshow all over the world without cracking a joke. And it was wildly effective at alerting the world to the looming disaster of climate change. But Al Gore's truth was merely inconvenient; his catastrophe preventable; his message, ultimately hopeful. Dan's, not so much. Dan's news is impossible; his catastrophe unavoidable; his message lacking any objective basis for hope. What else but gallows humor could carry such a show?

On an average day, laughter can lighten our load, poke fun at the officious, point up the sweet absurdities of life. In a dark time such as ours, it can be an existential survival strategy. You don't have to buy into the worst-case climate scenario, as Dan does, to realize the wisdom of Oscar Wilde's observation: "If you're going to tell people the truth, you'd better make them laugh, or they'll want to kill you." Or as my good friend Dave Mitchell[27] puts it: "Without gallows humor, all we have are gallows."

Grief. Dark laughter. A twisted kind of hope. It was evident I was going to need all three to get through my portion of the 21st century.

Each of us has to walk our own path through the Five Stages of Climate Grief. Mine seemed to demand, first, that I admit that "My name is Andrew and I too am a climate denier," then bargain with the terror of it finally being "now," until, cleansed by despair, I can awake to my burdens. Along the way, the strange alchemy of gallows humor, that runt-of-the-litter "Sixth Stage," helps me find my way.

It would be a lie to say that this lets me completely "get over" anything. But, reeling and settling and reeling again, I slowly sober up for the tasks before us.

And hope? Knowing the depth of horror awaiting us later in the century, Guy McPherson and Tim DeChristopher, as well as my own early stumblings, were telling me we need a different, maybe darker, way to hope—or failing that, at least a way to remain compassionate amidst our hopelessness.

DR. MARGARET
"MEG" WHEATLEY

"Give in without giving up."

In search of answers, I now pointed my quest west.

Every quest needs a mythic arc, and here, I suppose was mine. "Gone West" was WWII slang for someone dying. And with all this talk of extinction and "death eyes," it seemed right that I was now following the setting sun.

From New York, I headed through the flash droughts of the Midwest (through the other side of the increasingly wobbly hydrological cycle that had dumped all that Manhattan rain on Tim DeChristopher and me), across the Plains that I'd criss-crossed as a youth by thumb and freight, landing in the mountains of Boulder, Colorado.

Here, Meg Wheatley, a Buddhist wisdom teacher and community facilitator, was leading a three-day session at the Shambhala Center on "Moving Beyond Hope and Fear," specifically tailored to folks alarmed by the prospect of climate catastrophe.

This would be my third meeting, with five more to follow it. Unlike any of the other meetings, it would not be a direct interview. I'd be part of a group of a hundred or so receiving her teaching.[1]

I'd been following Dr. Wheatley's work for a while. I'd read her books *Perseverance* and *So Far from Home*, as well as a smattering of her many articles and poems. She was asking some of the tough questions: "How can we do our work without hope that we will succeed?" "If we don't have hope, who will save the world?" And: "What if we can't save the world?"

In response, she was proposing a philosophy and practice of "spiritual warriorship." Reworking the Tibetan Buddhist tradition of the Shambhala warrior in light of our intractable contemporary predicaments, she'd crafted an "engaged Buddhist" approach that included such tenets as:

Resisting the illusory comfort of certainty and stability.
Letting go of needing to impact the future.
Not placing blame on any one person or cause.
Confidence that humans can get through anything as long as we're together.

Which all sounded serene, and high-minded, and wise, but I wasn't sure I agreed with any of it.

Also attending the workshop was my good friend, colleague, and "kayaktivist" Lois Canright. She is a mainstay of the Mosquito Fleet, an amphibious direct-action team, using swarms of kayaks and other small watercraft to blockade fossil-fuel infrastructure in the Puget Sound and other waterways in the Pacific Northwest. Over the years, we'd shared many a grief-laced conversation about our climate future, so it felt good and right to be going through this process together.

In the morning, Lois and I borrowed two bicycles and cycled through the well-heeled eco-streets of Boulder (median income $57,000; population 88% Caucasian) to the Shambhala Center, located a couple blocks from the parks and shops and restaurants of Boulder's pedestrian-friendly downtown.

After registering, we proceeded upstairs to a large second-floor meditation hall, took off our shoes in the foyer, and along with maybe a hundred others, found a cushion and mat to sit on. There were chairs in the back for those unable to sit meditation-style. Tangka tapestries depicting various deities graced the walls, and a Tibetan shrine with a large brass gong dominated the space. Meg, in black robes, sat on a little dais at the front-center of the room not far from the brass gong. She was maybe 70, reading glasses draped around her neck on a chain, her hair curly and reddish-brown. She took in the room, nodding to a few familiar faces.

I'm no stranger to Buddhism. As a half-Jew, functional atheist, and card-carrying member of the counterculture, I'm probably more Buddhist than any other religion. I've read my Alan Watts and D. T. Suzuki and done stints at meditation retreats from New Mexico to Thailand. But I have a strong contrarian and anti-authoritarian streak that is decidedly unhelpful when it comes to following a teacher or specific lineage, a situation not improved by the meditation instructor who, when I had asked him how I could fight the power and stay serene, had said, "the government is all in your head."

Anyhow, here I was again, shoes off, sitting on a meditation cushion looking up at yet another wisdom teacher. After a short welcome and silent meditation, we jumped into the deep end.

"In this time of ecological devastation," Meg began. "Many of us are exhausted, despairing, and overwhelmed by grief and by everything we're losing. How can we keep going in our good work for the planet? How can we persevere without being ambushed by hope, which always comes with fear? In this time of profound disappointment, we can't help but wonder, 'What's the point?' and 'Why should we bother?' Given our expectation that things *will* get worse, how do we stay motivated?"

Bam. She wasn't pulling any punches here. The stage was set.

Then, as if to answer her own questions, she read a long passage—a parable or prophecy—which I recognized from her book *Perseverance*, where it is attributed to "Elders of the Hopi Nation" (though, after much googling, it'd be most accurate to declare its true provenance unverified[2]). It went like this:

Here is a river flowing now very fast.
It is so great and swift that there are those who will be afraid,
    who will try to hold on to the shore.
They are being torn apart, and they will suffer greatly.

Know the river has its destination.
The elders say we must let go of the shore.
Push off into the middle of the river, and keep our heads above
    water.

And I say see who is there with you and celebrate.
At this time in history, we are to take nothing personally, least
    of all ourselves, for the moment we do, our spiritual growth
    and journey comes to a halt.

The time of the lonely wolf is over.
Gather yourselves.
Banish the word struggle from your attitude and vocabulary.
All that we do now must be done in a sacred manner and in
    celebration.
For we are the ones we have been waiting for.

Over the next three days, she would re-recite this parable—maybe a line here, or a stanza there—to ground the various elements of her teaching, which she proceeded to lay out, strand by strand.

We began with openheartedness. "We must open our hearts," she told us, "not just to difficulties in our daily life, but to the larger reality and great grief of our time." In her soft, weary voice, she then stepped us through all that being truly open brings into our hearts: mass extinctions, scorching heat around the planet, incalculable devastation and human suffering, destruction of our beloved places, loss of so much we've counted on.... There were sighs in the room, slumping shoulders, shaking of heads.

"We're numb to this. It's too abstract, too enormous," she continued. "We can only feel it for a moment. We know it intellectually, but it's almost impossible to be alive to it in our hearts." She asked everyone to raise their hand if we feel guilty or complicit in this destruction. And again, if we feel rage and blame. "How many of us feel powerless or overwhelmed?" she asked. "How many of us can barely function?" I raised my hand each time, and I watched Lois raise hers.

"Welcome," Meg said, "to the whole range of emotions. Make space, be willing to feel it all. What's happening to the animals, the trees, the oceans. Feel our society reeling in confusion and uncertainty. Feel what's happening in our own body. In our own heart and mind. It takes courage, but we cannot skip over what's happening in our own hearts. We cannot skip over this."

Meg continued in this vein, enjoining us to use our meditation practice to open our hearts and feel it all. "Be all-accommodating," she told us. "Reject nothing. We need to include the pain and the chaos of our world in our practice. This takes gentleness and bravery."

I looked over at Lois. I saw her trying to take Meg's instruction seriously, and the pain it caused her. Afterwards Lois told me how she tries to "sustain the gaze." How she tries to not look away from the suffering and destruction, but rather let it break her heart open and help her build her capacity to stay in the game. Lois leads with her heart. For better or worse—and not untypical for my gender—I'm a bit more compartmentalized and hard headed. Together, I like to think that we make at least one whole human.

From this heartbroken place Meg invited us onto the warrior path. "We need to understand and accept all these feelings without needing to change them," she told us. "The sadness we feel is not actually a feeling." I'm looking at Lois and thinking, *Yeah, well, tell that to the people feeling it.*

"Sorrow is not a feeling we should try to escape from," she continued, "but a place of wisdom we can make a home in. The fragility of Mother Earth is written into our own human body, and that brokenheartedness is the fuel for our warriorship."

Having invited us to the warrior path, Meg attempted a definition. "A spiritual warrior protects life without adding to the aggression and fear in the world." She then shared a short passage from Chogyam Trungpa,[3] the controversial Tibetan monk and founder of the Shambhala lineage in America:

> The path of the warrior is very difficult and painful. We cannot change the way the world is, but by opening to the world the way it is, we can be of service with gentleness, decency, and bravery.

None of this sounded like a warrior to me. A warrior resists evil. A warrior identifies the problem, names the enemy, outlines a path to victory, and then fights like hell to make it happen, while trying not to break too much shit along the way. That's a warrior! In all this talk of spiritual warriorship, where is the *war*? As Meg continued, my doubt-filled monkey mind gibberred away.

"In a climate of increasing fear and destruction," she intoned, "the warrior steps forward with basic human goodness so that others may see what's possible." (*Yeah, because basic human goodness can declaw The Machine? Um, how exactly?*) "Potency, strength, brilliance, confidence, unconditionality—this is warriorship." (*Really? Are you sure it's not just a list of really, really excellent nouns?*) "We must follow a Middle Path..." (*ah, that mystical triangulation through extremes favored by Buddhist sages and Clinton Democrats—*) I spun around and slapped my monkey mind. *Shut up!* I hissed at it. *Show some respect! You could learn something here!*

But who can blame my monkey mind? None of this is easy. Meg is trying to walk a Middle Path between two psychological extremes, a numb, "quietist" checking-out and a blind, frantic "rushing" into the climate change fight. Neither of these will serve us well, she argues, and neither will bring forth our full humanity.

"Only when we're outside the grip of the push and pull of these feelings can we know 'right action,'" she said, and even my monkey mind, still sulking in the corner, gave a shrug of acceptance. "It's important to stay in the work we're doing," she concluded. "Don't abandon it. Stay in it, don't flee."

None of this was easy medicine, and to process through it, we would often break into twos and threes, or engage in small group exercises, then come back to the full group again to discuss.

One man spoke about his anger at what is happening in the country. He talked about Black Lives Matter and the clarifying potency of righteous anger. Meg nodded and replied: "The energy of anger is crucial to our activism once we are no longer angry." *Koan? Or quip?* I wondered. "Anger," she went on, "puts us on the attack. We're eating the poison and thinking it will kill the rat. It runs hot and deep, but it's superficial—and beneath it is fear. Anger cannot solve anything."

"But what about legitimate anger?" the man asks.

"Feel it, legitimize it, but don't use it as a strategy for action."

My own activist experience has borne this out, but I also know it's wildly difficult advice to follow, especially if you're at the receiving end of the abuse.

Lois also asked a question: "When I open my heart to the chaos of the world, it's not just grief that comes in. Right alongside it, I feel the urgency to act. And soon enough, urgency is driving, and, honestly, I'm not sure that's always a bad thing."

Meg responded with another koan: "The task before us is very urgent, so we must slow down." Which is about the most annoying thing anyone can ever say to you. Especially when you're feeling urgency—and the task is in fact urgent.

Meg continued: "Replicating the patterns—like urgency—that made this mess in the first place aren't going to solve it. As activists we're often reactive. We tend to leap into the next battle with an ungrounded, brittle urgency. When we meditate we learn to slow down, we learn to pay attention. What we need to do is bring that same slow, open-hearted mindfulness into our work in the world."

But how do you resist at the speed of mindfulness when the problem is moving at the speed of capitalism?! *We need to get on Tim DeChristopher's bicycle and start pedaling. Hard! We don't have time to slow down and take another deeeeeeep breath, just to reconsider our lack of options!*

And then I caught myself. *Hold on. You know better. This crisis is not like the other battles. We're not dealing with a "solvable" problem, here. We're in a predicament, a Gordian-knot-level existential conundrum. And maybe slowing down and getting wise to ourselves is exactly what we need to do...*

I let all this churn inside of me, in a probably not very Buddhist way. I looked over at Lois, then around at the faces in the room, wondering what thoughts—Buddhist or otherwise—anyone else might be having. I'd find out soon enough.

On that first day we covered a lot of spiritual ground. We invited grief into the room, tried to ground our anger and even threw some cold water on our urgency. But it was all just prelude, because on day two we were invited to demolish our hopes. "Moving Beyond Hope and Fear" was the title of the workshop, after all.

Meg began by acknowledging the central place of hope in our lives and culture. "The idea that history moves forward, the idea of progress, is one of the building blocks of Western civilization. Part of the bedrock we stand on. But we made it all up. We think our technology will save us, but it actually traps us. Our activism is fueled by similar notions and is also a trap," she said, ticking off the myths we tend to tell ourselves: "'If we just work harder,' we say, 'and learn a few new tricks, we *will* finally be able to create the change we need.' 'We can figure this out,' we say. 'I believe in people,' we say. 'Yes we can,' we say. 'We know what to do, all we need is the political will to do it.' These are the mantras we tell ourselves to keep our heads and hearts in the game. But this attitude comes at a cost." And from there she proceeded to lay out a Buddhist perspective on hope, which went something like this:

As soon as the human mind is focused on goal-oriented grasping, we're taken away from the now and into the goal. Hope shrinks our attention and limits our imagination because it confirms our bias that we already know the right way. Hope even fuels our aggression. It gives us the self-destructive permission to keep pushing our agenda and ramming our heads against the wall. Hope fuels our need for something external to make things right, and sets us to measuring and quantifying our effectiveness. We trap ourselves on a hamster wheel powered equally by our manic hope-driven activity and our fear of that those hopes will be dashed.

"And so," she continued, "for those of us who want to be wholeheartedly engaged, hope is an insufficient source of inspiration and motivation. It is dangerously fragile. Only when we release our hope and fear can our most limitless capabilities be awoken. Only then can we 'actualize our own being.' In fact, our work on climate chaos has the potential—exactly because it is such a near-insurmountable predicament—'hopeless' in the classic sense—to be the vehicle to awaken us. But 'How can one live without hope for change?' we worry. 'Who am I if I give up hope?'"

Having named the anxieties on so many of our minds, she attempted an answer.

"Something different is being asked of us this time. The regular frantic reactions won't solve our situation. We must move beyond hope and fear to something deeper. It's not just resilience, but something deeper still," she said.

Yet it was hard to tell exactly what this deeper thing was; over the course of the rest of the workshop she sometimes called it "patience," sometimes "perseverance," sometimes "faith."

Whereas hope hangs on the future, she said, this other deeper quality resides in the present. Hope is an expectation, and a set-up for disappointment; what we need to do is cultivate a confidence independent of circumstance. She encouraged us to be warriors of lovingkindness, and insisted that beyond our hopes and fears there was a "timeless and primordial confidence" available to us. "It's not hard work to achieve," she promised. "It's just us settling into our full humanity."

But then she told us that this "settling into our full humanity" included "abandoning the idea of 'struggle' entirely." And also somehow accepting that multinational corporations had already crushed any chances we had for a livable future.

Her argument went as follows: Corporations are slaves to the logic of profit, growth, and expansion. Their bylaws are inescapable. The entire thrust of their existence causes harm. So far, this was not a controversial view to most of us in attendance.

As an organizational consultant she'd been hired by a host of Fortune 500 corporations to try to integrate an ecological consciousness and a "second bottom line" approach into their decision making. She'd seen things up close, and it wasn't pretty. She told us a story about one company she'd consulted for, and how they'd hunted down some of the last members of an endangered species and then deep-froze them so that *after* they went extinct—which in this company's analysis, they surely would—the company could still profit. There were groans from the audience.

Then she turned the knife. There is no longer the promise of a better future. Corporations already have the future of the planet in their unbreakable grip, she said. They are going to take us down. It's the inescapable momentum of the lifestyle we all crave. Our task is no longer to try to stop it. That is futile. Our task is simply to recognize the enormity of what is happening right now, and take better care of each other. "We can't give up," she said, "But we must give in." Another koan.

I saw Lois involuntarily shaking her head. I could see her lips and eyes scrunched up with skepticism. And she wasn't the only one.

The current system, Meg continued, is unchangeable. Many factors have landed us here, but now that it has emerged, it cannot be reversed. What we *can* do, however, is change our own expectations and choose a different way to do our work: not on behalf of "saving the world" but for the people around us. The planet is weeping, yes, but ultimately, the planet will be fine. The question is: How will *we* be? Our bravery is to continue to trust in human goodness and be present for people as things get more difficult. We can do local mitigation work but we are ultimately powerless to stop what's happening. The best we can offer each other—at this time of despair, loss, and collapse—is togetherness and compassion.

There was a pause. Then a revolt from the floor.

"What do I tell the young people I work with?" asked one woman, trembling. "That's there's no hope? So they shouldn't bother? We leave them this world, and tell them that?!" From the other side of the room: "Maybe that is just the path *you* needed to take, and it is not for us." "What about magic?" asked a woman wearing three jade necklaces. "What about all that can't be seen?"

From the back a nervous middle-aged fellow piped up: "We can create all the local alternatives we want, but if we don't blunt the ravages of the fossil fuel industry, what world will there be for these alternatives to exist in?"

Meg didn't interrupt or contradict these speakers. Up at the dais her face remained impassive. It was hard to read her thoughts. Was this all part of the teaching—an expected, even necessary, reaction that she'd encountered before? Or was the vehemence of the response catching her by surprise? She did seem a little pained, a little weary. More people spoke up, challenging Meg, maybe seven in all.

People's reactions were understandable. Meg had basically told us that she'd given up on changing the world and so should we. She'd just told a hundred people who were primed to fight for a better world, but who were looking to her for wisdom on how to stay whole and grounded *while* they fought, that there was no reason to fight.

She was counseling strategic surrender. Corporate America was locked on course, she'd said, and there was nothing we could do to change it. I have been to the mountain top (aka the boardroom), she was saying, and there is no promised land. Those who are wrecking the world won't change, she was saying, so the only thing we can do is take care of each other as it all gets inevitably worse. *Give in, but don't give up.*

Was this more hard medicine? Was she laying down some deep mystic wisdom that I just couldn't quite wrap my head and heart around yet because I was too attached to the fight? Or was it toxic advice?

I stood up and asked for the mic. I didn't know where I was going at first, but as I spoke I slowly found my footing. I commended her for her bravery in putting things out there so starkly, and for trying to walk a warrior path through our dark times. Yes, I think things will be getting worse. Yes, our number could be up. But we don't know that for sure. *You* don't know that for sure. Everything is impermanent, as the Buddha says. That includes capitalism: It didn't exist 400 years ago, and one day it will cease to exist, just like everything else in the Universe. Sure, it may take us down along with it before then, but we don't know that for sure. Yes, we must move away from a manic, fragile, optimism-driven hope, but that doesn't mean we withdraw from the field. That doesn't mean we relegate ourselves to a kind of social work for climate victims.

No. We've still got to fight ecocide at its source. We still have to take on the problem at the highest level. Because even if we ultimately lose—how we lose, how badly we lose, and how quickly we lose, matters. Even if we're just slowing down the speed at which things get worse. Even if all we get for our efforts is a bad catastrophe instead a worse one. It all matters. So, yes, maybe we should proceed like spiritual warriors, without hope, and knowing that we will lose—and thank you for your teaching—but even so, maybe we still have to fight *as if* we could win, which probably means fighting *as if* there was hope, even if, um, there isn't, because that's, um, the best way to, um…I trailed off.

I was sputtering a bit by the end, but I'd said my piece as best I could. I was a little surprised at myself: I'd been convening "hopelessness workshops" in my own community, but here I was standing up for hope in someone else's. Meg acknowledged the emotions in the room, and thanked us all for speaking up. As we broke for the day, it occurred to me that my gallows-humor notion of a "better catastrophe" had just made an unplanned—and not particularly funny—public debut.

Participants reconvened the next morning. Meg noted how the previous day had been a little rough for some folks, and acknowledged that she'd been intentionally knocking out our props. More than a few folks nodded in recognition. After some breakout-group activities and a debrief, she closed the workshop with a blessing, both beautiful and ominous. "If you feel you're falling into the abyss," she said, "welcome to reality. If you feel yourself falling without a parachute, the consolation is that there is no ground. But fear not because there is a bottomless source—a vast state of being unhooked from any attachment to outcome—which can keep us going even as we fall."

Afterwards, Lois and I headed to a nearby park. We mulled over the days' teachings.

"Here is a person," Lois said, "telling me to slow down and not act out of urgency. To be compassionate, to face the dirge music with an open heart, to care for those around me..."

"...to not try to solve our problems with the same mindsets that made them..." I chimed in.

"Exactly."

"But...?" I asked her.

"But it's making me realize I am more addicted to hope than I care to fess up to."

"Ah."

"This is a fragile moment in human history," Lois continued. "We're in a giant hinge-time as the unraveling occurs. How much do we fight, and how much do we accept it?" And as she launched into all her heartbreak and confusion and resistance, I remembered why Lois is one of my dearest friends. "Are we being called to warriorship?" she asked. "Or just to do hospice work in the shadow of corporate rape and pillage?"

Meg's resignation dovetailed with all the worst things both of us had been reading about our climate prospects. "I don't disagree with the facts, or the analysis," Lois said. "I just refuse to accept that our doom is so sealed." She wiped away a tear. "They're basically saying that the beautiful human qualities that I believe in—our essential power to do the right and good thing—can't hold a candle to the degree of fuckedness we're in. I feel like they're beating me with it, that their misery demands my company, and they won't be satisfied till they get it."

I was glad to be with a friend who shared my conflicted response to Buddhist no-hope. Neither of us were ready yet to become hospice workers—not yet, maybe not ever. Lois withdrew from the struggle once, holing up in her farm in Washington State, and it was a dark period for her. With all the frustrations and challenges that come with politics, she still feels better being in the fight, paddling her kayak into the path of oil tankers, blocking the gates of refineries, sounding the alarm and gumming up the works.

She admitted she might just be throwing her life at the wall, hoping something sticks, but said she's chosen "to do my best for what matters most." Why? I asked her, though I already know the answer. "Simple," she said. "It's a better life." It aligns her ethically, it helps her feel not quite as guilty as she otherwise would.

"Yes, it's goal-oriented work," she said, feeling that strange need one often feels after a long weekend of Buddhist instruction to justify that you even have goals. "I want to save what we can. I can't just mop up behind the conflagration. Maybe when I must, I'll find that pool of complete, detached-from-results love, and become a service wench. But until then, I don't want to give up on us pulling a rabbit out of the hat in the last three seconds. I want to be a good Buddhist, but maybe not thaaaat good."

In the following weeks, we continued this conversation online. In one email, Lois wondered how to take action and offer hospice at the same time. In another, she wondered whether it's even helpful to ponder how impossible our task is. In a third, as if answering that, she quoted Andrea Dworkin: "I found it was better to fight, always, no matter what."

Can I get my Buddhism
with a side of strategy, please?

If you don't have a strategy, you're part of someone else's strategy.
— Alvin Toffler

"Give in, but don't give up," Meg Wheatley had counseled us. But what exactly did she mean? If she means: Wave the white flag and let Exterminator Capitalism have its way with the planet, then count me out. If, however, her koan-like teaching is more akin to a Buddhist restatement of the Serenity Prayer, a middle path between radical acceptance and right action, then color me a convert.

"There's no hope," she said over and over. If by this she means things are going to get worse, even catastrophically worse, alas, I agree. But does this mean there's no hope? Only if you equate hope with optimism and the expectation of a good outcome. If, instead, we take, say, Tim DeChristopher's deeper definition—"hope is the will to hold on to our values in the face of difficulty"—then, so long as we can keep rising to that spiritual challenge, there will always be hope.

How much of this disconnect is just semantics? While both Meg and Tim seem to agree that an outcome-dependent hope is not going to serve us well, Meg calls this weak sauce "hope" whereas Tim calls it "optimism." Likewise, they both agree we need a more robust kind of soul-force to face our difficult circumstances, however this is what Tim calls "hope" whereas Meg refers to it via her quartet of "faith," "patience," "perseverance," and "confidence."

If giving up hope means accepting that we won't be achieving our grandest utopian dreams, and the world will always be rife with injustice and violence and abuses of power; if it means the world will always be imperfect, just as we ourselves are imperfect, well, sure, fine. But if it means: There's nothing we can do that'll make any difference; nothing we can do that's strategic or programmatic or at the scale of the problem; nothing we can do other than patch up the victims of climate chaos as the Machine rolls over us—then I'd rather keep hoping and fighting.

It's one thing to counsel not being attached to outcomes; it's another to say that outcomes don't matter, or that we have absolutely zero power over which outcomes occur. Because outcomes *do* matter, and we have at least a marginal influence over those outcomes. And in our histori-cal moment, a marginal difference in outcomes could be the difference between, say, extinction and survival—in other words, all the difference in the world.

So, can I get my Buddhism with a side of strategy, please? I may not be smart enough to "be here now" at the exact same time that I'm mak-ing plans for the future, but I'm at least stupid enough to try. I figure, as long as I don't get too attached to results, it's fine to be results-oriented. Because, well, we absolutely must be.

Buddhism 101 tells us that "attachment" is the source of all suffering, but what is often translated from the original Pali into English as "attach-ment," is better understood as "acquisition" or "grasping" or "craving."[4] Could it be that, while craving is problematic and leads to suffering, some degree of attachment is natural and necessary? It's going to have to be because I am quite attached to what is beautiful about the world, and I want as much of it as possible to survive into the future.

I can try to not fixate on outcomes, and I can try to not hang my ego and expectations on one particular future coming true, but I'm sure as hell going to fight for the future(s) I want. And if you're going to fight, you need a strategy, and strategy requires planning, and it's hard to plan without some kind of map.

Tim had talked about the need to think "beyond the cliff" (because otherwise our fears and doubts and all the bad climate news get knotted up in one big undifferentiated "we're fucked"). The blow-up at Meg's workshop made clear the need for better tools to help us think and feel and plan beyond this cliff.

A map with scenarios and some specifics would help. Not sea-level-rise and carbon-parts-per-million specifics, but the existential specifics of what choices our hearts must make, what world we might get, and what we can do to get the least terrible outcome. Less what's going to happen, than how to be human regardless of what happens.

Could you map all that out in one place? Could you squeeze the com-plexity of our predicament into a single diagram and (end)game it all out? Told to abandon Western notions of hope, fear, attachment, and the illusion of control, I made a big flowchart.

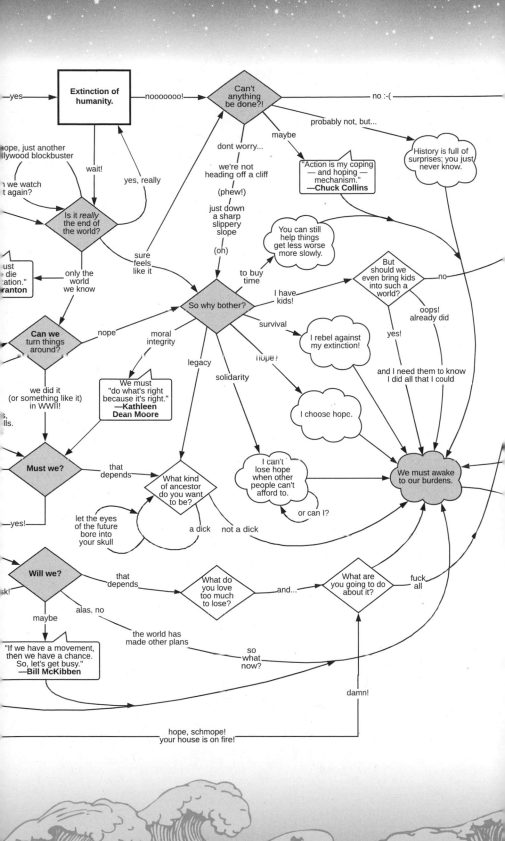

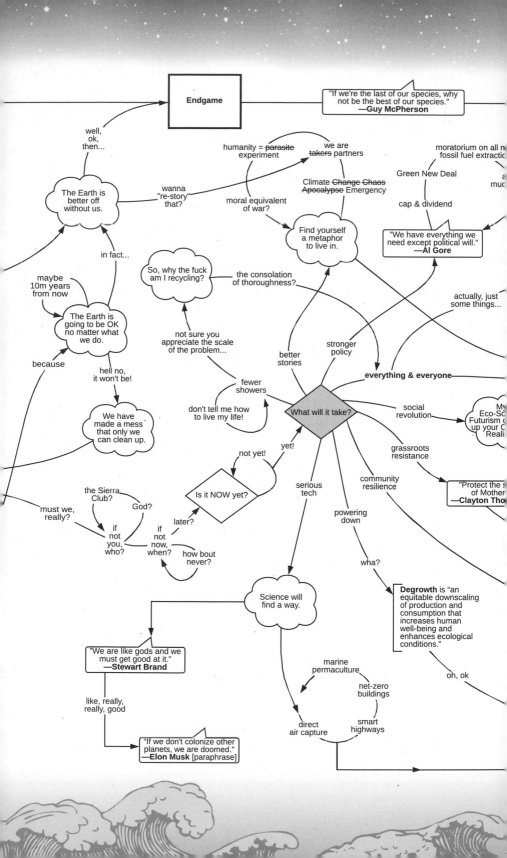

EXISTENTIAL CRISIS SCENARIO PLANNING

Mann Tracht, Un Gott Lacht.
(Man Plans, and God Laughs.)

— old Yiddish proverb

At some think tank somewhere, guys with pocket protectors and knotted brows are asking their computer models questions like: "If global carbon emissions peak by 2030 and sea-level rise is 1.6 meters by 2050, how many people along the Eastern Seaboard are likely to drown in storm surges, and what will the effect be on national GDP to relocate the rest?" Unfortunately, such crisis scenario planning is increasingly necessary. But no matter how accurately it captures some of the gruesome realities before us, this big-picture quantitative approach to gaming out our future will only ever do half the job we need. Because the climate crisis is also a personal existential crisis for each one of us.

Where are the tools to help us regular Joes and Janes game out our next 50 years? Because the questions most of us are asking ourselves are quite different from the ones the think tanks are asking their models. We want to know: *Can I handle the truth? Is there any hope? To whom am I responsible? How bad is it going to be? What is still worth doing? And how can I survive whatever is to come with my humanity intact?* What we need is something more akin to *existential* crisis scenario planning.

I'm not a scientist or policy wonk or game theorist, so I don't have much to contribute to standard scenario planning. But as a long-time climate activist, angst-ridden student of social collapse, and a generalist with a warped sense of humor willing to go there, I'll take a crack at the more existential kind.

It's not the end of *the* world;
it's only the end of *our* world.

People are always calling me a pessimist but if I write a lot
about the fall of civilization, it's because I want a better one.

— Robinson Jeffers

In our nightmare imagination we toss around "the end of the world" like a hand grenade, but what do we mean by "end," and what "world" is it exactly that we think is ending? Are we talking about the literal, physical extinction of humanity? Or are we using "end of the world" as a vague shorthand for civilizational collapse? Or maybe we have a specific empirical threshold—say, the tipping point of runaway climate change—in mind? Or simply that there'll be nothing left that we recognize or feel at home in?

Understandably unwilling to dig in and distinguish between these different flavors of absolutely fucking terrible, we tend to conflate all these scenarios together, and just throw our arms up in despair. Or numb ourselves into passivity. *Oh, well, shucks. I guess that's it then. I'll just putter away here in the garden till it's time to go.* We imagine a single cataclysmic black wall of doom that somehow unmakes the world. There's nothing to be done, no choices to be made; the lights simply go out.

But that's not it at all. Because, like the "I'm only partly dead" peasant in *Monty Python's Holy Grail*, it's probably not the end of *the* world, it's only the end of a *particular* world. Or as Paul Kingsnorth and Dougald Hine put in their 2007 *Uncivilization* Manifesto, "The end of the world as we know it is not the end of the world full stop."[1]

It is therefore essential to identify what world we think is ending, how much agency we think we have, and what our options are. If we don't do that, it's hard to make plans. And if you're not planning your own apocalypse, you're probably gonna end up in someone else's.

Once we accept that we are on track for some kind of catastrophe, the vital question is "How bad?" Because the scenarios are *differently* bad. And which one (or which mix of them) you think we're on track for matters. And, because, well, here we are. And we're not going to stop trying to be here—or making plans to keep being here—just because things are bad and getting worse. In fact, the worse we expect things to get, the more important it probably is to make plans.

So, what kind of worse do we think we're in for? :

1. a slightly rockier version of *business-as-usual* ("bad-ish");
2. a combustible mix of unravelling, upheaval, and revolutionary *transition* ("pretty bad");
3. a horrendous civilizational *collapse* with billions dead, but some humans will make it through over the long haul ("really bad"); or
4. human *extinction* within the next few generations ("really, really bad").

First, let's take *business-as-usual* off the table right now, as nothing is going to be "as usual" in the future. (In fact, things are not even as usual in the present anymore.) Next, let's turn our attention to the other outlier scenario, *extinction*.

Guy McPherson is not the only prognosticator armed with reams of analytical data and hockey stick-shaped graphs who claims we are literally on the path to extinction. More mainstream analysts, among them James Lovelock and James Hanson, have also acknowledged that such an outcome is at least theoretically possible.

What if they're proven right? Well, there's good news and bad news. The bad news: this bad news is as bad as it gets. The good news: we as a society, and each of us as individuals, still have a lot of very, very bad options.

We could throw in the towel early and just "party like it's 2099!" We could descend into chaos and tribalism, and turn on our neighbors. Or we could try to follow the counsel of eco-Buddhist Joanna Macy: "If we are going to go out," she says, let's "do it with some nobility, generosity and beauty."[2]

When I try to place myself at the human Endgame, when I truly step into the chilling eeriness of that possibility, a host of voices tug at me. Sometimes the artist in me steps forward and I imagine myself creating a sad, beautiful epitaph to the species. An odd gesture, I suppose, and at some level futile, but nonetheless an attempt to snatch a final meaning from the darkness.

Or, I might set my sights on simply trying to be as present and as kind as possible. Here the idea of Hospice Earth—that at the end of the world we're all in hospice together, both to and for each other—becomes strangely comforting. Like the dinner and prayer scene at the end of *Don't Look Up*. These perspectives give me a little more courage, a little more compassion, as I imagine trying to face the End, however it comes.

But I soon rebel at the whole exercise. Why plumb this darkness? Why rehearse an Endgame that isn't yet upon us?

"History is full of surprises," said an attendee at one of my hopelessness workshops. "You just never know." His voice, and a host of others like it, pull me out of my Endgame nightmares. They carry me back to life, hope and humanity's ongoing story. Because extinction is only one of our possible futures and, according to most prognosticators,[3] not the most likely one. With luck, we still have a chance at a really bad future, and even—fingers crossed!—just a pretty bad one.

Which leaves Collapse (really bad), or Transition (pretty bad—and then later maybe not quite as bad). Most likely, we're going to get a tumultuous combination of the two. So, how do we prepare ourselves? How do we navigate this future? How do we make plans as the world we know unravels?

To help us find our bearings, we're going to look at several frameworks, including Richard Heinberg's "Four Ways to Decline," David Fleming's *Surviving the Future*, Peter Frase's *Four Futures*, Dmitry Orlov's *Five Stages of Collapse*, John Michael Greer's *Long Descent*, as well as the broad notion that climate change is an unsolvable "super-wicked problem."

But before we can follow any of these maps, we must first learn how to die.

We have to learn how to die as a civilization.

In a 2013 *New York Times* Op-Ed piece, "Learning to Die in the Anthropocene,"[4] and his subsequent 2016 manifesto of the same name, Roy Scranton—Iraq War vet turned professor of writing and environmental humanities at Notre Dame—diagnosed our civilizational crisis through the lens of individual mortality:

> What does human existence mean against 100,000 years of climate change? What does one life mean in the face of species death or the collapse of global civilization? How do we make meaningful choices in the shadow of our inevitable end?

While acknowledging that "these questions have no logical or empirical answers," he notes, "They are philosophical problems par excellence." If it is true, "that studying philosophy is learning how to die," he argues, "then we have entered humanity's most philosophical age.... The rub is that now we have to learn how to die not as individuals, but as a civilization."

Scranton experienced his tour of duty in Iraq and the subsequent military policing of the streets of New Orleans during Hurricane Katrina as prefigurations of civilizational collapse. He knows a thing or two about dying, and acknowledges that "learning how to die isn't easy."

In Iraq, at the beginning of his tour, the everyday possibility of death paralyzed him. Scranton found his footing with the help of an 18th-century Samurai manual, which commanded: "Meditation on inevitable death should be performed daily."[5]

Instead of fearing his end, he owned it. Every morning, he would imagine the many different ways—"blown up by an I.E.D., shot by a sniper..."—that he could die. Then, before heading out that day, he told himself he didn't need to worry because he was already dead. "The only thing that mattered was that I did my best to make sure everyone else came back alive."

Scranton got through his tour in Iraq one day at a time, meditating each morning on his inevitable end. Our doomed civilization, he argues, must do something similar:

> The human psyche naturally rebels against the idea of its end. Likewise, civilizations have throughout history marched blindly

toward disaster, because humans are wired to believe that tomorrow will be much like today.… Yet the reality of global climate change is going to keep intruding on our fantasies of perpetual growth, permanent innovation and endless energy, just as the reality of mortality shocks our casual faith in permanence.

The biggest problem climate change poses…won't be addressed by buying a Prius, signing a treaty, or turning off the air-conditioning. The biggest problem we face is a philosophical one: understanding that this civilization is already dead. The sooner we confront this problem, and the sooner we realize there's nothing we can do to save ourselves, the sooner we can get down to the hard work of adapting, with mortal humility, to our new reality.[6]

I read—and re-read—Scranton's op-ed the year it was published. It forced my first deep reckoning with the reality that we were not going to be able to "prevent" climate catastrophe, or "undoom" our civilization. It's taken me years to accept it. In fact, I'm still trying to, because accepting it requires me to murder my hopes, or at least those hopes that depend on a good-ish outcome.

We saw with Tim DeChristopher how this kind of realization, after a period of mourning, could eventually be powerfully liberating.

Environmentalist Derrick Jensen went through a similar experience. "When you give up on hope, you turn away from fear," he writes. "You quit relying on hope, and instead begin to protect the people, things, and places you love."[7] For Jensen, losing hope, accepting that we are doomed (or at least doomed-ish), didn't demobilize him; it made him more active, more responsible, more courageous, more free.

"It's as if giving up on saving the world," writes eco-visionary Charles Eisenstein, "opens us up to doing the things that will save the world."[8]

For Scranton, "The choice is a clear one": We can get "more and more desperately invested in a life we can't sustain. Or we can learn to see each day as the death of what came before, freeing ourselves to deal with whatever problems the present offers without attachment or fear."[9]

Which begs the question: Exactly how? What's the next move?

If we can't prevent climate catastrophe and we can't ultimately stop the doom-bound trajectory of our civilization, then how do we move forward? Or at least downward, in as wise and just a manner possible?

Many people, it turns out, have already begun to game this out.

How do you want to decline?

In the dark times
Will there also be singing?
Yes, there will also be singing.
About the dark times.

— Bertolt Brecht

Scranton counsels us to accept that our civilization is already doomed. For him, this is the only honest and responsible attitude. And until we're able to do that, we'll remain stuck in denial and the frenzy of false solutions.

If Scranton is right, then the basic motivation of activists like me—to "make things better"—would also need to change. But what would our new role be: Save what we can? Help things get less worse more slowly?

Most folks—especially us Americans—are not used to thinking this way. We're good at progress, bad at decline. But in our era of civilizational unraveling, it seems we'll need to get good at decline.

How do we do that?

Richard Heinberg and his colleagues at the Post-Carbon Institute have outlined Four Ways to Decline:

1. Waiting for a Magic Elixir: wishful thinking, false hopes, and denial
2. Last One Standing: the path of competition for remaining resources
3. Powerdown: the path of cooperation, conservation, and sharing
4. Building Lifeboats: the path of community solidarity and preservation[10]

Here we have four choices: We can sit on the sofa and, paraphrasing Rebecca Solnit, clutch our hope like a lottery ticket (Path 1). Or we can duke it out with resource wars and Walls and enclaves and gated communities and sacrifice zones, further endangering the web of life, wasting the little time we have left to act, and reducing our overall collective chances of survival (Path 2).

Or, if dog-eat-dog terminator capitalism is not your cup of tea, we can advance policies to partially power down our richest economies, and better share what we have left (Path 3). Compared to the first two, this third path sounds like a no-brainer, right? Just try getting it through a Congress that hasn't yet accepted that our civilization is doomed.

We can also foster greater resilience and cooperation at the community level (Path 4). Millions are already doing this. Witness the worldwide Transition Movement; the hundred-plus initiatives in the New Economy Coalition here in the US; Zapatista base communities in Chiapas in Mexico; the decades-long anarchist co-op movement in Catalonia, Spain; or the Kurdish, eco-feminist inspired, stateless state of Rojava, in northern Syria, just to take a few examples.

In the worst case scenario, Paths 1 and 2 conspire to sentence us to something like the hellscape of *The Road*. In the best-case scenario, Paths 3 and 4 work in synergy to head off the worst and prepare us to survive the now thankfully not-quite-so-worst.

"Our strategy must change," Heinberg says, "from crisis prevention to crisis management."[11]

This best-case scenario of PowerDown plus Lifeboats is, broadly speaking, the path of Descent. In his magisterial *Lean Logic: A Dictionary of the Future and How to Survive It*, British economist and anthropologist David Fleming describes it as the only "path that has both reality and hope."[12]

Like Heinberg, he views climate breakdown as one element in a matrix of breakdowns that will inevitably cause a systemic failure of civilization as we know it. "Our economy and society depend on a lot of things working right, all at the same time," he says, noting that many of these things—from cheap, and reliable flows of energy to a stable climate to a cohesive culture—"are all in trouble."[13]

He identifies four broad possible responses to this situation: Growth, Continuity, Descent, and Collapse. Noting the inevitable double-bind that a market economy only works when it's growing but "growth destroys the foundation on which it relies," he rejects the possibility that large-scale reforms to the economy could achieve Continuity. He sees zero evidence that growth and ecological repair can be twinned, and any attempt to continue in a "no-growth" mode, would not only "do nothing to reduce the accumulated damage, which is already more than the Earth can bear," but "the lack of growth would break the economy." For all these reasons, he sees Descent as inevitable. Distinguishing it from the more sudden, involuntary, and unmanaged Collapse, he describes it thusly:

...a steep winding-down of the size of the industrial economy. It strips away its burdens and complications, nurses the human ecology back to health, builds local competence and discovers a sense of place.... This is managed descent.... The shock is as gentle and as survivable as foresight can make it.[14]

Fleming's survival strategy rests on local communities, which, "as the industrial economy descends," will have to provide many basic necessities, such as food and water for themselves, or do without. "They will need to rediscover their locality and local skills, rebuild a culture, and apply the power of lean thinking."[15]

Fleming notes that, "When long-established systems break down they often do so in many different ways at the same time." He acknowledges that these converging shocks "will leave nothing in our lives unchanged," but believes this inevitable descent "can be managed, mitigated, made survivable," and even "recognised as our species' toughest, but greatest, opportunity."[16]

Civilization as we've known it is doomed. What comes next depends on how we respond. We can think of our task as "learning to die as a civilization" or "existential crisis management" or "helping things get less worse more slowly" or "our species' toughest, but greatest, opportunity." It's not what any of us imagined we were signing up for (myself included), but here we are. We have our work cut out for us.

What kind of future social system might we get if we decline poorly? Or if we decline well? Or if we pull off a Green-New-Deal-level transformation that lets our civilization "continue" in some deeply reformed way?

Economist Peter Frase lays out four possible futures in his appropriately titled 2016 book *Four Futures*.

Frase argues that we are facing a paradoxical crisis of both scarcity (ecological catastrophe) and abundance (AI & automation) at the very same time. Under these twin pressures, he argues, capitalism is going to end. The question is what will replace it. On intersecting axes of scarcity/abundance and hierarchy/equality, he maps out a 2 × 2 grid of our possible futures:

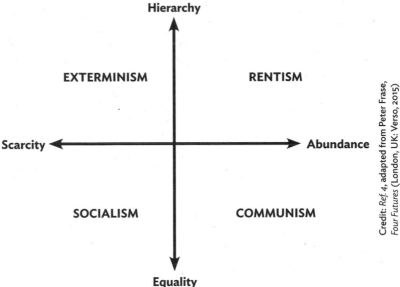

Credit: Ref. 4, adapted from Peter Frase, *Four Futures* (London, UK: Verso, 2015)

According to Frase, if we get clobbered by climate change and fail to harness the power of automation for the common good, we're going to end up with "exterminism." A society that's the worst of all worlds: where a redundant (and highly policed) mass of permanently unemployed poor are forced to live on a planet whose systems of life support are severely degraded, while the benefits of automated production go to an elite few. Think Matt Damon in *Elysium*.

If, however, we manage to stave off the worst of climate change but still fail to evolve to a more equal society, we end up with what Frase calls "rentism," a system in which the economic elite tries to maintain their power and wealth in a context of large-scale automation. Here, a new class of "rentier-capitalists," by owning the robots of the future—and especially by controlling the code, patents, DNA, 3D-printing patterns, and other intellectual property of the future—and charging at every point of transaction for the right to use them—could turn the potential of universal abundance into artificial scarcity. Imagine the *Fifteen Million Merits* episode of *Black Mirror*.

But what if, partly by harnessing the immense potential of automation, we manage to build a more equal society, yet still get clobbered by climate change? What then? We end up with "socialism." Here, as we adapt to a changed planet and dig our way out of our ecological mess, at least we're in it together, sharing the burdens as well as the joys. Think of almost any story out of solar punk sci fi, or Cuba in the 90s, adapting to the fall of the Soviet Union by, among other things, shifting most of its agricultural system off of fossil-fuel based fertilizers to organic cultivation.[17]

Finally, if we manage to both stave off ecological catastrophe and harness automation and AI to benefit society as a whole, we have the best of all worlds. A society of both abundance and equality. Frase calls this "communism." Think here not of the grim scarcity and gray concrete of the old East Bloc, but the "from each according to their ability; to each according to their needs" of Marx's ideal. And if that (even with a small "c") still makes you uncomfortable, just think *Star Trek* minus the Klingons.

And—minus the Klingons and some very stilted ("He's dead, Jim") dialog—who wouldn't want to live in a society that has achieved the abundance and egalitarianism of *Star Trek*? Unfortunately, it's pretty clear that's not the future we're going to be getting anytime soon. With wide-scale ecological disruption already baked into the atmosphere, we're headed for one of the two scarcity scenarios—either *exterminism* or *socialism*.

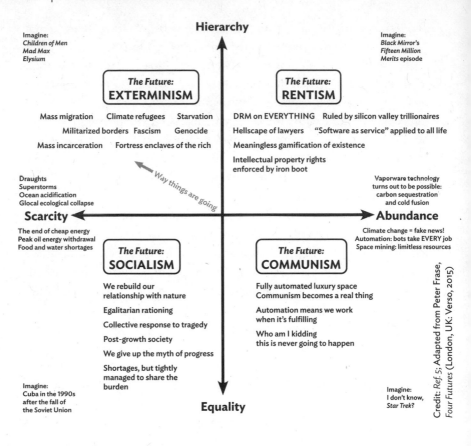

Hierarchy

Imagine:
Children of Men
Mad Max
Elysium

Imagine:
Black Mirror's
Fifteen Million
Merits episode

The Future:
EXTERMINISM

Mass migration   Climate refugees   Starvation

Militarized borders Fascism   Genocide

Mass incarceration   Fortress enclaves of the rich

The Future:
RENTISM

DRM on EVERYTHING   Ruled by silicon valley trillionaires

Hellscape of lawyers   "Software as service" applied to all life

Meaningless gamification of existence

Intellectual property rights
enforced by iron boot

Draughts
Superstorms
Ocean acidification
Glocal ecological collapse

Way things are going

Vaporware technology
turns out to be possible:
carbon sequestration
and cold fusion

Scarcity ◄─────────────────────────────► **Abundance**

The end of cheap energy
Peak oil energy withdrawal
Food and water shortages

Climate change = fake news!
Automation: bots take EVERY job
Space mining: limitless resources

The Future:
SOCIALISM

We rebuild our
relationship with nature

Egalitarian rationing

Collective response to tragedy

Post-growth society

We give up the myth of progress

Shortages, but tightly
managed to share the
burden

The Future:
COMMUNISM

Fully automated luxury space
Communism becomes a real thing

Automation means we work
when it's fulfilling

Who am I kidding
this is never going to happen

Imagine:
Cuba in the 1990s
after the fall of
the Soviet Union

Imagine:
I don't know,
Star Trek?

Equality

Credit: *Ref. 5*; Adapted from Peter Frase,
Four Futures (London, UK: Verso, 2015)

Given those choices, you'd think even the most die-hard capitalist would choose socialism. However, you'd be wrong. We don't just get a future, we have to make our future. And, for most of us to not die hard too, it'll take something like the Green New Deal or better, backed by powerful social movements and serious political muscle, along with a culture-wide ecological awakening and the flourishing of resilient communities described by Heinberg and Fleming, in order to bend the needle towards a livable future.

The beginnings of ecological catastrophe are already upon us, but we still have a choice. Between the truly horrific catastrophe of an eco-cidal and genocidal *exterminism*, and the not-so-bad (in fact, in many ways actually good) catastrophe of a post-carbon *socialism* of shared burdens and ecological restoration, let's set our sights on the not-so-bad catastrophe.

Don't worry, we're not heading off a cliff, just down a sharp slippery slope.

You can find helpful guides for our challenging times in the strangest of places.

One of my favorite books when I was a kid was *Fortunately* by Remy Charlip. It was the story of a little boy who would stumble into danger and difficulty and then get out of it, and then get into it again. It was a big format picture book, and each pair of facing pages went something like this: Left side: "Unfortunately...he fell in a big, big hole." Right side: "Fortunately...there was a tunnel he could walk through." Turn page: "Unfortunately...there was a tiger in the tunnel. Fortunately...there was a passageway he could squeeze through too small for the tiger to follow. Unfortunately..." And on it went.

Poised on the cusp of climate catastrophe, this basic plot structure seems to fit our collective predicament quite well: Unfortunately...we're going down. Collapse is coming. We're heading for the cliff at full speed. Extinction seems quite possible. Fortunately...it's not a cliff. Unfortunately...it's cliff-ish—and steep, and sharp, and slippery. Fortunately... there's a bit of level ground in between the various jagged drop-offs. Unfortunately...the momentum of our crash is likely to carry us off these ledges each time. Fortunately...we will eventually hit bottom and we won't all be dead. Unfortunately...and so it goes.

This is not just a cute metaphor. The stages and mechanisms of collapse have been mapped out in compelling detail by some of our most prescient analysts. In *The Five Stages of Collapse*, Dmitry Orlov lays out five increasingly dire stages of collapse—financial, commercial, political, social, and cultural. In the first stage, central governments have used up all their magic bullets and currencies implode. This triggers commercial collapse: commodities are hoarded, import and retail chains break down, and widespread shortages of survival necessities become the norm. As official attempts to mitigate the situation fail, the political establishment loses legitimacy and relevance. People lose faith that the government will take care of them. Local social institutions rush in to fill the power vacuum, but as they run out of resources or fail through internal conflict, people lose faith that their own communities can take care of them. Finally, we reach stage five, cultural collapse, where faith in the goodness of humanity itself is lost. People lose their capacity for kindness, generosity, and reciprocity. Even families disband and compete as individuals for scarce resources.

So, how is this good news? Fortunately…it's not one stage. That would be immediately terminal. Poof! Done. Game over. We'd be all collapsed before we even got started collapsing. Instead, we get five stages. Five big chances to try to figure our shit out. Because there's things we can do at each stage to survive, stabilize, and brace ourselves for the next descent. And if we miraculously manage to do enough things right at any one stage, we're not necessarily fated to descend through all five. The Soviet Union, for example, whose late-20th-century collapse informed Orlov's model, went through partial versions of the first three stages, then stabilized and bounced back (though in a fairly ugly way).

So: Unfortunately…we're going to collapse. But: Fortunately…it's not all at once. This is an incredible boon. It's like finding a magic fossil-fueled lamp, rubbing it, and having an all-powerful genie come out, and instead of ripping you to shreds on the spot (I mean he is all-powerful and you, in your fossil-fueled lust and greed, did rouse him from his cosy little lamp-home to ravage the Earth) he grants you five wishes. It just happens that all five wishes are going to hurt you really really really bad.

In *The Long Descent: A User's Guide to the End of the Industrial Age*, John Michael Greer argues that climate catastrophe and resource depletion will lead to "staircase descent" over the next 100+ years.[18] There will be 25-year periods of extreme difficulty, population contraction, and public health deterioration—but there will also be plateaus, as society responds and alters its behavior. Middle-aged folks like me won't see the worst of it. Millennials and their children will, while the next generation after that will experience it like a "new normal." In other words, we adapt, change our expectations, readjust to conditions, and keep on. The ancient Mayan who had to walk out of her drought-stricken city in the Guatemalan highlands suffered, but her children had a different life experience. Now, that might look like doom to most people, but to me, well, it also looks like doom. But not *total* doom.

So, yes, there's a whole lot that's horribly unfortunate about our situation. But we're incredibly fortunate, too. So, let's try to say it with a smile: we're not heading off a cliff, just down a sharp slippery slope that will wreck our world.

Houston, we have a super-wicked problem.

Whenever I run into a problem I can't solve, I always make it bigger... I can never solve it by trying to make it smaller, but if I make it big enough, I can begin to see the outlines of a solution.

— Dwight. D. Eisenhower

Roy Scranton's 2013 pronouncement of inevitable civilizational doom notwithstanding, the decades-long debate over how to "fix" the problem of climate change continues. But is climate change—even in the best case scenario—a problem we can ever really fix? The answer depends on what kind of problem you think climate change is. Wait, what do you mean, what *kind* of problem?

Well, for example, Rex Tillerson, CEO of Exxon from 2006 to 2016, and Donald Trump's long-suffering-for-a-short-while Secretary of State, believes climate change is just "an engineering problem," and as such, "it has engineering solutions."[19] Rex Tillerson is an engineer by training, and he's half-right.

Climate change is partly an engineering problem, and there are helpful ways to address the problem at the engineering level (more efficient solar, electric-powered passenger jets, carbon capture technology, etc.). But climate change is not *just* an engineering problem. As we've seen, it's also a cultural, psychological, political, social, economic, spiritual, philosophical, and indeed *civilizational* problem. It can't be "fixed" in any simple, straightforward engineering-y way.

Philosophers like Timothy Morton call it a "hyper-object";[20] policy analysts, a "super-wicked problem." Climate change has earned a reputation as "the most complicated problem humanity has ever had to face"[21] for a reason. It's non-linear; it's laced with uncertainties, complex interdependencies, and unintended consequences; it's also a moving target.

The constraints that the problem is subject to and the resources needed to solve it change over time. The effort to solve one aspect of the problem may reveal or even create other problems. And even the best-intentioned attempts to address the problem are dicey experiments done in real time at the volatile intersection of industrial civilization and our planet's ecosystem, two of the most complex systems we know.

Also, the various stakeholders (all of us) have wildly divergent points of view, with little agreement on the solutions, the nature of the problem, or whether there even is a problem. Just think how unable we are, in our pluralistic society, to arrive at an indisputable notion of the public good, and multiply that by denial, existential terror, and melting Arctic ice.

For all these reasons and more, climate change is a problem that doesn't have a "solution" in any normal sense of the term. And, as John Michael Greer already taught us, when you have a problem without a solution, instead of having a problem, you're in a predicament;

Not wanting to be outmaneuvered by such a simple, understandable word as "predicament," academics who study social complexity have a special term for this kind of problem. They call it a "wicked problem." Yep, even the ones that aren't from Boston call it that. What makes a problem wicked? The following ten characteristics:[22]

1. There is no definitive formulation of a wicked problem.
   Oh, great, you're probably thinking. We can't even define the problem. Right out of the gate, and we're already lost. To be fair, though, what they mean here is that there's no "scientific" basis for describing the problem. Climate change is simply too complex, fickle, and subjective to lock down and definitively describe.

2. Wicked problems have no stopping rule.
   "No stopping rule" is management-speak for a problem that never ends. So, it seems that climate change—whatever it is, since we've just been told we can't properly define it—is nonetheless going to be with us for a long, long time.

3. Solutions to wicked problems are not true-or-false, but better or worse.
   Climate change has no right answer, no agreed-upon optimal solution. (Should we go with a cap and trade system that lets the market decide whether poor communities drown? What about nuclear power—it's zero-carbon [yay!] but extremely dangerous, expensive, and central-ized [boo!]. Or how about an immediate moratorium on all new fossil fuel infrastructure? The scientists tell us it's necessary, but it'll likely wreck any political coalition that attempts to implement it. You choose.) There's no right answer here, just a range of suboptimal solutions, and a whole lot of trade-offs, some worse than others.

4. There is no immediate and no ultimate test of a solution to a wicked problem.

   Before you throw up your hands in despair, this doesn't mean there's no way to compare the relative merits of competing potential solutions, just that there's no (good) way to do so before you've actually tried them out—and then, no way to be absolutely sure that they're working the way you think they're working. OK, now you can throw your hands up in despair.

5. Every solution to a wicked problem is a "one-shot operation"; because there is no opportunity to learn by trial and error, every attempt counts significantly.

   Not only do we not know exactly what problem we're trying to solve, or whether our solutions will work, but every shot we take has got to count. We're conducting improvisational surgery on the only home we have. And we have to get it right. No pressure.

6. Wicked problems do not have an enumerable (or an exhaustively describable) set of potential solutions, nor is there a well-described set of permissible operations that may be incorporated into the plan.

   Even though the solutions we come up with can be neither true nor verifiable (see #3 & #4), at least we get a potentially infinite number of them. Not sure if this is good news exactly, but as far as climate change goes, it's certainly an invitation for everyone to pitch in and get creative.

7. Every wicked problem is essentially unique.

   If climate change was anything like problems we'd had before—and ideally solved before—we'd have data and templates and methods for solving them. But it is a problem we've never had before, and certainly never solved before; it's terribly, terribly unique.

8. Every wicked problem can be considered to be a symptom of another problem.

   Just when you thought things couldn't get more complicated, they get even more complicated. The causes of climate change have been variously ascribed to industrial civilization, short-termism, our alienation from Nature, regulatory failure, and a host of other causes, and if any of these are the true cause (and most likely all of them are), it means

*that the problem that our wicked problem is just a symptom of is not
just a problem, but itself a wicked problem.*

9. The existence of a discrepancy representing a wicked problem can be
explained in numerous ways. The choice of explanation determines
the nature of the problem's resolution.
*Wait, wha?! Okay, OK, maaaaybe what they're saying here is that even
though the same problem is messing with all of us, that mess is going to
look and feel different to each of us. And not only will we be unable to
agree on how to explain how it's messing with us differently, but all our
efforts to stop it from messing with us will be similarly misunderstood.
In any case, this does not bode well.*

10. The social planner has no right to be wrong (i.e., planners are liable
for the consequences of the actions they generate).
Don't fuck it up or else we all die.

So, let's sum up. Not only is climate change a problem that will never be
definitively solved, it can never even be fully described. The good news:
Healthcare provision, drug trafficking, poverty, and nuclear weapons are
also wicked problems They haven't been "solved," and they haven't totally
wrecked the world yet, so maybe climate change won't either.

Unfortunately, climate change combines all the apocalyptic global
complexity of the nuclear weapons problem (requiring intricate, hard-
to-enforce, international agreements like the Paris Accords) with all the
treacherous psychology of the drug problem (multiplied by the fact that
all eight billion of us are addicted), plus, time is running out.

And so, in due recognition of the uniquely pernicious dynamics of the
climate change problem, some even cutting-edgier academics felt moved
to create a category of problem for problems even more messed-up than
wicked problems.[23] Yes, they did this to us. These they dubbed "super-
wicked problems," and include four additional defining characteristics:
1. Time is running out.
2. No central authority.
3. Those seeking to solve the problem are also causing it.
4. Policies discount the future irrationally.

"Irrational future discounting" might sound like a new-fangled Wall Street investment vehicle, but it's actually behavioral-psych speak for how decision makers, in spite of overwhelming evidence of significant impacts (for example, that most of southern Florida will probably—but not definitely—be underwater by 2050), tend to make decisions that "discount" those impacts. It's "irrational" in the same way that smokers, knowing the risks, but abetted by some level of uncertainty, keep smoking.

So, where does all this leave us? Is the problem of climate change so impossibly tangled and riven that we should just give up all hope? Not so fast. While there are no strategies for "solving" super-wicked problems, there are strategies for beginning to solve them, or at least beginning to address them, or at least "intervene" in them. Which is actually one of the main strategies: to stop looking for one big "Solution," and instead get busy at every level with a host of "interventions." All the while knowing that even if those interventions are "successful," we'll still need to follow them up (and ideally scale them up) with further interventions. Um, forever.

Basically, if you've got a super-wicked problem, you're gonna need super-wicked strategies. Luckily, the policy ninjas have come up with a host of such strategies, from "path-dependent incrementalism" (making the most of how seemingly small steps can accumulate to produce significant results—the way many small "nudges" like preferential parking for electric vehicles in downtown Los Angles in the 1990s[24] slowly gets us to many cities and nations[25] banning new fossil fuel vehicles entirely by 2030), to "collaborative dialog mapping" (actively involving all impacted stakeholders in addressing the problem—as Canada's LEAP Manifesto process is attempting to do), to "constraining our future selves" (policies that operate like a reformed smoker who, knowing that severe cravings will make her unable to trust her future self, pays someone on Monday to hide her cigarettes on Thursday).[26]

But even the cleverest policy ninjas aren't going to make much headway unless the people involved are able to open their hearts. Or so E. F. Schumacher would have us believe. The author of *Small is Beautiful* and *A Guide for the Perplexed* distinguished between "convergent problems" (where attempted solutions gradually converge on one answer) and more super-wickedesque "divergent problems" (where different answers appear to increasingly contradict each other the more they are elaborated).[27]

According to Schumacher, divergent problems like climate change can't be solved with engineering or policy tools alone, but require a more human approach involving "higher order faculties" like love and empathy. And where best to start, but with yourself? After all, the hardest part of solving a problem—even a super-wicked problem—is to first admit that you have one.

To take that first step, we have to admit that we have a problem we can't "fix." We have to accept that we're in a predicament—a predicament we must fight, manage, mitigate, intervene in, adapt to, learn to live with, and, ultimately, survive. Further, not only will we remain in some version of this predicament effectively forever, but all the while we'll never agree on exactly what in fact it is or how best to resolve it. Finally, we have to accept that our predicament is a moving target that we are simultaneously causing even as we try to disentangle ourselves from it, and all our attempts to resolve it are irrevocable experiments done in real time—something we're desperately running out of. And if we fuck up, it's on us.

Ready to get to work?

What's a meta for?

In our "time of unravelling," it's not just the economic order and the stability of our climate that is unravelling, but the stories we've been telling ourselves. The good news: we can tell ourselves new ones. The bad news: who cares! Do you really think something as flimsy and made of air as a story can hold its own against the raw momentum of capital and ecocide? Yes, sort of.

On the cusp of climate catastrophe, in the face of a super-wicked problem of existential significance, the stories we tell ourselves matter— philosophically, ethically, even strategically. They tell us who we are responsible to, what agency we have, what our options are, what kind of world we think we are living in —and, possibly, to what fate we feel we are doomed. Some of these stories (man is separate from and master over Nature) are lurking deep in our psyche and civilization, the mythic root structure of our ideas and lifeways; others occupy the middle ground, taking shape around a powerful metaphor or allegory (we're on a Train heading for a Cliff); yet others are as superficial or as arbitrary as the phrase (Climate Change or Global Warming or The Apocalypse) by which we choose to name our predicament. After all, what is a name but a story in concentrated form?

In a time like ours, choosing a new story, a different metaphor, or simply better nomenclature—far from being "just semantics," can actually make the difference between human extinction and survival. How terrifying! But also: how thrilling to know that stories matter, and better ones can help us find our way.

If a huge chunk of ice and rock were hurtling towards the Earth about to wipe us all out in one shuddering blow, it wouldn't matter whether we called it an asteroid, a meteor, or a comet. (In *Don't Look Up*, it was a comet.) If it was big enough, it'd still kill us all no matter what we called it. But the name we give to the large-scale disruption of Earth's ecosystems that *we're doing to ourselves whether we admit it or not* does matter. Because "Climate Change," which sounds so neutral and innocuous, is anything but. "Global Warming" at least feels eerie and wrong. Which is exactly why, in trying to downplay the issue, Republican pollster Frank Luntz encouraged[28] the George W. Bush administration *not* to use it.

But "warming" doesn't even capture the whole spectrum of climate dislocations,[29] some of which, like Boston's record 110.6 inches of snowfall in the winter of 2015,[30] actually weren't warm at all. "Climate Chaos" does a better job of covering that range of extreme weather, but came off poorly when informally focus-grouped, with one attendee criticizing it for being too vague and catch-all, "as if, like, whoa, it's just cray cray out there!"[31] Meanwhile, those wishing to emphasize the complex, "unsolvable," super-wickedesque nature of the problem, favor "Climate Predicament." One climate communicator (me), in an attempt to bring a clear Good v. Evil narrative to the issue, even proposed "Climate Hitler." It never caught on.

In 2019, in belated recognition of rising temperatures and the wider social dimension of the problem, *The Guardian* officially changed their style guide,[32] replacing "climate change" with "climate crisis," and "global warming" with "global heating." The rising prevalence of "Climate Catastrophe" is an indication of how much worse people are expecting things to get. However, neither it nor its full-monty version, "Climate Apocalypse," leaves much of a role for human agency.

Enter "Climate Emergency."[33,34] While it doesn't bring out the thicket of complexity we're mired in the way Climate Predicament does, and it isn't quite as visceral and mustachioed as Climate Hitler, it does seem to strike the right mix of alarm and action. Recently, its aptness for evoking the notion of a planetary emergency that demands a WWII-level mobilization of our whole economy and society has made it a go-to favorite for everyone from Chuck Schumer[35] to Extinction Rebellion.

Across this decades-long stumble to give a name to the crisis—and the clarifying, galvanizing effect of the Emergency frame—we see the power of metaphor at work. There's a reason Naomi Klein begins her climate masterwork, *This Changes Everything*, with a search for "the stories that got us into this mess."[36]

Australian science writer Gillian King collects climate metaphors on her blog, craftily wielding them to reframe key climate debates and combat misinformation. To bring out the futility of trying to outsmart Mother Nature, she compares it to "trying to box with a glacier."[37] To highlight the polluting crimes of fossil fuel companies, she makes it relatable and human scale by likening it to the bad manners of "pissing in the pool"[38] we all have to swim in. To highlight the absurdity of people who refuse to take the reality of our fixed carbon budget seriously, she compares it to the limited air in a scuba diver's tank.[39]

King's delightful interventions are "tactical" metaphors, deliberately devised and deployed to make a specific point about a particular aspect of our climate situation. But the metaphors that most deeply rule our climate thinking are more foundational, and often more unconscious. We grab hold of them to try to make sense of our predicament, often without realizing the full consequences of the story they're telling, or the alternate stories we could be telling instead. Let's look at four very prominent ones.

Humanity Is a Virus

Here humanity plays the role of a toxic, out-of-control virus overwhelming the Earth's ecosystems. In an attempt to eliminate the threat and restabilize herself, goes the logic of the story, Gaia marshals her antibodies (droughts, superstorms, global pandemics), ratcheting up her vitals into a defensive fever of rising temperature and sea levels. And who can blame her? She's just doing what any other organism would do to rid itself of a pesky parasite.

This story-frame—let's call it "Humanity Is a Virus" or "Gaia Has a Fever"—corrals us into an extremely brutal non-choice: All we can do is continue to be the parasites we are...until the worst happens. In effect, it's a big, resigned thumbs up to cynicism and the status quo. To some, it might feel cosmically pro-Earth, and even democratic in a "Kill us all. Let Gaia sort it out."[40] kind of way, but given who is most likely to be hurt first and worst, we're basically saying: "Kill them all. Let racism and inequality and the brutal logic of the market sort it out."

We're Addicted to Oil

The metaphor of addiction quickly becomes an explanatory framework: If our civilization is "addicted" to oil, well, it can't be reasoned with as one might reason with a healthy person. If it's addicted, it's going to lie, beg, borrow, steal—and if necessary kill—to get its fix. Which goes a long way to explaining climate denial, the Iraq War, etc.

And who are *we* in this metaphorical regime? Codependents of our civilization's bad habits? Or recovering addicts trying to drag our civilization to AA? And what is our best strategy? Can we step up our game and become DEA agents, only the drug we're trying to police is oil?[41] Can we try to cut off the toxic supply ("keep it in the ground!") so our society does less self-harm? Or maybe: instead of trying to scare ourselves eco-straight (with more melting-glacier disaster porn), we should gather our civilization's loved ones (citizens, stakeholders, etc.) and stage an intervention (election, pipeline blockade, Green New Deal, Extinction Rebellion, etc.) that can finally get our civilization into treatment (a fast and just transition) even though the clinic is still so unfortunately low on methadone (wind and solar).

With every metaphor we use, we inevitably lock in (and lock out) a host of meanings. As *Virus* and *Addiction* amply demonstrate, every metaphor has a set of built-in assumptions and a half-hidden politics lurking inside it. The next one, *The Train*, shows how controversial and contested these assumptions can get.

We're on a Train Headed for a Cliff

In this all-too-common metaphor, the train of fossil-fueled civilization is heading full speed for the cliff of climate and ecological catastrophe. In the standard interpretation, we're locked on course and doomed. But consider how much the story calls out to be questioned, and how natural it feels to do that questioning by tweaking and extending the metaphor itself:

Has the train already gone over the edge, or is it still speeding towards the cliff? If it's still approaching, is there any way to stop or slow it down? Can we coax the driver to ease up on the accelerator? Can we break into the engine room and pull the brakes? Or maybe there's a switch we can throw to fork us off onto a different track? In any case, is it right that some of us are in the champagne car while others of us have to shovel coal? Can any of us unhook our car from the doomed train? Up ahead folks can vaguely make out an advance team building a bridge to a post-carbon 22nd century. It's only half-finished (because we started too late), but maybe, just maybe, it'll hold when we arrive at full speed....

Here the metaphor becomes a creative narrative space in which we can probe the nature of our predicament, argue out some of our politics, and explore questions of agency, power, justice, solidarity, and hope.

We can also do this by challenging the metaphor itself. "There's no climate 'cliff' that we go off," says climate scientist Michael Mann. He prefers the analogy of a highway: "We want to get off the earliest exit we can. But if we missed the 1.5°C exit, we still work like heck to get off at the 2°C exit."[42]

Or maybe, it's not a cliff or a highway, but a sharp slippery slope (see page 155–57). What we're in for is a "staged-descent," and the better metaphor (both more accurate and offering more agency) is not a train going over a cliff, but a canoe going down a treacherous rapids (see page 12). In this new story, we can still get our paddles in at key moments to avoid the very worst.

Here, a slight shift in metaphor can radically reinterpret the world.

Or we can reject this framing whole hog. "The end of this road is *not* a cliff," Tim DeChristopher said to me. "And I think that's one of the biggest things holding people back." He reached for other stories: A pilgrimage. A doomed love affair. A world lost and reborn.

"We need a new mythic-scale story to tell about climate change," contends astrophysicist Adam Frank.[43] Instead of the "we suck" story, or even the "we must save the planet" story, which posit humanity as either the villain or hero of a story that fundamentally has neither, he suggests we look at our species' story as an "experiment." (Counseling humility, he reminds us that we're not the first species to have transformed the climate. Blue-green algae, for example, did so three billion[44] years ago.)

But what if you feel that "experiment" is irrevocably failing? What story do you reach for then? You could do worse than to follow collapse psychologist Jamey Hecht's lead and carve out a tragically aware story— call it Cosmic Witness or Hospice Earth or the Beautiful Sunset—that enjoins us to take care of each other as things fall apart, and continue to honor the beauty and nobility of life even as the sun sets on our species.

We're at War

For those not ready to throw in the towel, War is an increasingly common lens. Naomi Klein often reaches for the war metaphor, sometimes to describe the problem: "Our economy is at war with many forms of life on earth, including human life,"[45] and sometimes the solution: we need a "Marshall Plan for the Earth."[46]

Others imagine social collapse leading to a Mad Max-style War of All against All; yet others, a War of Us against Them, where the Them are climate refugees, and the Us had better armor ourselves up into a Fortress Amerikkka or Europa, or else.

But, as metaphor-blogger Gillian King says, what we want is a war, not *between* civilizations but *for* civilization.[47] For her, as for so many others, WWII's full-on mobilization is the story we should model our moment on. It took only a few weeks[48] after Pearl Harbor for the US government to ban the production of personal automobiles, King points out. Invoking the heroism of the Battle of Britain, she imagines how stepping up to meet the crisis could be our finest hour.[49]

Bill McKibben, in his 2016 *New Republic* piece, "A World at War,"[50] takes the war analogy about as far as it can go, stating, "we're under attack from climate change—and our only hope is to mobilize like we did in WWII." His narration reads like a battlefield report: "In the Pacific this spring, the enemy staged a daring breakout across thousands of miles of ocean, waging a full-scale assault on the region's coral reefs." Raging forest fires in the Western US are the work of "saboteurs behind our lines." And in a deft repurposing of the "Remember Munich!" arguments used by war hawks to browbeat us into unnecessary wars, McKibben extends the metaphor to describe the kind of no-appeasement "political realism"[51] we need in order to meet the climate threat.

Others question the "we're at war" framing. Ecovisionary Charles Eisenstein argues that the war metaphor straitjackets us into "a linear response to a non-linear problem,"[52] while Paul Kingsnorth wonders whether an arduous trek or "trial might be a better metaphor and guide than a war." He sees our challenge more as a long, complicated, intergenerational test "of patience and hard work and attention to nature."[53]

After years of searching for new stories of "Uncivilization," drawn from myth, forgotten voices, and from wild nature itself, Kingsnorth concluded, "actually, we don't need new stories, the old ones will do just fine." Whether new or old, many of these stories—like Joanna Macy's version of the Fisher King (see page 200), or Robin Wall Kimmerer's interpretation of the Skywoman creation story (see page 286), or ecofeminist Sharon Blackie's retelling[54] of the Celtic myth Voices of the Wells—tend to be commandments to care for sacred Creation, often delivered as cautionary tales of what happens when we don't.

Choosing a story to live in is an existential and political choice, an assertion of one's world view and values, equal parts willful act of imagination and a squaring of our narrative container with the data.

Can we ignore the facts and just tell ourselves the stories we want? No. We must understand the inertia of the machinery we're dealing with, both chemical and political. We must make a sober judgment about the degrees of freedom we have left. But once we've done all that, we can choose for ourselves a more honest, more courageous, and maybe more hopeful story than the ones we tend to be given.

Metaphor is not destiny but it can become destiny. If we believe there are no lifeboats on our Titanic, if we believe Humanity = Virus, if we believe our Train is barreling forward with no way to change course, that's probably how it will turn out. Likewise, if we see ourselves as the guardians of sacred Creation (or even DEA agents interrupting our society's oil addiction); if we think, no, we're not actually heading off a Cliff of Doom but rather down a treacherous rapids we can partly navigate; if we believe that we're in an all-hands-on-deck War to save ourselves from the evils of fossil-fueled civilization, then maybe we will.

The road to catastrophe is paved with other catastrophes.

We can see why the climate crisis is so often viewed through the lens of war—and WWII in particular—but such analogies can also create expectations that don't pan out. For years, I kept waiting for a "Pearl Harbor of the Climate"—a spectacular, inarguable calamity that would finally shock us all into an emergency footing. In spite of biblical-level fires and floods, it never came. Eventually I realized the climate crisis is simply too diffuse, amoral, complex, and "super-wicked" to deliver such a single moment of awakening. The best we were going to get was a slow accretion of awareness that was unlikely to achieve critical mass until our most vital time windows had closed.

Then along came Covid. No, it wasn't a climate event per se, but it was a global cataclysm that we all experienced together, viscerally, in our bodies and in our daily lives. Many of us lost loved ones. Millions died. (Over a million in the United States alone.) Billions of us had our lives disrupted in both terrible and trivial ways, and as of this writing, it is not over yet. It was also a test run for how we might respond to the larger climate cataclysm awaiting us in the years ahead.

For some, it confirmed their worst fears. For Richard Heinberg, "The pandemic was a stress test on our social cohesion," he told me via email. A preview—and a terrifying one, in his opinion—of what is to come. "If we aren't able to set aside some personal freedoms to keep a million or more people from dying needlessly from a novel virus, and to keep our health care system from imploding, then heaven help us when climate change hits and the economy crumbles due to our hitting a series of environmental limits." [55]

For others it was a global "teachable moment" of how regular people can step up in an emergency, take care of each other, and get the job done. The pandemic and our response to it may not have "changed everything," but it changed many things, and taught us a host of Pearl-Harbor-sized lessons: We learned that governments can move at the speed and scale of a social problem if there's enough pressure and political will. We learned that once we decide we are in an Emergency, we (mostly) know how to act like it. We learned there is power in mutual aid and self-reliance. We learned that if science is communicated well, it can move political majorities. We learned that crisis is also opportunity, and it can be leveraged by organized people power to deliver everything from more bike lanes to the end of austerity politics. We learned that a "solution" is not really a solution unless it is also a just solution. We learned that hyper-productivity and economic outputs and good-for-nothing jobs are not what life is all about. Simplicity can be beautiful. Staying close to home (though not locked up in it) is not a bad way to live. And as Covid-related economic stoppages bent the 2020 emissions curve down (for the first time since the 2008 global recession), we learned how beautiful it was to see the blue sky again over Beijing and New Delhi.

Ultimately, we learned (or at least some of us did) that if we look after each other, we can get through these kinds of things together. We learned the lesson that humans learn again and again about ourselves (and seem to *need* to learn again and again): that "the worst of times can often bring out the best in us."[56]

We learned that resilience isn't just about gritting our teeth and powering through the dark times. Because the dark times might last. Resilience is about finding our footing for the long haul. It's about accepting that terrible things are going to happen and trying to make the best of them. It's about trusting that amidst these terrible things, beautiful things will also happen. It's about friendship, and community, and bitter laughter, and faking-it-till-you-make-it, and mourning (and moaning), and, yes, fighting back, but also being flexible—flexible enough to adjust expectations and plans (even your apocalypse plans), and still stay in the game.

Covid stopped the world for a while. (Something revolutions also do.) And when it started up again, even though we tragically let that emissions curve bend right back up,[57] things were not the same. We were not the same. Alongside our losses and exhaustion, we were stronger, wiser, more serious, more prepared. Yes, this unexpected (and unwelcome) apocalypse was a distraction from the one we really need to focus on, but it was also a revelation, and a rehearsal.

Extreme Sisyphus.

When Twitter isn't raging on about how Bill Gates is using the Covid pandemic as a pretext to implant trackable microchips in everyone[58] or how Hillary Clinton is sex-trafficking children in the basement of DC's Comet Ping Pong Pizzeria,[59] it can sometimes offer the most exquisite and soulful window into the human condition and our climate predicament. Like it did on November 3, 2021, when acclaimed novelist Michael Chabon spun out the following twitter thread on nihilism and existentialism:[60]

Today I was reading over one of my kid's college application essays, responding to a prompt that asked him to imagine his future.

I was moved, watching him fight to swim against the rising floodwaters of nihilism all around.

Nihilism is a valid response to the randomness of existence. It's free of the willed illusion and wishful thinking that haunt traditional methods of ascribing meaning and purpose to life. But it can't be sustained. Its logic leads inflexibly to suicide, individual or societal.

Worse, in negating the purpose of life, nihilism invalidates agency, conscience, self-determination. It's a surrender, and like any form of surrender immediately raises the question: To whom?

Here, capitalism steps in, unparalleled in its ability to profit from the standard nihilistic pursuits: hedonism, false nostalgia, cocoon- and bubble-construction, and the malaise that can be treated, but never cured, by "retail therapy."

Nihilism is the ultimate product line. Increasingly our most despairing and antisocial thoughts bear the logos of the corporations that bring them to market.

That is reason enough, to me, to reject nihilism.

The class of 2026 lives in a fraught, even overwhelming world, different in important ways from the world faced by the class of 1984. And yet I remember, and can understand, the viability of nihilism, its ease and allure.

When I was applying to college, we were staring into the Cold War nuclear abyss; it was one minute to midnight. I went to bed

128

every night with the thought that the missiles would be launched while I slept.

It was around that time that I first encountered an alternative to nihilism, equally valid, equally clear-sighted and unillusioned, yet compatible with an acceptance, even an embrace, of life.

This approach had evolved in response to the terrible absurdities and burgeoning sense of meaninglessness that arose in the wake of the 20th century's cataclysmic wars, totalitarianism, rationalized slaughter, Auschwitz, Hiroshima.

It was called existentialism. It argued—put roughly—that life's only meaning is the one we bring to it, that its purpose is for us to determine, each for ourselves.

And most importantly, it argued that in this absurd universe without purpose, meaning, or objective morality, in a world where nothing matters, the only principled alternative to suicide is to behave as if it all *does* matter. As if we and the consequences of our actions matter.

Which, in a tentative, small way, was the modestly defiant conclusion my kid was feeling his way toward in his essay, looking ahead to the rest of his life.

Chabon concludes the twitter thread with a promise to get "some Camus" for his son, saying, "It helped me, way back the last time the world was falling apart."

I stumbled across *The Myth of Sisyphus*, Albert Camus' mid-century essay on suicide and the absurd, when I was about the same age as Chabon's son. In the original myth, Sisyphus defies the Gods by lingering in the world of men beyond his appointed time. For his impudence (basically, for the sin of loving life too much), he is hauled down to the skyless Underworld and sentenced to push a boulder up a mountain only to have it roll back down again and again, forever. It sucks to be Sisyphus. And since he represents all of us, doomed as we are to strive meaninglessly in an absurd universe, it sucks to be us, too.

In the middle of the essay Camus states with startling simplicity: "I want to know whether I can live with what I know." What Camus "knows" is his own irredeemably absurd predicament: a creature who cannot help but seek meaning and wholeness in a world whose irrationality and opaqueness will forever thwart him. He wonders "whether

I can live"—and he means it literally. He wants to know, given the circumstances of our condition, whether suicide is a defensible act. And if not, then, how—without God, without hope, without any premature resolution or mystical leap—he can live.

Like most adolescents, I knew everything: Farrah Fawcett was hot; God was dead. I didn't need religion's false comforts, and I was pretty smug about it—which is often the cue for nihilism or Ayn Rand to walk into a young man's life. Luckily for me, it was Camus. His ideas captured my post-pubescent imagination; his example even more so. Here was a man of not just words but action (he'd fought with the French Resistance and assisted—or at least tried to assist—the Algerian Revolution), who said it plainly: life had no meaning or purpose; there was no hope; and yet, with "passion, courage and lucidity," and with a great humanity, he chose life. He had his boulder-pushing rebel hero choose life as well. In Camus' telling:

> Each atom of that stone, each mineral flake of that night-filled mountain, in itself forms a world. The struggle itself towards the heights is enough to fill a man's heart. One must imagine Sisyphus happy.

Sisyphus is sentenced to his task forever. There is no exit. There is no hope. Crucially, Sisyphus gets this. He understands his predicament. Knowing "the whole extent of his wretched condition," he pauses at the top of the mountain and smiles. "The lucidity that was to constitute his torture," says Camus, "at the same time crowns his victory." "His fate belongs to him. His rock is his thing." He descends to find his burden again.

Sisyphus sees his fate plainly, and chooses life. What about us?

Even in normal times, it's hard enough to push a boulder up a mountain only to watch it tumble back down again and again forever and try to be happy about it. We, however, also know that air temperatures of 130°F will soon clobber us with heat exhaustion after just a few steps. Ragged mobs of climate refugees will moan into the night as we toil. Climate change will so grotesquely ravage the world (the Underworld, too) that our mountain will be unrecognizable; the Gods themselves will lose their bearings.

Can such an extreme fate still belong to us? Can this new, terrible rock be our thing? (Consider that it will be our additional curse to slog away always knowing that with a bit of long-term planning and some basic decency we could have preserved our original absurdly hopeless fate.)

In the end, whether we are destined for that ancient skyless Underworld or the extremis of some new self-made climate hell, we must find a way to love our fate. So: *Bring on the rock! Gimme that boulder! Let me push it up that treacherous slope and watch it fall back down again and again forever! Can we start now with the endless, meaningless striving? I can find a way to be happy about it. I know I can.*

Credit: Diane McIntosh

I want a better catastrophe.

Climate catastrophe is coming. We know this. What we don't know is how bad it will be.

In the best case scenario, an unprecedented global Green New Deal rapidly transitions the world economy off of carbon, holding global temperature rise under 3°C.[61] This causes large-scale polar ice melt, 12 inches of sea level rise by 2050,[62] and major habitat disruption. We lose Miami, Shanghai, and many of our greatest cities.[63] Tens of thousands of species become extinct. Coastal flooding and systemic crop failure lead to hundreds of millions of climate refuges,[64] global resource wars, and partial social breakdown, but enough of us survive, and a chastened version of civilization—one that's learned to live within sustainable limits—stumbles through. That's our best case scenario, if we're being realistic.

In the worst case scenario, runaway global warming of 6°C+ superheats the planet, wiping out humanity and most complex life forms.

In the worst case scenario, catastrophe is total. But in the best case scenario, what we do matters. If projections tell us that 1,000,000 of the Earth's species are on track to die off,[65] and we can do something to help make that "only" 500,000, shouldn't we try? Indeed. We must protect all that we can. We must do everything possible to limit the damage, including gracefully "powering-down" our civilization, and becoming lean and resilient enough to survive in a broken ecosphere with our humanity intact. What we want here is a better catastrophe.

Imagine the protest rallies: "What do we want? A better catastrophe! When do we want it? As late in the century as possible!" Door-to-door recruitment: "Excuse me, Ma'am, care to sign this petition to only half-fuck over the planet?"

A defeatist attitude, you say? Hardly. It's hard-nosed, courageous, and full of hope for the future. What was Winston Churchill's rallying cry in the darkest moments of WWII? "I have nothing to offer but blood, toil, tears, and sweat!" Not exactly an upbeat message. He wasn't one to overpromise, and nor should we. He too had a choice of catastrophes: Europe as a Nazi-occupied Death Camp. Or Europe bombed to pieces and split down the middle, with some takeaway wisdom about the nature of evil. He chose a better catastrophe and so should we. If catastrophe is where we're headed, let's fight hard to get the best catastrophe we can.

GOPAL DAYANENI

"We're going to suffer, so let's distribute that suffering equitably."

"When is the best time to plant a tree?" one of the protagonists in Richard Powers' 2018 novel, *The Overstory*, asks. "Thirty years ago," comes the reply. "When is the next best time? Right now."

Same with us. Because of our failure to act at the scale of the problem 20 or 30 years ago, we're locked in for a catastrophe worse than any other in human history.

But here we are. And it is now.

My self-appointed mission: to think and feel my way into our darkening future, trace out the pathways before us, game out our options, and along the way—amidst my own splutterings of hope and despair—take good notes and draw a few maps that might be useful to all of us.

It made sense, then, that I'd arrived now in California, the place where American Manifest Destiny, both murderous and visionary, had come up against the vast Pacific and built its utopia/dystopia. The whole country, with a mix of fascination and unease, still looks to California as a laboratory of the future.

Here, Elon Musk designs his electric cars and plots his way to Mars. Here, Apple, Meta, and Google spin their webs of data and silicon to entrap us in their next-gen thing. Here, Richard Heinberg and his team at the Post-Carbon Institute return from their deep dives into economic trends and energy metrics to tell us "the Game is over" and we must "Powerdown" our civilization.

But professional futurists and CEOs in gleaming California offices aren't the only ones inventing or mapping out the future. The future is also being invented on the ground, in neighborhoods and backyard think tanks.

One such think tank is Oakland-based Movement Generation, a grassroots laboratory that has devised an influential Just Transition organizing model and is thinking through (and, maybe more importantly, living out) some of the "intentional pathways towards local, living, loving economies" that might help us make it through the bottleneck of ecological collapse.

Over the years, I'd crossed paths with one of its co-founders, Gopal Dayaneni, a leading voice in the Climate Justice movement, and now I was bringing my dilemmas and rough-drawn maps to his door. What would he say?

Gopal lived in the middle of Oakland on a leafy street in a three-building commune. I arrived to find him in his coveralls, moving scrap wood from his yard into the trunk of his car. I hopped off my borrowed bike. "Perfect timing," he said, "just wrapping up." We loaded the last of the wood, and he gave me a tour of the place.

There was a sizeable garden with vegetables and herbs and chest-high corn. There was a hammock, a play area, and a chicken coop. The coop contained fewer chickens than it had a few weeks ago, Gopal said, but whether the now-missing chickens had been donated to a home for at-risk youth, or a farm for at-risk chickens, I couldn't quite tell.

A total of eight kids inhabited this strange wonderland, the oldest now headed off to college. The adults in the community—nine in total—were long-time friends who'd known each other since college in the 80s. They'd gutted and rebuilt one of the houses, creating a common space in the basement. They were now working on the second building, getting ready to pour concrete that weekend.

Gopal emphasized how no one in the collective owned any of the land or buildings individually (we don't own it "separately," but "in common," he said, noting how all of them together own it equally and indivisibly, and collectively manage its governance). When they began the venture, none of them were particularly skilled builders, but over the years, they'd taught themselves what they needed to learn, here and there hiring an odd electrician or plumber or two to help them out.

I was impressed at the intentionality, dedication, and politics behind it all. And I said so. After the tour, Gopal invited me inside to meet his partner Martha—a teacher, labor organizer, and activist—and his two pre-teen kids, Kavi and Ila. Eventually Gopal and I stepped out onto the back porch, cracked open a beer, and settled into our interview.[1]

Gopal: Before we get started, let me just say you're a damned fool.

Andrew: Why, um, thank you.

Gopal: I would never, ever write a book. I would much rather be in a chapter of someone else's book than actually have to write one myself.

Andrew: I guess it's your lucky day, then, isn't it?

Gopal: Ha, I guess so.

Andrew: So, let's get into it. Our civilization is at war with the ecosphere it depends on. Some kind of catastrophe is coming, but there's so many unknowns. "We don't have a future. We have *futures*." Or so argues much of the scenario-planning literature I've been reading. We have to think of the future in the plural sense. Given all the research you and the Movement Generation collective have done in this realm, what does our future(s) look like?

Gopal: Again, perfect timing, Andrew. We are just now beginning to put down on paper the ideas we've been developing for the last several years, so this is a great chance for me to talk through some of them with you.[2]

Back in 2010 our collective decided to take a break from our usual retreats and campaigns and do a deep dive into scenarios. We looked at food, water, waste, energy, capital flows, globalization, deglobalization, expansion, and contraction. We looked at population, at culture and cosmology. Then, using a simplified scenario planning mechanism, we drew out trend lines for each of these categories, and mapped out three future scenarios, which we named Gray, Green, and Gaia. We crafted them as "news feeds from the future," and included a guiding strategy document.

Andrew: OK, let's take it from the top. Gray?

Gopal: Gray is more or less business as usual extrapolated out. It's the mostly collapsitarian view of how the social and ecological implications of climate will play out if power relations don't change.

Andrew: And what does that look like, exactly?

Gopal: Fortress America. A wall to keep climate refugees out. Increasing militarization of police. Continued corporate concentration, increasing criminalization of the less powerful, and a more and more inequitable distribution of and access to resources. Not just peak oil, but peak soil and peak water. Which all result in peak incrementalism.[3]

Andrew: "Peak incrementalism"? Say more.

Gopal: The ability of our economy to accommodate change—which is a dimension of resilience—is shrinking as the resource basis for its existence shrinks. The purpose of the economy is still the accumulation of wealth and power—that isn't changing (unless *we* change it)—but as resources dwindle, and it tries to maintain that purpose, it's less and less able to accommodate the pressures from mobilized folks demanding change—whether that's for clean water, for $15/hr, or what have you. The capacity of the system to accommodate change is shrinking because, with massive resource depletion, it's running up against its limits.

Andrew: There's not a lot of give. It's harder for the powers-that-be to share the pie in order to head off revolt.

Gopal: As long as they're desperately trying to hold onto the purpose of the economy as a profit engine, yes. People think peak is a material thing, but it is more than that.

Andrew: It's more like a ratio; a rate of return.

Gopal: It's the amount of energy you put in for the amount of energy you get out. Once you've passed "peak," it doesn't matter how much energy you put in, you'll get less and less out—until you reach the point at which no matter how much energy you put in, there is no return, because there is nothing left to extract. How you measure energy invested and energy returned matters. If energy invested is in "dollars" and energy returned is in "profits" you can always subsidize the peak. If you live in Appalachia, Peak Coal happened a long time ago because we just had to blow up your whole mountain and destroy your watershed to get it. If you live in Iraq, Peak Oil happened a long time ago. We just occupied your country to exert pressure on the price of oil. You can call the easiest to extract oil "light sweet crude" if you want, but the truth of the matter is we decided we would subsidize it with a massive military intervention in Iraq, a 60-year-long petro-dictatorship in Nigeria, etc.

Andrew: By some estimates, the Iraq War was a 1.7 trillion-dollar intervention aka "subsidy."

Gopal: Exactly. We're way past Peak Oil and Peak Coal, and it's driving the more extreme—and highly subsidized—energy extraction we're seeing these days.

Andrew: So, that's "Gray." How about the other two scenarios?

Gopal: The Green scenario is the world we're in today, which is really the festival of false solutions. It's typified by the techno-fix, the idea that the solution to the climate crisis is simply clean energy, like solar panels. Or the notion that we can mitigate climate change through hi-tech engineering of our planet or ecosystems, whether that's synthetic biology, geo-engineering, or gene drives. Gene drives didn't even exist 20 years ago and now we're talking about driving mosquito populations to extinction, bringing back extinct species, daughterless mice, and the like.

At the heart of the Green scenario is carbon fundamentalism. It's the idea that the climate crisis is just about how much greenhouse gasses we are emitting into the atmosphere. You only understand the climate crisis by looking up at the atmosphere and counting carbon, as opposed to actually understanding that the enabling resource of the climate crisis is the exploitation of human labor, which, when concentrated and controlled, can be wielded like a chainsaw against the rest of the living world.

The pool of carbon that really matters is life itself, as our friends at ETC Group say.[4] The larger crisis is actually the erosion of biological and cultural diversity, and climate is just one particular dimension of that. It's both a cause and consequence, but it's not actually the thing we should focus on. What we really should focus on is the fundamental organization of the economy.

Andrew: You're giving "Green" a fairly negative characterization here.

Gopal: The Green scenario is incredibly terrifying to me because it assumes we can "fix" the climate, without questioning the underlying assumptions of the economy or the world we're in. When we do that, instead of actually addressing the root cause of the problem, we just exacerbate the inequitable distribution of suffering. There is no question that the climate is going to change, and bring with it a lot of suffering. The question is how we navigate it.

Andrew: Once we acknowledge we're heading for a catastrophe, the question becomes: how do we shape that catastrophe for the better, not the worse.

Gopal: Instead of waiting for the crisis to come to us, we should be shaping—even provoking—the crisis we need. It should be a crisis of jurisdiction over who has the right to define the economy and govern our communities. Now, people tell me not to say this because it sounds sucky, but the measure of success of a just transition is, in many ways, the equitable distribution of all that suffering. And while that sounds awful it really isn't. The suffering is *in* the inequity. The coming changes are going to be hard, but the more we can shift our economies and cosmologies towards greater equity, cooperation, deep democracy, sacredness, and caring—then, the more *how* we experience those changes, while being hard, may not be so bad. I find that pretending we can "fix" the climate is a fantasy rooted in a deep desire to preserve the kind of economy that has never worked, and that has always been dependent on endless frontiers of extractivism, from land to the labor of the living world—including our own labor.

All economies are nested within an ecosystem.[5] Our economy is now running up against the hard limits of the ecosystem in which it is nested. It's impossible for everyone to have a Western style of living. If everyone can't have a Western style of living, no one should, right? But how do we elegantly backpedal out of our expectations? That's the hard part.

Andrew: Very hard.

Gopal: The Green scenario allows us to indulge the fiction that we can technologically innovate our way out of the crisis; that progress is inevitable; that wealth generation can happen forever infinitely; that we can convert the living world into shit forever if we just do it greener, nicer, and cleaner. That's the Green scenario.

Andrew: And the Gaia scenario?

Gopal: The Gaia scenario is built around the idea of "earth democracy." It's a transition toward bioregionalism, toward boundaries instead of borders. Borders are rigid, and arbitrary, and always enforced through violence; while boundaries are permeable, flexible, and socially and ecologically defined. It's reorganizing our economic systems around the living systems of which we are a part—not just around foodsheds and watersheds, but also around energy sheds and trade sheds. Here in Northern California, prior to colonization, the trade shed was the Salmon Nation. That's a name Indigenous peoples in this area have given it. It's an eco-regional identity defined by all the Pacific salmon runs that stretch from Alaska to the Bay Area. It's possible to reorganize our ideas and living systems along these kinds of lines. And we can do it without huge death-dependent energy systems like fossil fuels. And when I say death-dependent, I mean it literally—it is both past life and requires violence to access. We can chant "no blood for oil," but where there is oil, there will always be bloodshed.

Andrew: And how do we unhook ourselves from that death-dependency?
Gopal: In the Gaia scenario, we're aiming for a dynamic balance so that the amount of energy you consume not only provides for our collective well-being, but also returns value to the living systems upon which we depend. That's the idea of a regenerative economy, a regenerative society. It'll require a profound transition to get there, composed of many smaller transitions.

Andrew: For example?
Gopal: Well, take, for example, all those deep-water wells in the Gulf. They're an ecological catastrophe right now even when they're functioning properly. Imagine when we're not using them. We're going to have to cap every single one of those wells in such a way that living systems can recuperate over the next thousand years. We have to have a 1,000-year plan, and some of the early years in that plan involve capping those wells—and we are not going to do that with compost! We're going to need some fossil energy to do that.

Andrew: Bill McKibben has put some very precise numbers on our remaining carbon budget. You're saying we're going to need to safeguard some portion of that budget for some of these critical transitional steps?
Gopal: We need to be very smart about how we use our carbon budget. We should absolutely not be expending any fossil energy exploring for more fossil energy, let alone burning everything we can reach now—but it's naïve to think we can just shut everything down. We shouldn't build any new fossil-fuel infrastructure. We must phase out everything we can as fast as we can. But we need to use some high-powered fossil fuels (though as little as possible) for the back-pedal—like capping wells.

Andrew: So…how's it going to play out? I know which scenario you're fighting for, but which one do you think is going to dominate our future?
Gopal: All of them. Our future is going to be a contest between Gray, Green, and Gaia, between those three paradigms, those three sets of forces. Which is exactly what we're experiencing today. The Right vs. Left, climate-is-a-hoax vs. climate-is-real argument, which unfortunately still absorbs so much of our energy, is an irrelevant conversation to what is actually happening in the world right now—which is about who is going to define the solutions. The battle over whether climate change is real or a hoax is the easy argument to have because it's polarizing; it's binary; it's factional on both sides. But the complexity you want to get into is the contest over solutions: who gets to define them, how they

get defined, and how we actually assert the solutions that we believe are right. [Movement Generation's Three Circles Strategy Chart[6] is a tool to help you do that.] And the only way to win is to actually organize around our solutions at a big enough scale that we become ungovernable through our own loving self-governance.

Andrew: We have to self-govern better than the systems that are trying to impose their larger will upon us?

Gopal: Exactly.

Andrew: Okay, so we're three future scenarios in. And what I want to know—and what I think a lot of people want to know—is: Are we in for Collapse? Or Transition?

Gopal: I'm probably sounding like a broken record here, but I think it's going to be both. I think we'll be toggling between the two.

Andrew: Say more.

Gopal: Originally, the idea of a "just transition" had a focus on transitioning workers out of polluting facilities and into green jobs. Folks in labor and environmental justice got together to talk about what it takes to transition out of dirty energy, petrochemicals, and more, and do it in a way that doesn't sacrifice workers or communities. Growing in, through, and out of that tradition, and the long tradition of EJ and other movements, we are now using Just Transition to talk about transitioning entire economies and entire communities into a whole new way of being in relationship to each other and the Earth. And we at Movement Generation have been big promoters of this framing. But the thing that often frustrates me about the phrase "transition," is that it makes people think that it's going to be some kind of smooth path. No. There will be harm, and hurting, and struggle. It's absolutely not going to be smooth. Intentional, bottom-up reorganization of the economy through grassroots community organizing will play a role, but that's only one kind of transition. Collapse itself is also a form of transition. Collapse also creates opportunities—sometimes painfully—for us to make a transition. So, in our work, we talk about Shocks, Slides, and Shifts.

Andrew: Try to say *that* three-times-fast.

Gopal: Shocks are those acute moments of disruption that seem unpredictable but are, in fact, now the reasonably predictable consequences of the system. Slides are slower, more incremental changes, which have actually been set in motion long before you experience them and happen gradually over time. If you have a really steep slide, once you get on it,

A STRATEGY FRAMEWORK FOR JUST TRANSITION

Extractive Economy

Living Economy

STOP THE BAD
SOLUTIONS THAT ARE VISIONARY AND OPPOSITIONAL
BUILD THE NEW

CHANGE THE RULES
DRAW DOWN MONEY AND POWER

International

National

Local

WORLDVIEW
Consumerism & Colonial Mindset

WORK
Exploitation

RESOURCES
Extraction
Dig, Burn, Dump

PURPOSE
Enclosure of Wealth & Power

GOVERNANCE
Militarism

DIVEST FROM THEIR POWER
STARVE & STOP

INVEST IN OUR POWER
FEED & GROW

WORLDVIEW
Caring & Sacredness

WORK
Cooperation

RESOURCES
Regeneration

PURPOSE
Ecological & Social Well-being

GOVERNANCE
Deep Democracy

VALUES FILTER

A JUST TRANSITION MUST:
▸ Drive racial justice and social equity
▸ Shift economic control to communities
▸ Democratize wealth and the workplace
▸ Advance ecological restoration
▸ Relocalize most production and consumption
▸ Retain and restore cultures and traditions

Developed by Movement Generation
with OUR POWER CAMPAIGN
COMMUNITIES UNITED FOR A JUST TRANSITION

UPDATED MAY 2017

Credit: Developed by Movement Generation with Our Power Campaign; Updated May 2017

you can't stop. Of course, slides can be bumpy, and shocks can set slides in motion. These processes are extremely intertwined and complex, but as a framework, we find it can be a helpful way to think about how we need to organize differently for shocks and slides. We talk about harnessing the shocks and navigating the slides towards the shifts we want. We're all going to travel the same shocks and slides; the question is how do we experience it and who gets to decide how it lands. That's the contest.

When folks say, "Catastrophe is inevitable," I think: well, yeah, but change is also inevitable; transition is inevitable. Part of the catastrophism mentality is the idea that we're giving something up. In a just transition, what we'll be giving up is mass incarceration. What we'll be giving up is precarious housing. We can imagine a transition—and actually organize toward one—that both reduces our consumption of resources, redistributes them equitably, and creates greater democracy. Since we absolutely need to power-down our civilization, why not embrace the peak. Let's celebrate the brown-out. In one of our GAIA scenario's "news stories from 2030" we embrace the brown-out. There is a future in which we shut off non-essential power from, say, 4 p.m. to 7 p.m., so you have to go outside and play with your kids, or be with your neighbors, or whatever. There's all kinds of ways we could socially construct the transition in a way that actually adds value to our lives, so long as we consider equity, justice, care, and cooperation. No solution will work if it doesn't center justice.

Andrew: We're not falling off a cliff; if we do it half-right, it'll be more like a stepped descent with plateaus of adaptation.

Gopal: There'll be both expansions and contractions. There'll be periods of deglobalization and reglobalization over time. Again, some of it is going to be how capital navigates the contradictions it creates, whether it makes concessions, whether it loses or gains ground in terms of power.

Andrew: They say "artists are the antennae of the species." Who are the antennae for this unprecedented transition?

Gopal: In the 90s we learned how environmental justice communities—people on the front lines of environmental pollution—have unique insight into what it takes to survive at the sharp end of injustice, and thus unique insight into what the solutions should be. This is still true. And in much the same way, today I think queer, trans, and gender-nonconforming folks have an enormous contribution to make to the just transition movement. I've been learning so much and been really inspired by radical queer/trans organizers who see the interconnectedness of systems and complexity—rather than over-simplified notions of how change happens.

Many queer and trans folks have shared with me how their daily embodied acts of rebellion against narrow conceptions of normal offer unique insights into the complexities and realities of a just transition. This is also true for the Disability Justice movements, Food Sovereignty movements, Indigenous movements, and more. All are fighting against the eradication of diversity—whether that's cultural diversity or biological diversity; these movements embrace the idea that "another world is possible." The transgender journey is the literal embodiment of transition—changing states, traveling through states, and dealing with that complexity. There is wisdom in the trans experience that our other movements need if we are to reimagine our way forward.

Andrew: You're saying that trans folks have a lot to teach the Just Transition movement because both experiences involve transcending norms and neat labels, embracing complexity, and getting comfortable with unknowns and the "in-between"?

Gopal: Queer and trans folks are also part of defining the Just Transition movement. I've come to understand that the lived experience of trans folks is not like, "Oh, I'm a man and I want to be a woman, and so I'm going to be a woman now, and boom." It's not one or the other. There's a level of complexity, and emergence, and mystery in the journey itself. It's the same with living systems; it's the same with a just transition. You can't know what's coming. You just have to be in the journey of it. But you still have to have a moral compass and you still have to have a vision of where you're going. You're navigating your way through, all the time asking yourself: Is this right? Am I in right relationship with myself? Am I in right relationship with others?

Andrew: It's difficult to stay open to that level of uncertainty. There's a part of us that wants to short-circuit the process into hard answers and linear strategies. We tend to think of climate denial as an exclusively right-wing problem, but folks who accept the science, even folks in the climate movement, have their own kinds of denial.

Gopal: When we first started bringing folks together for the Movement Generation retreats, the first three days were deliberately intense. *The shit is real*, we tell folks. *It's more real than most of us ever thought it was, and it is worse than most of us ever thought it was, and we have not been thinking about it.* And every single time, the folks in the room—most of them grassroots organizers of one kind or another—are like, "So, what do we do about it?" But we say, "Look, you cannot understand what you are up against if you do not first sit with the grief." So, strategy can be a unique form of denial for folks on the Left. Because we're not going to come up with an answer of how to deal with it without actually sitting through the pain of what it means and the pain that it took to get us here.

Carbon fundamentalism is another form of denial. Energy obviously matters; it's an absolutely strategic point of intervention for shifting us to a more just society, but when people talk about renewable energy and clean energy as if the only thing that matters is reducing carbon emissions and it doesn't matter how we get there. We say, *No, it's not just about that.* Clean energy is not a solution to the climate crisis. *The reorganization of the very purpose of the economy is the solution to the climate crisis.* Sure, energy needs to be renewable, but as much, if not more, it needs to be democratized, decentralized, distributed, and decolonized. We need renewable energy, but as part of a full program of Energy Democracy.

Andrew: There's also a measure of denial in people's obsession—mine included—with their "carbon footprint." I fly, but I try to fly as little as possible, and I never want to fly casually. But at the same time, it feels almost foolish to even worry about it.

Gopal: It's good to take it seriously, but if you think your personal air travel is a strategic point of intervention, you completely misunderstand the scale of the problem. I too try to limit my carbon footprint, but less for its direct impact—which is minimal—and more for the purpose of reminding myself what it's going to take in the bigger picture. And, of course, none of us are doing it very well. We are all struggling with the contradictions of living in the world we are in. Contradictions are inevitable.

Andrew: Say more.

Gopal: Let me back up for a moment. At Movement Generation, we have four strategic principles of Just Transition. The first is, *What the hands do, the heart learns.*[7]

If all we do is fight against what we don't want, we learn to love the fight, but we'll have nothing left for our positive vision but longing, and longing is not good enough. Instead, you must directly apply your labor to meeting your needs in such a way that you become invested in that solution, in the very literal sense that you depend on it, so that when the powers-that-be threaten it, you're prepared to defend it. This is not about individuals. This is about organizing. This is about community.

The second principle is, *If it's the right thing to do, we have every right to do it.* Because the basis of revolution is not the struggle for power; the basis of revolution is rights. If people are organized enough to assert their rights—whether these are new rights or existing rights that are being infringed upon with violence—then they can contest the legitimacy of existing authority. The only way to assert a right is to exercise it. We can say we have the right to free speech, but until we actually exercise it, we have no idea where that right begins and ends, and how existing authority will try to limit it. Same with the right to housing; same with the rights of Mother Earth. Again, I'm using "free speech" as an example, but what we are concerned about is not individual liberties or entitlements, but fundamental, collective rights, such as Rights of Mother Earth, Food Sovereignty, and New Economic Rights.[8]

The third principle is, *If we're not prepared to govern, we're not prepared to win.* And it's in this context that I care about my footprint. I don't care about my footprint in terms of its impact on the world. I don't care about my footprint in terms of how much carbon I'm consuming when I drive my car. I do care about my footprint in terms of whether or not I am self-governing in my community in a way that demonstrates to the world what we want, and that we're willing to fight for the world that we want.

The fourth principle is, *If it is not soulful it is not strategic.* This principle we learned through our work with BlackOut Collective and the Black Land and Liberation Initiative. We have always wanted to uplift art and culture as essential to organizing. It was in that process, through songs and cultural organizing—led by amazing organizers such as Charlene Corruthers—that we realized we needed to uplift art, culture, spirit, ancestry, in our strategies. We have to remember our ancestral ways and sometimes create new culture that embodies the spirit of the transition and the vision.

Andrew: So, how do these principles play out in your personal life?

Gopal: When they were younger, I used to take my kids to school in a little trailer bike with another trailer attached. This wasn't so much about my individual footprint or somehow convincing 300 million Americans to ride more bikes. But every time I got on my bike and took a lane of traffic, I was reminded that the rules were designed to privilege cars, and it motivated me to change things. As I biked, I'd say to myself, "Oh, yeah, we need transit justice. I want to connect my community's struggle around transit/housing/gentrification/etc., with a larger climate and energy action plan." I'm not trying to change peoples' personal behavior one at a time; I'm interested in changing structures, so that people have better options, and different paths of least resistance. We want it to be more convenient for people to, say, hop on their bike to do the groceries, than to jump in their car. And, of course, the more people change, the more folks there are who are committed to changing the path. This can become a virtuous circle. In the end, it is always and only ever about relationships and organizing.

Now, these principles are not intended to be an individual thing, but more of a community practice, a social movement practice. In our little community here, we have solar panels and we try to be responsible. We try not to waste. We reuse what we can. Sometimes we are better, sometimes not. We do all this because we believe it's a useful model and somewhat because everybody is looking at us because we're nine adults and eight kids, and we're all activists, and we're living in a community together. We've been doing it for a long time, and our kids have been born and raised in it. Everyone is like, "Oh my God, that's amazing."

Amazing or not, we're demonstrating that it is, in fact, possible to do. And the struggles that we go through are the struggles of learning how to navigate the difficulties of self-governance. In fact, in part, we're doing it for the *practice* of self-governance.

Andrew: Achieving a smaller footprint is only one small part of this.

Gopal: It's not that I want my footprint to be *smaller* as much as I want it to be in a *different direction.* As a species, we may have once had a chance to shrink our footprint as a way of dealing with the crisis, but that moment has passed. We must now increase our impact on the planet—by orders of magnitude over the next 100 years compared to the last 500— just in an entirely different direction. We've been the keystone species of ecological erosion, and now we must become the keystone species of ecological restoration. Nothing is going to rip out concrete faster. We have to act in balance with other species, but we are the key ingredient to back-pedaling our way out of this situation.

Andrew: Now, when most people talk about "ecological restoration," they're not thinking about justice, but you are. In your work, they're inextricably linked.

Gopal: Absolutely. Our "Justice and Ecology" retreats are six days long. The first three days are about the scale, pace, and implications of the ecological crisis and our politics around it. The second three days focus on strategy. The retreats are really, really intense. And they can profoundly shift people's perspective. Many racial and economic justice organizers who thought climate and environmental issues were just for white people, or only for environmental justice organizers, realize that, actually, everybody in their communities are on the frontlines and that climate change will impact *all* the things they care about.

Andrew: You've worked for years to bring a deeper understanding of race and class into the climate movement. Why is this essential to the success of any just transition?

Gopal: Trying to bridge the gap between folks focused on climate and the folks focused on race and class, is a challenge, to put it mildly. There's a tendency for the folks focused on race and class to think—uncharitably— that most climate folks are white, middle-class, self-interested, environmental geeks. On the other side, there's a tendency by the folks focused on climate to be dismissive, to be like, *Hey, given the scale and pace of the catastrophe, we don't have time for all this justice stuff.*

Andrew: How do you bridge that gap?

Gopal: At one point in the workshop we make the case that the trans-Atlantic slave trade was an ecological catastrophe. And we tell people: *Sit with that for a second and understand that it was not fossil fuels that enabled American capitalism and the empire; it was enslavement. Sit with it, understand it, feel it. Understand that enslavement is, in fact, deeply connected to the fact that there are now islands of trash in the ocean, that whole countries are disappearing under the Pacific, that entire cultures are under assault, and species are being wiped off the planet at a catastrophic rate. For all of us, the histories of colonization, enslavement, migrations, globalization that run through our families and stories are what brought us to the brink of ecological collapse. Do not separate what is happening from your own history, and your own ancestry, and your own experience. And grieve it, so that you can then step into the power that it takes to fight back and build something different and lead the transition.*

Folks grappling with climate seem to fall into one of two camps. Either: the scale, pace, and implications of the crisis demand action that requires us to set aside our critiques of capitalism and race, gender and class issues, because we just don't have the time. Or: the scale and pace and implications of the climate crisis require us to reorganize the economy towards justice, because we can't afford to do anything else. I'm in that second camp.

Another way to put it is to see that Climate Justice is not a "win-win." It is not, "we can tackle climate and win on social justice." Climate Justice is the *only* way to win, because it is from the injustice that the crisis has emerged.

The traditional climate perspective takes a 30,000-foot view, looks at global tipping points, worries that once we lose 30% of the Amazon, then we have runaway climate change, that kind of thing. But if you're an Indigenous person in the Amazon, the very first bulldozer getting ready to take down the very first tree to put in the very first road to access the very first oil well is potentially the end of your way of life. *That* is the tipping point.

Andrew: Nonetheless, those global tipping points are still critical.

Gopal: Sure, but the way to stop a tipping point is not to freak out about 30% of the Amazon. It's to devolve power and control to the people who are on the frontlines of that struggle and to return the power to them to self-govern their relationships to land and living systems and to fight back against the desecration of the sacred. Whether that's in the Amazon or at

Standing Rock or here in Oakland. Fighting for our rights—particularly the rights of nature and our right to create commons of land and capital—are actually fundamental to the fight for the survival of our species. And, again, the only way to fight for these rights is to actually exercise them and to inevitably provoke crises around them. And, ideally, to provoke those crises on our terms.

The catastrophe I want, to use your phrase, is one provoked when the people demand—and the system can't deliver on—two really important, very pragmatic sets of rights. The rights of Mother Earth, and the right of people to have access to the resources required to create productive, dignified, and ecologically sustainable livelihoods. These are collective rights that we can organize around, and in fact, are organizing around. So, yeah, that's the better catastrophe—as you say—that I want.

I want the catastrophe of becoming ungovernable because of our own self-governance. I want us to be meeting our own needs in such a way that when the powers-that-be say, "Hey, you don't have the right to do that!" We get to say, "You misunderstand the meaning of the term 'right.' You and some army are going to have to come stop us." That will not be pretty, and that will not be a smooth transition, but that is what is required if we are to actually disrupt the very purpose of the economy and reorganize our relationships to each other and all of the living world that we are a part of.

This is the resistance of Indigenous peoples as Water Protectors and Sky Protectors. This is peasants creating Food Sovereignty and Agro-Ecology. This is the movement for Energy Democracy. This is houseless folks taking land and creating homes. This is re-matriation of land by Ohlone women through the Segorea Te Land Trust. The movement is everywhere all around us.

Andrew: In one sense, then, the frontlines are everywhere

Gopal: Yes, but the most strategic site of struggle for me might still be throwing down for other folks. True solidarity isn't about helping others out of charity or sympathy. It is recognizing that our liberation is connected. People ask, "Why do you say Black Lives Matter and not All Lives Matter." Apart from that being just lazy, and assuming that one precludes the other, it's because when Black lives matter, all lives will matter—because in this economy and society Black lives have always mattered the least. When Black lives matter, then my life will matter because that's how it's going to play out. I see a shared struggle for liberation. When I hear "Black Lives Matter," I hear a call for collective liberation.

Andrew: So, amidst all our intersecting justice struggles, and the inevitable shocks and slides and partial collapses and possible catastrophes that are going to be part of this transition, where does hope come in?

Gopal: Some people take the attitude, "Oh, we have to talk about hope because otherwise people won't follow us." But they themselves don't actually have hope.

Andrew: Right! That's more or less the attitude that launched me on this whole investigation—when I realized folks were doing that—when I realized *I* was doing that.

Gopal: Yah.

Andrew: So how do you *do* hope?

Gopal: That's a great question. I don't peddle in the usual discourse of hope much because I find that it too often is used either as a way to look away from the scale of suffering and inequality, or a way of hoisting upon future generations responsibility for dealing with the crises we've created. Instead, I seek out daily reminders that we not only have the capacity to transform the world towards greater equity, justice, diversity, and integrity, but that, if you look around, you'll see that we are actually exercising that capacity everywhere.

I'm a believer in action. And not just individual action, but, more importantly, collective action. Unlike the collapsitarians or catastrophists, whose whole frame and identity is centered around how they *conceive* of the problem, I'm all about how we *navigate* the problem. Yes, I'm resigned to the inevitability of ecological system shift. But the question is what will the specific human experience of that shift be. A collapse based on total inequity, violence, brutality, and disorder? Or an elegant backpedal where we come into right relationship with the ecosystems upon which we depend?

Andrew: It could get really ugly.

Gopal: Yeah, but at any given time, in any given place, in any given shock or slide, we will look up and be like, *Holy shit, if we do not take care of each other, we are not going to make it.*

Andrew: I worry that those same gut-check crisis moments might just as easily kick folks into an even darker me-mine, go-it-alone, might-makes-right attitude.

Gopal: This is another great Western lie. We're actually evolved for co-operation, not individualism. In moments of stress and crisis we really see this come out. We've been around for a couple hundred thousand years or so, pretty much the way we are today. For only a very tiny sliver of that time, we've been organized in empires, doing this large-scale fucked up shit to the Earth and to each other. The vast majority of our wisdom, of our deep ecological knowledge, our ancestral wisdom, our ways of being in the world, is to care for each other in community. I have an exceedingly high level of certainty that we as a species are going to survive. Honestly, it's almost axiomatic for me.

But to survive, we have to understand our relationship with all other living things. The survival of my species depends on understanding the salmon's relationship to the ecosystem upon which I depend.

The question for me is, what is the process? How do we get that gut-check moment to happen sooner rather than later? How do we ensure that the greatest number of us get to participate in the reorganization of our communities and our lives so that we can navigate the transition better together? We are an incredibly resilient species with unique capacities; we have deep instincts for cooperation. I think we've still got a lot of future left. Even during the darkest periods and the most violent and brutal periods of human history, we have shared food and sung songs. We celebrate birth, we honor death, we tell stories, we sit around the fire. People navigated the unprecedented brutality of chattel slavery, and found ways to be resilient, to survive, to resist, to rebel, to reorganize.

Andrew: And slavery was not actually very long ago.

Gopal: Right. It's just yesterday. And out of that very recent history we have so much powerful knowledge and wisdom about how to resist and be resilient. Look, the transition is inevitable. The question is whether there will be justice. If there is justice, then we will navigate it well together. But if the suffering continues to be racialized and inequitably distributed along lines of colonial legacies, the worse the crisis will get. The real indicator of how we are dealing with the climate crisis is not CO_2—it is justice and well-being. Because regardless of what is coming, that is what will determine how we experience it.

If we manage to shift towards a more and more equitable distribution of the consequences, then we will be able to be in right relationship with each other and the living world upon which we depend. We will reorganize our cultures and cosmologies. We will "remember our way forward."

If you stick your fingers in the soil and you attempt to actually figure out how to be in the world, you will find your way home. If you want to call that hope, you can call that hope.

Andrew: Fair enough.

Gopal: So, I don't mess with despair and I'm not interested in the kind of hope that lacks agency. When I talk to grassroots organizers who work on gentrification, it doesn't matter how bad the situation is, no one ever says, *Oh, San Francisco is lost.* But when climate folks take that 30,000-foot perspective—and make it all about carbon in the atmosphere—it's easy to feel defeated. But it's the way that we're understanding the problem itself that is creating the despair.

First of all, the scale of the problem does not dictate the scale of the solutions. Because we can actually aggregate to scale. No question the scale of the solutions has to address the scale of the problem, but that doesn't mean we're trying to create One Big Answer. We should start at whatever scale we have access to, and scale up from there. The way we are going to mitigate the worst impacts of climate change is not by stopping the climate from changing, but by reorganizing our relationships to each other.

Andrew: If we're too narrowly focused on preventing climate change, we're identifying the wrong problem, and we're going to come up with the wrong solutions. In fact, we're missing the real site of struggle.

Gopal: Climate catastrophe is going to happen. I have no doubts about that. But to then say, *it's too late, all is lost,* is to stay stuck at 30,000 feet and fundamentally misunderstand the issue. It presumes that the way things are is immutable, that the only solutions we can imagine are those that are politically possible. My argument, in fact, is that the solutions we need are going to happen through a contest over the very question of what is possible.

Andrew: As they often do.

Gopal: As they *always* do.

My head was a riot of possibilities and impossibilities as we came in from the backyard to join the rest of Gopal's family. Dinner was veggie curry with some extra-hot pickle chutney. The five of us ate around a big wooden table off the kitchen, then watched Monty Python YouTube bits together on Kavi's iPad.

At one point, I turned to Ila, age 11, and asked her whether she was excited about her future.

"Yeah," she replied.

"Why?"

Nonchalantly, she answered: "I want to grow up and die." Which brought me up short.

Later, as we were doing the dishes, Gopal explained: "Ila sat through one of our Justice and Ecology retreats. She was in the session on biological and cultural diversity, which is framed around the web of life, which, obviously, involves a lot of dying, too. After it's over, she's like, 'It's not really a web. It's more of a tangle.' And everybody laughed because it was cute. But I was, like, 'Wait a minute, say more.' And she said, 'It's too knotted up to ever come undone.'" Gopal and I smiled at each other, and for a moment, the future—a future of growing up and dying—which is every lucky person's future—didn't seem so grim. In fact, it almost seemed bright.

Fortunately, the web of life is a complex and resilient "tangle." Unfortunately, our civilizational predicament is also a tangle. Gopal had helped me see how tangled, sometimes how horribly tangled, it was.

I integrated many of his insights—how the Transition would unfold in the teeth of Collapse; how race, class, power, injustice, and carbon are inextricably intertwined; how the contest would be over solutions and who gets to define what is possible; and all the rest—into subsequent iterations of the flowchart, and the broader notion of a "better catastrophe."

What stood out so strongly in talking to Gopal was his sense of agency. He had little patience for the kind of hope that just made you feel better about a bad situation. For him, as for Grace Paley, "The only recognizable feature of hope is action."

In that spirit, Gopal chose to face our circumstances squarely, map out exactly how things were likely to fall apart, and—crucially—how social movements could respond so that things fell apart in ways that favored communities over capital. For Gopal, each Shock and Slide of our civilizational unravelling is a chance to remake the world, a chance at liberation, a chance to redistribute power—and suffering—more equitably.

In our interview, Tim DeChristopher and I had spoken about the need to "strategize the inconceivable." And here was Gopal, doing it.

Tim felt that because the consequences of our looming catastrophe are so unthinkable, people "don't have a context for understanding it in any other way besides complete annihilation" (see page 52). Climate catastrophe is uncanny. Without adequate words or emotional containers for it, we end up frozen in a state of "hypocognition"—we don't know how to think or feel about it, so we don't. Meg Wheatley courageously steps into that void, but without hope or a plan, she can only guide us to "give in." But Gopal is very clear: the future is not a void, it has a terrain. It's a treacherous terrain, but it has its points of intervention where collective action can gain leverage and make a difference.

In the decades ahead, we're heading into the roughest of waters, but Gopal is fully confident we will survive. He offers us principles to set our compass by, and strategies that can help us navigate our way through.

Sartre is my whitewater rafting guide.

"Freedom," according to Jean-Paul Sartre, "is what you do with what's been done to you." This is true at the individual level—say, someone climbing their way out of a traumatic childhood into full adulthood—as well as at the collective level—think, for example, of the generations-long liberation struggle of Black people in America.

What about us, as we slip into the darker waters of the 21st century?

Consider "what's been done" to us,[9] as Gopal might describe it: in spite of knowing for decades about the devastating consequences of climate change, our Lords of Carbon locked us further into a fossil fuel-addicted economy that has systematically plundered and traumatized billions of people, while concentrating the vast majority of wealth and power in the hands of an elite <1% and enmeshing us all into dependency on a system that is wrecking the planet.

What can we possibly "do with" that?!

It might feel like we've got no room to maneuver, here. That we've been put in a straightjacket by our past greed and folly, if not handed a death sentence.

No, says Gopal. We're in for a very rough ride, but reality is always a tangled interplay of circumstances we can't do anything about and circumstances we can do something about. As Karl Marx wrote over 150 years ago:

> Men [sic] make their own history, but they do not make it as they please; they do not make it under self-selected circumstances, but under circumstances existing already, given and transmitted from the past. The tradition of all dead generations weighs like a nightmare on the brains of the living.[10]

The past weighed like a nightmare on the men and women of Marx's time just as it weighs like a nightmare on us now. Unlike the nightmare of Marx's time, however, our nightmare is currently wrecking the very foundations of life itself and will inevitably haunt us over not just historic, but geologic time. Even so, we can (and must) make our own history. Our history making might feel different than the history making of Marx's time. Rather than, say, "storming the heavens," it might be more like navigating down a treacherous river.

In 2019, a cross-sector team of scholars, activists, and system-designers met to explore strategies for "Navigating the Great Unraveling."[11] They projected out best- and worst-case scenarios for the year 2040 across four key drivers: available energy, degree of climate change, level of economic activity, and amount of accessible freshwater. For climate, for example, their best case had warming staying under 1.5°C, with a worst case of runaway warming. For water, their best case was current per capita freshwater consumption, with a worst case of only 25% of current per capita consumption. Across this range of possible scenarios, they then gamed out what could be done—by communities, movements, and governments—to shape outcomes in a more sustainable, just, and democratic direction.

While acknowledging that there was "no single, magical leverage point," they identified a host of points where organized efforts could make a significant difference. These included a prohibition on privatized water extraction; land redistribution and other policies to create pathways for "re-ruralization"; "more explicit and systemic disaster planning"; a Green New Deal that would also transform agricultural and water practices; as well as being ready in moments of crisis to move radical ideas from politically impossible to politically inevitable in a "Bottom-Up Shock Doctrine."

Amongst all these proposals, they paid special attention to what they called "no regrets" strategies, efforts worth doing "no matter what specific scenario transpires." (How comforting to know there's a host of "next right things" to do regardless of how badly or not so badly it all turns out!)

Possibly more instructive than all this strategizing and scenario-planning, however, was the way they began to think about the freedom we can exercise in the 21st century:

> The canoe of our civilization is caught in a treacherous rapids that's carrying us downwards, the current is fierce and unpredictable, we can't control it, but there are places we can get a paddle in, to pivot, to steer clear of the choppiest water, till the current grabs us again…

This perspective has much in common with the Shocks and Slides on Gopal's map. Unlike Guy's cliff of extinction, or Meg Wheatley's go-with-the-flow prophecy, there's a role for collective action here. There's stuff we can do that matters. Yes, we're looking at a catastrophe, but as fair economy campaigner Chuck Collins, one of the conveners of the gathering, put it, "We have some agency around the margins to make things less catastrophic." He elaborated: "Even if we can only make a 1% difference in the outcome, we must try. That marginal difference sounds small, but it's actually huge. It could be the difference between survival of the species and complete extinction."

We are riding down the dark rapids of Collapse. Earth chemistry and the whole historical complex of Sartre's "what's been done to us" have a vast say over our destiny, but we too have a say. The stories we choose, the choices we make, the fights we step up to, the solutions we put forward, matter. As Marx said, we make our own history, but not in circumstances of our own choosing. As Gopal said, "We're all going to travel the same shocks and slides; the question is who gets to decide how it lands." Some of our possible futures are pretty terrible, but even in those, we have a say. And even if the very worst happens, we are still free, existential psychologist Victor Frankel reminds us, to choose the attitude by which we face those most terrible of circumstances.[12] At every juncture, and at every moment, we don't just "get" a future, we "make" our future.

Paddles in!

Same storm; different boats.

"If you have come here to help me you are wasting your time," says Indigenous Australian artist Lilla Watson, "but if you have come because your liberation is bound up with mine, then let us work together."[13] Part of what I understand Watson to be saying here is that just being an advocate, just being an ally, falls short. Everyone has to locate their own skin in the game.

Watson is also pointing to the notion of "intersectionality," the idea that all forms of oppression—including racism, sexism, ableism, homophobia, transphobia, ageism, classism, colonialism, and speciesism—can compound one upon the other to deepen injustice. By the same token, and because these oppressions may share similar root causes, our many liberation efforts can intersect, too. Simply put: if we can find the points of intersection in our struggles, and leverage them, we can be powerful together.

The "together" part can get real complicated real fast, however. Just consider the challenges of uniting around a Green New Deal.

While the Green New Deal (GND) is a positive, action-oriented framework that evokes a heroic era of civic accomplishment, it also brings with it a difficult legacy. While the original New Deal created jobs for the unemployed, those jobs nearly always went to white workers first. And while the 1937 Federal Housing Act opened new home loan opportunities to white Americans, it simultaneously institutionalized redlining policies designed to segregate African Americans into urban ghettos. The New Deal was also part and parcel of the long, grim history of US infrastructure projects violating Indigenous sovereignty, with massive losses of homelands and sacred sites due to freeway construction in direct violation of treaties. With that kind of history, Indigenous peoples and others who've been historically left behind have an understandable distrust that the Green New Deal will be any different.

A collaboratively-written memo I received in 2019[14] mapped out some of the complex tensions at the heart of the GND, including:

- The GND must happen ASAP—AND—A just transition of livelihoods and infrastructure cannot be rushed.
- The US Government cannot be trusted to craft policy that supports

158

climate justice—AND—The US Government *has to* craft policy that supports climate justice.

- "Green" is snappy, broad, and easily understood—AND—"Green" is a consumerist term that waves away real impacts.
- If done wrong, the GND will perpetuate inequities—AND—If done right, the GND shifts the balance of power.

If the central imagery for selling the GND ends up being burly white dudes erecting windmills, a lot of people are going to feel left out. We'd be wise to bring in a more intersectional-feminist perspective less centered on metaphors of production and building and more on service and care, which will be a core component of the GND economy in any case.

If we can cast the GND as "a futurism for our time" that celebrates the regional particulars of place, and as a "doing" framework that's making a new society possible (as opposed to an "abstinence" framework of *no, you can't have your fossil fuels anymore*), we're more likely to create the big tent we need.

But creating—and sustaining—such a big tent will be fraught, especially one that can sustain the deeper, *transformational* vision of the GND (beyond the watered-down jobs and energy package currently stuck in legislative limbo in Washington) whose broader aim is to restructure the economy (not just rewire it), redistribute power, and deliver on the historic aspirations of oppressed groups.

Our transition from fossil fuels to renewables, must be both fast *and* just—not the easiest pairing. Achieving collective liberation is a complicated, centuries-long project. No one should expect smooth sailing when flawed people (that'd be all of us) try to take on the legacy of genocide, slavery, pillage, and faulty cultural software while trying to fundamentally transform the economy in a decade.

Guided by twin pole stars, we are doing the best we can. On the one hand, we know the "Frontlines must be in Front." That's a core tenet of climate justice, one that Gopal had invoked time and again in our interview. But it's also going to "Take Everyone to Change Everything." An "everyone" that includes Indigenous Water Protectors, Black and Brown leaders bringing community solar to their neighborhoods, enlightened bankers, anxious coal miners, solar-friendly Tea Partiers, as well as clueless, semi-clueless, and less-clueless-than-we-used-to-be white dudes "just here to help."

We're all tangled up together in a brave new intersectional hairball of justice. In any Green New Deal-sized coalition historic resentments will surface, nerves will fray, meetings will go long. Those of us accustomed to privilege need to get comfortable being uncomfortable. The more all of us can lean in to what connects us, while staying aware of all that separates us, the better we'll do. We may be in different boats, but we're all caught in the same storm, and we need to find a way to sail together.

4

HOW TO BE WHITE AT THE END OF THE WORLD

The moment we cease to hold each other,
the sea engulfs us and the light goes out.

— James Baldwin

Sobered by Gopal's notion of a "more equitable distribution of suffering," I took a more honest look at my own tangle of race and privilege. As fossil-fueled civilization collapses around our ears, my privilege—white and otherwise—is going to distribute a whole lot less suffering to me. The end of the world is going to be bad for everyone, but instead of drowning, starving, or getting rounded up in a refugee camp, so far my allotment of suffering has mostly been to feel uncomfortable.

Something I've been doing for decades now.

"When you're accustomed to privilege," the saying goes, "equality can feel like oppression."[1] It isn't oppression, but it can feel that way. It can be profoundly disorienting to peel back all your layers of privilege. It can be a real shock to learn your own history, to go from obliviously enjoying your unearned privilege to being crushingly aware of it.

I was in college when that first happened to me. My classes were no longer the reassuring civics of middle school or the march-of-American-Manifest-Destiny-minus-a-strategic-blunder-in-Vietnam of high school. Now, I was learning about rape culture and the long history of CIA-engineered coups and torture in Latin America and how corporate monocrop agribusiness was destroying not just family farms but whole ecosystems.

Meanwhile, college friends (as college friends are wont to do) were challenging me to see how I was complicit in colonialism and patriarchy and white supremacy and anthropocentrism and...a whole host of systems of power so diffuse and pernicious that they seemed to be, well, everywhere! And before I knew I even had a phallocentric-postcolonial-cis-normative-ish-white-male ego, that ego was in full meltdown. Suddenly the world was a crime scene and I was the prime suspect. Everything felt upside down; everything I thought I knew was wrong.

When my privilege was invisible to me, I glided through the world. I ruled from a pedestal I couldn't even see. In this new upside-down world, I didn't know what to say—except "sorry." Or where to stand, except on this narrow, ugly pedestal of my privilege. The spotlight had once felt good, but now I just felt naked and ugly. I wished I could disappear. Or melt invisibly in among my new heroes, the oh so nobly oppressed. (Until I remembered that they were oh so nobly oppressed by…me!) I wished I was a righteous victim, but I was just a sheepish perp; I wanted to be a cool code switcher, but I was just the dominant code. (I was dull and flat and white and male and storyless. *Wait*, I thought, *I do have a story!* Yeah? Don't even.)

I had two options: blame the people making me uncomfortable, or find a way to get comfortable being uncomfortable. I chose the latter. Why, you may ask, would I choose to spend my limited time on this earth grappling with something that's just going to make me uncomfortable? Well, because justice, respect, and healing? (And not wanting to get called out for *not* grappling with it, to be honest.) And over time, this decision has grown into a lifelong commitment. What's in it for me? Maybe nothing except the satisfaction of trying to be a decent person and take responsibility for my own mess. And also get my full humanness back.

Privilege can be a "disconnection drug" that separates us from our natural reciprocity and solidarity with others; that cuts us off from the full gamut of human feeling, both its sufferings and joys. Even in those early days, I had the sense that acknowledging and relinquishing privilege could help me along my path of liberation, towards a deeper connection with my fellow humans and the Earth.

How to do this as the climate crisis increasingly becomes one of the key arenas in which the consequences of privilege and justice play out? Let's investigate via a series of inner monologues, story-songs, and thought experiments.

We're all in this together. Not!

Given the existential nature of our climate predicament, you might think we're all in this together. And, in some ultimate way, you'd be right. But in every other way, you'd be so fucking wrong.

Take, for example, white, well-off me. Am I "all in this together" with black kids in the South Bronx getting clobbered by asthma? Or Honduran climate refugees fleeing three seasons of crop failures? Or Indigenous Amazonians being hounded out of their villages by ranchers and mining companies?

Climate catastrophe isn't neutral. With disproportionate force it hurts exactly those people—poor folks, communities of color, and residents of the Global South—who have historically done the least to cause the crisis.

Wait, if they did the least to cause the crisis, who did the most?, asks my inner-fragile-white-person.

You did! You forced the Indigenous to dig the gold and silver out of the New World that made the Old World rich that brought black bodies across the Atlantic to extract the surplus value that built the capital that dug the coal that drilled the oil that made empires in the North that colonized and neocolonized the South and poisoned the skies with carbon before the rest of the world had a chance. You did it! I saw you there!

I wasn't even born yet.

OK, maybe you weren't *there* there, but your ancestors were there and they did it!

My ancestors were poor tailors in the shtetl.

OK, maybe not your ancestors per se, but folks who looked like you, and you benefited! And you're still benefiting! Odds are your family has ten times[2] the wealth of some black family across town, and that gap is growing—and climate change is no different. You've run up a huge carbon debt over the last couple centuries, and it's payback time.

Look, I just want to help.

Me too, but this struggle isn't just about recycling and putting solar on your roof, it's not even just about blocking pipelines and curbing emissions, it's about Climate *Justice*, it's about not always putting toxic dumps in the poorest Black and Brown neighborhoods, it's about making sure the people most impacted have a say over how the problems get fixed, it's about righting big historic wrongs.

Wrongs that, er, I—or my ancestors, or people who looked like my ancestors—did, huh?

Alas, yes. And if you're still thinking the world is a beautiful, fragile blue-green orb floating out there self-evidently worth saving and why can't we all just get along and stay focused, don't be surprised when someone comes back at you with: "It's a pretty picture those white astronauts took from space, but why should I trust you to come into my neighborhood with your solutions to save a world that's just gonna keep fucking me and my people over?!"

I don't want that to happen.

Right, me either.

So…?

So, let's agree: Yes, we're all in this together, and…

And…and…No, we're not in this together?

Right! Yes *and* no.

Wait, how can both of these opposite things be true at the same time?

Ah, grasshopper, now we're getting somewhere.

How do I sing my story?

We say again and again that statistics don't move people, stories do. We're told to tell our story—our "story of self." But the story of my self is white, male, and relatively well-off. If I try to celebrate myself and my "people," I quickly find myself in a tricky spot. If I try to tell my people's story as anything but an *apologia*, it's often seen as an act of power by someone who already has too much power.

Everyone else gets to sing their song. How do I sing mine? And what, pray tell, would that song even be?—"For the love of corn flakes and pop-tarts and oatmeal, and all the breakfasts mom made, come let us recycle!"?

I get it: We white folks already have 44 out of 45 Presidents, 450 out of the top 500 CEOs[3] and 0% of Milli Vanilli. And, hell, Whiteness doesn't even really exist (it was socially constructed sometime around the 1790 Naturalization Act, and revised in the 20th century when they let us Jews, Irish, and Sicilians be white, too). Whiteness is not an ethnicity; it's the way power confers status on those it deems worthy, and excludes—sometimes brutally—those it does not. Whiteness is the norm that doesn't need to announce itself; it just gets to be the norm. But even though there's no real white "we," I know that somewhere out there, beyond being guilty or a good ally or an annoying liberal scold or a raging nationalist or an overly precious celebrator of ethnic heritage (Scottish tartans, anyone?) or an ironic self-loather of "things white people like,"[4] I have a story to tell.

Yes, I showed up 500 years late to the apocalypse. I'm not on the sharp end of Climate Chaos. My island home isn't disappearing under the Pacific; the cornfields in my sub-Saharan village aren't being withered by freakish heat; my family isn't living in a crowded New Orleans FEMA trailer. But I'm going to have to pay for my civilization's carbon sins sooner or later, and I want to bring everything I've got to this beautiful movement for climate justice.

I never have to drive while Black, or walk home at night while woman, or wake up while poor. There's no lynchings or Trail of Tears or being left to die in the desert by a *coyote* in my family history. And no one is going to abandon me to fend for myself in public housing zero feet above sea level in the path of a climate-chaos-fueled Frankenstorm.

But I have a story, and it's a story that shapes my commitment to justice. And it's as formative and sacred and necessary to me as anyone else's story is to them.

My dad was born in Chicago more than a hundred years ago, landed a full scholarship in the middle of the Depression and had his Ph.D. in chemistry by the age of 25. He broke with the faith of his Presbyterian minister father about the same time he finally came round to voting for FDR. His ancestors were some distant mix of Scot and other Northern European tribes. While the sound of bagpipes makes my heart swell with a strange bloodlust (don't they do that for everyone?), my Scottish heritage is more disconnected curiosity than pride per se.

My grandparents on my mom's side were poor Romanian Jews who landed in Ellis Island in 1914. They settled in Brooklyn. My mom's father worked as a tailor in the Lower East Side for 50 years. My mom went to City College tuition-free in the 1950s and became a librarian. Some of my mom's extended family were murdered in the Holocaust and others escaped to Israel, but I don't root my identity in either Auschwitz or Jerusalem. I could, I suppose, but I don't. I even resented the Jewish kids in high school who, once they found out my mom was Jewish, tried to claim me as one of their own. I'm a WASP-Jew mutt, and as proud of the "mutt" part as either of the halves. I lean into my Jewish side (the humor, the food, the Brooklyn attitude), but I don't pretend this somehow makes me magically not-white, or part of some oh-so-persecuted minority. Alt-right conspiracy theorists might think I personally own several international banks and media companies, but even without them, I'm still sitting pretty: my papers are in order. Cops aren't gunning for me in my hoodie. I'm not a gay or trans teenager getting bullied to the point of harming myself. I don't have to work some dangerous industrial job where I might pick up cancer or lose a limb.

My dad landed in New York in the 1950s for an NYU professorship, went looking for a book one day and met my mom the librarian instead. I come from a stable upper-middle-class family. No divorce, no incest, no broken home, no evictions, no violence. (Except maybe for the time I took an adolescent swing at my mom and luckily missed.) My younger brother got into drugs in high school and never quite got out of them. He OD'd when he was 32 years old, a few weeks after he'd sent me his six-month recovery badge from NA. I lost my dad a few years later, and my mom a decade after that. I have no immediate family left alive. With each death, I softened up and toughened up, and inherited some money—yet another "intergenerational transmission of advantage"[5] that I and older, whiter folks like me in the top 20% disproportionately benefit from. In my case, it has allowed me the freedom to write, and to give more of my life-force to causes—climate justice and others—which, on and off for the 30 years since college, I've been throwing down for.

I was the kid at recess who would talk to the kid everyone else shunned. That early, inchoate sense of justice and compassion—that something was wrong and needed to be set right—was bolstered, as I grew up, by my parents' liberal politics. But it was two experiences in particular that shaped my adult self and made me the activist I am.

At 19, with the Cold War accelerating, I was waking up in sweats from nightmares of nuclear apocalypse. In one dream—and I can still recall it vividly—I'm running James Bond-style through a vast techno-complex desperately trying, and failing, to stop the missiles from launching WWIII. Instead of continuing to live in terror, I finally screwed up my courage, and joined a growing grassroots effort to shut down the largest nuclear weapons laboratory in the US, in Livermore, California, where some of the most destabilizing new weapons systems were being developed. The morning of the action, in the crisp 6 a.m. air, there I was—law-abiding, good-student me—sitting down in the middle of the road, linked arm in arm in a human blockade in front of the gates of the sprawling facility, until police eventually arrested all 1,300 of us.

We stopped business as usual for just one morning, but this tiny act of defiance changed me forever. To do what was right regardless of fear or the law was cathartic and empowering, releasing me from a life-long trance of goodie-two-shoes no-questions-asked deference to expertise and authority. It was more than a shift in political orientation; it changed how I perceived reality. The status quo was no longer "natural" or inherently deserving (in fact, it was putting us all in mortal danger); the world was suddenly alive in a way it hadn't been before; it was fluid, up for grabs, "political" if you will. Not only did I realize I had the right to challenge power, but I realized it was my responsibility as well. I realized—in my bones—that all of us together were responsible for the fate of the Earth. It was a life-changing awakening and liberation; and the shouldering of a life-long burden.

That same year, in the California desert, I had a run-in with death and the sacred. Trekking off alone through Death Valley I was soul-struck by the desolate beauty of the place, the silence, the huge sky, the haunting severity of the whole surround. I'd found a terrain that matched the terrain inside me. As twilight fell, I had to hurry back towards camp. Scrambling down a crease of broken sandstone, loose rock spilled alongside me. Turning my back to face the dimming sky, I worked my way down the steepening ravine, only to notice my error too late.

The rock wall dropped down and away. Suddenly, there was nothing beneath me but empty air. Death was real in a way it had never been; my life was completely in my own hands. Slowly, I inched my way, cranny by cranny, hold by hold, across the rock face. I was half-way across when one foothold gave way. My knee banged against the rock wall, my foot forced down violently, jangling in the air, weightless. *I'm falling. I'm dead.* I thought.

But I did not fall. I clutched even harder to the rock face, clenching it—hugging it—and held on. And within the flow of that single motion, a remarkable thing happened: my face also reached closer to the rock and kissed it—in farewell or in thankfulness, I cannot say. It was a pure bodily reaction, yet it was sacred; it was an act of instinctive reverence. I was kissing my fate, kissing god, kissing nature, kissing the desert, kissing the moment, kissing the particular piece of rock that held my life and chose to spare me. I felt eyes on me, then, voices whispering my name. Whether these were my ancestors, or the ancestors of that particular place, again I cannot say. But they helped me carry on, and after a second near-fall, I managed to scramble down to the desert floor and eventually back to camp.

I died up there that day—in the best kind of way. Ever since, life has felt like a gift, an act of grace, a second chance bestowed upon me by the desert itself. In a single shuddering insight, I understood how life and death are twinned and tangled presences. It birthed in me a kind of existential humility, a sense of gratitude, a reverence for nature and the wild that has never left me. I became bonded to the desert that day, a bond that later in life has extended to other wild places on this beautiful planet.

And now all these places that I love—that unforgiving (and yet so forgiving) shard of California desert; the back canyons of Southeastern Utah; a special spot along Drummer Cove on Cape Cod, where the carpet of pine needles meets the salt marsh; the shifting confluence of the Sauk and Skagit rivers as they flow out of the North Cascades; and even that hometown ribbon of Manhattan coastline studded with urban parks and bike lanes; as well as the serene, open-to-all oasis of the Suffolk Street Community Garden in my increasingly treeless Lower East Side neighborhood—are threatened by climate catastrophe. And not just these places, but all places; and not just places, but everything and everyone; and not just everyone, but some of us more than others, in ways that are savagely unfair, so it's not just a matter of repairing our relationship with the Earth, but also with each other. It's a question of justice as well as survival. And here we are, and it is now. Now or never. And I choose Now.

And that choice ushers directly from the crucial moment we are all now in, as well as everything that has led me here: from the strivings of my immigrant grandparents to the truth-telling rationality of my scientist dad to the "choose which rules to break" M.O. of my feisty Brooklyn mom; from the patch of California desert that taught me the meaning of life and death and gratitude, to the courage and steadfastness of comrades who've had my back across many decades of social justice struggles.

You see, I have a story, too. I'm a New Yorker god-smacked by the wild desert; a radical, contrarian, troublemaker, author-artist-activist, WASP-Jew mutt; a skeptical-mystic, compassionate-nihilist, democratic-socialist, lapsed-bisexual straight white cisgender upper-middle-class adult-orphan man; and I'm angry and heartsick at what is happening to our one and only planet and all the people on it (I don't have "a people," so can "people" be my people?), and this is my song.

Sing yours.

"None of us are free, until all of us are free" is a beautiful sentiment. It captures the sense that we are all connected, that my liberation is bound up with your liberation, that an injustice to one is an injustice to all. At a spiritual level, it's the idea that I cannot rest, I am incomplete, I am wounded until the rest of the world is healed; on a material level, it's the credo that any system that makes my wealth dependent on your exploitation, my freedom dependent on your slavery, is unacceptable.

It's a beautiful sentiment, and I believe in it with my whole heart, but at another level, I'm also free to ignore it. Across the globe, millions of poor women are working in sweatshops. Here in the USA, millions of young black men are locked up in our sprawling prison system. And here I am: free. What I do with my freedom is another question, but I am wildly, terribly free to do nothing.

My climate change comrades often say, "We can't give up hope because the folks on the frontlines of this environmental shock or that brutal struggle against oppression don't have the option of giving up hope." Take, for example, Nabila Espanioly, a Palestinian human rights activist and one-time member of Israel's Knesset. For decades, she has strived to bring peace and justice to a conflict that year after year only seems to get more brutal and intractable. She's not only living in a war zone where two peoples vie for one homeland; but she must bear witness as her people's land is stolen, their homes bulldozed, friends and neighbors are harassed and sometimes shot by occupying soldiers; and on top of all that, she must also struggle to be a liberated woman in a conservative Palestinian society. She's triply oppressed.[6] If anyone has earned the right to feel a little hopeless, you'd think it'd be her. Instead she says, "I don't have the privilege of losing hope."

I try to mouth her words. There's a fuzzy taste of guilt. On a moral and intellectual level, yes, she's probably right. But here I am, not in a war zone, not oppressed, insulated from the worst impacts of the climate crisis, basking in the warmth of all my security and comfort, and I've come down with a touch of the hopelessness. Dearie me. Meanwhile, halfway around the world, because of my carbon profligacy and the carbon legacy of my Northern Hemispherian forebears, Africans are starving, Syrians are drowning, reefs are bleaching, and everything's getting hotter and meaner.

Nabila Espanioly can't afford to lose hope. Moms in Flint can't afford to lose hope. Sioux youth in North Dakota can't afford to lose hope. Pacific Islanders whose homes are sliding under the Pacific can't afford to lose hope. But, here's the thing: I can. I *do* have the privilege of losing hope—and I indulge that privilege all the time. I lose hope every day. Hell, I'm writing a whole book about losing hope. What happens to me when I give up hope? Nothing. At least nothing in any physical or immediate way. At a karmic level, who knows? But to me right now? Very little. One of the trickiest aspects of the climate crisis is exactly this: Many of us—at least for now—actually do have the privilege of losing hope.

So, what do we do with that privilege? Well, as I see it, we have three options: (1) Indulge it. Feel as hopeless as the facts seem to warrant. Give over to the nihilism and anomie all around us. At least you'll be in tune with the times. Or (2) Guilt trip yourself. Police your weak moments. Stay hard with hope, as an act of solidarity with those who have no choice in the matter. Educate yourself about the legacy of colonialism, and the climate debts we owe. Feel in your bones the asymmetries of power, privilege, and suffering, and our historic responsibility to reset them. Feel it all like a weight, soaked with the blood of centuries. Let it make you feel uncomfortable. Or (3) From a place of love, learn how to listen. Step back to make room for the voices of those who can't afford to be hopeless, and also step forward into your best, most hopeful self. Or (4) All of the above.

I choose "All of the above." Because, well, we're going to do them all anyway. We're human. That's how we do. Yes, we're going to be hopeless— some of the time. Yes, even if other people can't afford to be. Hell, these famous "resilient other people" are hopeless some of the time, too, even though *they* can't afford to be. And yes, we're going to guilt trip ourselves. After all, there's a lot to feel guilty about, and beating yourself up for sins real and imagined can be a uniquely educational experience—an electric shock to the moral senses. And love? "Justice," says Cornel West, "is what love looks like in public." So, in the name of love, let's dedicate ourselves to justice even if we can't always live up to our ideals. In fact, we're likely to fail miserably at justice (in the same way we often fail miserably at love).

And so, with a choice between hopelessness, guilt, and failing miserably, I say: let's do all three.

Yes, let's be hopeless, but, then—if and when we can—let's rally. Yes, let's guilt-trip ourselves, but then, lesson in hand, let's shake ourselves loose from our own grip and gripe. Yes, we'll fall short as allies, but there'll also be moments when we will feel our solidarity—know it in our bones—not just as a duty or an abstract idea, but in its concrete beauty; in its cathartic—and possibly cataclysmic—we-actually-*are*-all-in-this-together reality.

So yes, lose hope. Be as hopeless as you need to be, as often as you need to be, because it's your privilege, and part of who you are, and you have to be true to your whole self. Because if you can't be true to the part of you that is hopeless, and hopelessly privileged, how can you be true to the part of you that—eventually, as it always does—finds hope yet again, and chooses to wrestle yet again with the incredible privilege of being hopeless.

And, finally, don't do any of this because other people can't afford not to. Do it for your own sorry ass. Because the climate crisis that's raging hard in the "Tropics of Chaos"[7] is coming soon to a flooded city, forest fire, killer heat wave, disrupted habitat, or oil shock near you. And the community you'll need to save will soon be your own.

IS THERE HOPE?

There is hope but not for us.

— Franz Kafka

A short history of hope might go something like this: Luckily, when Pandora opened that box of human afflictions, hope was there to soften the blow. Marcus Aurelius, stoic that he was, had no use for it; Dante, Christian that he was, abandoned it at the gates of Hell; Emily Dickinson, poet that she was, wreathed it in feathers; Ernst Bloch, Marxist that he was, weaponized it for the revolution; and Henry Miller, nihilist that he was, likened it to the clap. The brutal logic of the Nazi death camps demonstrated its survival value, while the soft logic of Hallmark™ sentimentality demonstrated its commercial value. Rebecca Solnit helped keep it alive in a dark time, but the times are even darker today, as Paul "hope beyond hope" Kingsnorth has made all too plain. With late-capitalism hell-bent on ecocide, can hope still serve us? That would be a very short history of hope. In comparison, what follows is a marathon of historical detail and philosophical logic.

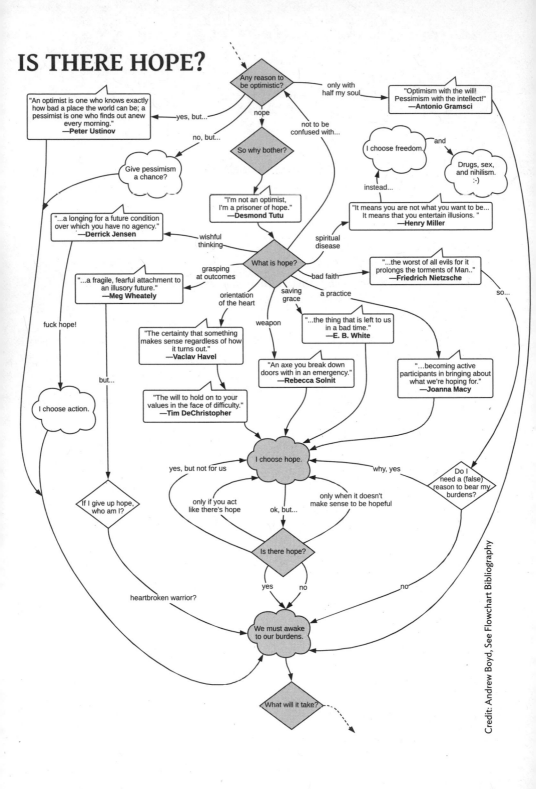

IS THERE HOPE?

Hope in the, like, *really* dark.

To be truly radical is to make hope possible,
rather than despair convincing.

— Raymond Williams

"Everything's coming together," says 350.org co-founder Jamie Henn, "while everything's falling apart."[1] Indeed it is, and we are all living on that crazy cusp. Except, most days, it's just a whole lot more obvious how things are falling apart, and not at all obvious whether we can get things together strongly enough and soon enough to avoid the very worst of our possible futures.

In the face of looming catastrophe—climate and otherwise—we don't know whether to double down on hope, or give up hope completely. We're not hopeful because things—like the facts—are pretty hopeless. But we're not hopeless either, because, well, we love life and have a heart that still beats and some part of us will always remain an irrepressible hope machine. It's a paradox, but that's how we do. And so, we need a strategy; we need a way to walk our paradoxical path, a way to twin our warring selves.

Over a decade ago, Rebecca Solnit showed us how to "hope in the dark," but things are darker now. These days we need a way to hope in the, like, *really* dark. What kind of hope can still serve us? (As there are many kinds.)

Per Espen Stoknes distinguishes four kinds of hope: passive hope, heroic hope, stoic hope, and grounded hope. Passive hope[3] is super-positive, almost Pollyanna-ish. It naively trusts that *technology will fix things,* or that *since the Earth's climate has changed before, we'll be fine.* The basic attitude here is *don't worry, be happy, because somehow it's all going to work out.* Which, though it gives you more peace of mind, leaves little reason to act.

Heroic hope, while also hyper-optimistic, is far more action oriented. It lives by the credo, "the best way to predict the future is to invent it." It takes a *Yes we can! There's no limit to human ingenuity! Just do it!* attitude. Despite their striking differences, passive and heroic hope share one important quality: they both depend on results. When actual outcomes turn sour and dark (or threaten to), this kind of optimism-based hope can quickly crumble and turn into pessimism.

"Optimism," Stoknes says, "has—scientifically—a weak case."[4] We should expect any hope that depends on results to get crushed by objective reality. Especially these days. So, now what? Fortunately, we have two other kinds of hope to turn to. Stoic hope says: *We can handle it. We've survived tough times before. Whatever happens, we can make it through, we can rebuild. (And, if worse really does come to worse, I'll drown with my boots on.)*

Unfortunately, stoic hope, though sturdy and resilient, is not particularly proactive or strategic—and we need to be both. Enter what Stoknes calls *grounded hope*. This kind of hope embraces the full paradox of our predicament. It says: "Yes, it's hopeless, and I'll give it my all anyway." This kind of hope is not dependent on outcomes, nor attached to optimism or pessimism; instead it's grounded in "our character and our calling." It recognizes the full difficulty of our situation yet still chooses to be hopeful.

Grounded hope channels the pivotal insight of Vaclav Havel: "Hope is an orientation of the spirit, an orientation of the heart. It is not the conviction that something will turn out well, but the certainty that something makes sense, regardless of how it turns out."[5] Grounded hope offers us no guarantee that we'll ever walk on out of the darkness, but it shows us how to walk through it. Here, one simply does what is right and what is necessary—and the doing and the walking are their own reward. It recalls Tim DeChristopher's understanding of hope as "the will to hold on to our values in the face of difficulty" (see page 97).

Embedded in all this is a crucial distinction between optimism and hope. Although we often conflate them in everyday speech ("She's an optimistic person." "I'm hopeful about our chances."), they're not the same at all. During a celebrated interview[6] with Archbishop Desmond Tutu, David Frost commented, "I always think of you as an optimist." Tutu replied: "I'm not an optimist, I'm a prisoner of hope." If they were people, optimism would be a very likable and somewhat overly caffeinated director of marketing; hope, a sailor caught in a storm. Optimism needs results and a rationale; hope is its own rationale.

Prominent non-optimist Richard Heinberg, bombarded at his day job at the Post-Carbon Institute by what he calls the "toxic knowledge" of our dark climate future, admits he's "not hopeful in the way that most people mean it." Instead he adopts an approach he dubs "strategic hope." "No matter how bad things get," he says, "and no matter how much worse they're likely to get, I know there's always something I can do to make things better."[7] In this way—and by playing his violin three hours every day—he's able to keep his head and heart in the game.

In the face of looming climate catastrophe, eco-philosopher Kathleen Dean Moore notes how we tend to polarize into one of two camps: either Blind Despair ("No matter what I do, it's not going to make a differ-ence") or Blind Hope ("I'm just going to trust that somehow it's all going to work out"). In either case, there's no reason to do anything. Both of these positions, argues Moore, are moral abdications, and together they suggest a false dichotomy. Instead, Moore suggests that "we respond to a lack of hope" by "do[ing] what's right because it's right, not because you will gain from it. There is freedom in that. There is joy in that. And, ultimately, there is social change in that," she says.[8]

Now, it's one thing to provide hopeless people with a way to act ethically, and quite another thing to accept that the world is objectively unsaveable. So, which is it? Rebecca Solnit weighs in on this question in her 2004 *cri de coeur Hope in the Dark*. Writing during the depths of the Iraq War and the Bush Presidency, she sees darkness all around, but it's darkness in the best sense of dark: unknown and full of possibility, a "darkness as much of the womb as the grave." She writes beautifully (uh, doesn't she always?) about how hope is a wild affirmation in this darkness; history an unpredictable trickster; activism a fluid, soulful, courageous project; and how revolutions are "days of Creation."[9]

For her, hope and despair are not simple opposites; one is not good and the other bad. "Despair," she says, "can also be liberating." To illustrate, she uses the metaphor of a door and a wall: "Blind hope faces a blank wall waiting for a door in it to open. Doors might be nearby, but blind hope keeps you from locating them; in this geography despair can be fruitful, can turn you away from the wall."[10] "False hope," says Solnit, "can be a Yes to deprivation, an acquiescence to a lie. Official hope can be the bullying that tells the marginalized to shut up because everything is fine or will be."[11] Meanwhile, "despair can lead to the location of alternatives, to the quest for doors, or to their creation." "The great liberation movements hacked doorways into walls, or the walls came tumbling down."[12]

"Hopefulness is risky," says Solnit, "since it is after all a form of trust, trust in the unknown and the possible."[13] But these days what exactly is still possible? Solnit was writing in 2004, when things were only dark. Now things are, like, *really* dark. Yes, Bush is gone, even Trump is gone, but we're 20 years deeper into the maw of climate chaos's relentless timeline. Up against its implacable math, what chance does Solnit's "Angel of Alternate History" really have? If we're basically past the threshold where we can prevent catastrophe, what kind of hope is there?

But Solnit's hope is not a naive kind of hope, far from it. It is a sober, hard-earned, long-game hopefulness, profoundly grounded in the complexities and uncertainties of how change happens. For her, "Hope just means another world might be possible, not promised, not guaranteed." "The planet will heat up," she acknowledges, "species will die out, but how many, how hot, and what survives depends on whether we act."

Pessimism of the intellect; Optimism of the will!

*I'm a pessimist,
but there's no point in being miserable about it.*

— Cormac McCarthy

*I am an optimist — it does not seem to be
much use being anything else.*

— Winston Churchill

In trying to distinguish optimists from pessimists, we often say that there's two kinds of people: Those who see the glass as half-empty and those who see the glass as half-full. How can two people see the same glass in such opposite ways? Well, because they're not just *seeing* it that way, they're *making* it that way.

"Pessimism," notes radical historian Howard Zinn, "becomes a self-fulfilling prophecy; it reproduces itself by crippling our willingness to act."[14] Does this make pessimism wrong? Not really. Optimism, after all, is a self-fulfilling prophecy, too.

"What I hope for is more hope," says James Richardson.[15] "Hope," says Jim Wallis, "is believing in spite of the evidence, then watching the evidence change."[16] So, really, the question isn't which view is more true, but rather, which self-fulfilling prophecy do you want to sign up for? The one where the world gets worse and confirms your worst opinions of it, and you get the thin satisfaction of being able to say, "I told you so!" Or the one where the world gets better and confirms your best opinions of it, and you still get the satisfaction of being able to say, "I told you so!"

Ha! If only it were so simple. Because, of course, the world doesn't always get better. In fact, given climate change, we know it's definitely going to get worse. So, if "I told you so!" is what you're in it for, then pessimism is clearly the way to go. Saddle up, Eeyore, time to ride!

But what if the choice between optimism and pessimism isn't about the likelihood of one outcome or another. What if "optimism is an ethic and an attitude, not a belief," as progressive blogger Josh Marshall wrote the morning after Trump's election. For Marshall, being optimistic in the face of adversity is a choice—an ethical choice, a choice that requires spiritual effort and a mustering of will.

When I was young, some author (Erich Fromm?) made the point that courage is not the absence of fear, but rather doing what you need to do in spite of being afraid. I soon learned this was a cliché, but I didn't care because it was a cliché that changed my life for the better. It made me feel courageous, or at least potentially courageous. Because I had fears. And there'd been a time or two when I'd stepped up in spite of my fears. Can we think of optimism and pessimism in a similar way?

We tend to gush "Oh, you're always so optimistic!" about someone who is naturally enthusiastic; for whom a positive outlook comes easily. And, sure, this kind of person might be good for the species, and sometimes fun to be around, but are they morally superior to the more depressive among us? I mean, how much credit, really, should an optimistic person get for being optimistic? On the other hand, consider someone who by nature trends more gloomy, yet—through an admirable act of will, discipline, and imagination—and in spite of how clearly they see the dark facts of the situation—still finds a way to hope, and act from that hope. These are the people, I think, who deserve our greatest respect and admiration. If the mayor is going to throw a ticker-tape parade, or give out keys to the city, forget the smiling astronauts and sports heroes, it's these can-do pessimists who should be honored.

"An optimist," says Peter Ustinov, turning our usual understanding upside down, "is one who knows exactly how bad a place the world can be; a pessimist is one who finds out anew every morning."[17] In other words, a pessimist is someone who feels *entitled* to a better world. *You did promise me a rose garden*, says the pessimist. *I keep looking around for it but it's not there.* This pessimist is actually a disappointed ex-optimist who wakes up every morning feeling betrayed by reality.

Meanwhile, an optimist is someone who knows—and, crucially, accepts—how bad things are, and still believes she can do something to make a difference. Every morning she asks herself what can I do to make this steaming pile of shit we call our world a little bit better because I know I can. In this reading, the optimist is actually a hard-headed realist, while the pessimist is a disappointed idealist who can't get over his disappointment.

So, is the glass half-full or half-empty? The answer, of course, is both/and. Stoknes suggests we think of optimism and pessimism not as personality traits or belief systems or moods, but as "tools"—tools we must choose wisely. Absolute pessimism and absolute optimism are faulty tools, he says. Both lead to "very poor scenarios," locking us into a fundamentalist storyline that can lead only to salvation or damnation (and contributing to the manic-depressive cycle activists are all too familiar with). Instead of letting our pessimism come up with ever more reasons to despair, then turning to optimism to repress that despair, Stoknes suggests we use both in parallel to help us imagine plausible futures and plan for them. Yes, we must feel our despair (after all, it is real, and one of our many antennae that tell us what is broken in the world), but we must also act. No easy task, of course. How do we set about it?

"My heart is on fire," goes the Zen Buddhist adage, "but my eyes are as cold as ashes." When I am despairing and need to act, this is the wisdom teaching I turn to. It spans my full soul. It names the contradictory qualities—commitment yet detachment; fierce engagement along with a letting go of results and ego—that I need to embrace. There's a pop-culture version of this: "Clear eyes, full hearts, can't lose!" Along with millions of grown men and women across America, Coach Taylor's locker-room pep-talk works me into a heart-swelling mess of tears, hope, and smashing shoulder-guards—even though the climate pragmatist in me knows that no matter how clear our eyes are with realism and how full our hearts are with idealism, we could still definitely lose.

In a similar spirit, Italian revolutionary Antonio Gramsci commands us to embrace "Pessimism of the intellect; Optimism of the will!"[18] Here he neatly and fiercely captures our twin tasks. (And you would be neat and fierce, too, if all your words had to be written on toilet paper and snuck out of a fascist prison.) Innovative leader of the 1920s-era Italian Communist Party, and Mussolini's most dangerous foe, Antonio Gramsci would die in that prison. He knew something about struggle and hope in a difficult time. He understood that we need to be lucid, critical, precise, and hard headed about the material reality of our circumstances, however dire it may be; and at the same time, we need to be passionately committed to our vision of a better (or at least less worse) world.

Taking his commandment to heart, I try with one hand to honor the truth of what is, and with the other, to reach for the dream of what could be. This is never easy. In Gramsci's era of rising fascism, it was devastatingly difficult. In our era of rising seas *and* rising fascism, we, too, will need both clear eyes and full hearts. (Especially if, instead of

"...can't lose!" the only pep-talk finale a climate-informed Coach Taylor could in good faith give us is "Clear eyes! Full hearts! We could definitely fucking lose!")

These disparate wisdom teachings—Zen spiritual instruction, high-school football locker-room spiel, and moral-strategic commandment from an Italian communist—are all telling me the same thing: The glass is not half-empty or half-full; the glass is half-empty *and* half-full. This is, of course, literally true (think about it), and also true in our souls: I for one need both halves (my optimism and pessimism both) in order to act in the world.

You don't need to "save the world"; it's already made other plans.

Imagine a job description that goes something like this:

> Ideal candidate cannot abide injustice. Is willing to put their shoulder to the wheel of History to make the world a better place. In lieu of pay will accept intangible rewards, including disappointment, defeat, and the crushing of all their hopes and dreams.

Would you answer such an ad? I did in my youth, though that last sentence was in fine print and I didn't read it till decades later.

Fine print or no, millions of us answer that ad every day, and who can blame us. The world is a wretched fucking mess. It needs to be saved from itself, so the folks who give a damn step up. We run for office; we march for justice; we volunteer in our community. We divest from the bad and invest in the good. We stand up for what's right and refuse to stand for what's wrong. We fight and sing and march and chant and organize and strategize and knock on doors and take to the streets and whatever else we need to do to try and bring about a better world. And underneath it all, powering it all, we hope.

What a strange burden, this hope; what a weight to carry around. Because maybe the world doesn't want to be saved. Maybe the world has already made other plans. In ways both terrible and beautiful, maybe it's just going to do what it will. The world is not our project; it is a mystery; a stubborn cosmic puzzle; a human-all-too-human mess, as unpredictable as it is tragically predictable. Maybe it's not in its nature to be saved. Least of all by you.

As this suspicion takes hold, it threatens to undermine all our visions of a better tomorrow. We recoil in horror and heartbreak. But why not embrace it as we would a welcome death? After all, the only one dying here is us-as-savior; the only world we're letting go of is one already neutered by our hopes for it.

This necessary humbling by reality leaves us mission-less and lost. But also free—terrifyingly free—of that strange burden of having to hope. Fuck hope! we say, because it's exactly this terror of losing hope that's been hounding us all along. And in saying this, we must now find the

183

courage for a new task: to be without hope and still true to ourselves. It's the culminating act in an already long and strange process of acceptance.

In Hegelian terms, we began with an urgent *thesis*: The world is fucked; I must—and can—save it. Hope will be my engine. But sooner or later we run into an immovable *antithesis*: The world will not be saved. It is what it is. And, as such, it's somehow perfect even in its imperfection. In any case, hope has no place here.

This paradox confounds us. Hope holds up a lantern of the possible, but marries us to a ghost world—a world in our heads that always chafes at the world around us. No-hope snuffs out that golden light while freeing us to love the world just as it is, in all its concrete suchness, both good and bad.

And, yet, we still find ourselves bidden to make the world better—and we know in our bones that it's right and good that we are so bidden. With our conscience a battlefield of these two truths, we struggle towards a *synthesis*: Isn't my attempt to save the world—no matter how riven with failure and disappointment—also part of the world, and thus part of what is perfect about the world? Yes. And so I must somehow accept all of it: the world and its fuckedness; how I'm called to save it; as well as the fundamental hopelessness of this task. It seems I must become a happy, hopeless warrior.

Which leaves each of us to wonder: What fool, having seen the fine print in the ad, would still choose to sign up? And also: What fool would not?

I dedicate myself to an impossible cause.

We are all incurable.

— Archbishop Oscar Romero, when asked why
he was attending to the sick at a hospital for incurables.

We have broken Nature. We have broken the world. Even the moral logic of struggle has been broken. Gandhi said "First they ignore you, then they ridicule you, then they fight you, and then you win." But in the shadow of climate catastrophe, we may have to update that to: "First they ignore you, then they ridicule you, then they fight you, *and then* a 6°C increase in the Earth's temperature wipes out all complex life forms." Martin Luther King said "The arc of the moral universe is long but it bends toward justice." But from where we stand now, it may be more accurate to say, "The arc of the moral universe *might* be long, and it *might* bend towards justice, but we're never gonna find out because: total ecosystem collapse." (Or as Bill McKibben put it in *Rolling Stone*, "The arc of the physical universe appears to be short, and it bends toward heat."[19])

I used to run on hope. I used to sign those petitions, show up at those demos, knock on my neighbors' doors—because I believed we could change things. I still show up, but the reasons have shifted. "So what if we're making progress on police brutality," a friend active in Black Lives Matter said to me recently. "Given where the climate is heading, the police might as well shoot us all now." She was joking, of course, but there are days when it seems our cause—maybe all our causes—are impossible.

Look across the full sweep of human history, with its wars and rebellions, its dark and shining moments: overall, it ain't pretty. Every revolution is replaced with the slime of a new bureaucracy. Every time you manage to overthrow slavery there's a Jim Crow 2.0 waiting for you. With the possible exception of indoor plumbing, smallpox vaccinations, contact lenses, and, OK, pretty much all modern medical technology, things don't seem to have changed too much for the better. At some point you just stop pretending that they will. And so, instead of fearing this loss of faith, I now welcome it as a revelation: Our situation *is* hopeless. Our cause *is* impossible.

Which leaves us with a stark choice: Do we dedicate ourselves to an impossible cause? Or do we pull back and just look after our own? The choice—once you've sat quietly with this question—is clear: You must dedicate yourself to an impossible cause. Why? Because, we are all incurable. Because solidarity is a form of tenderness. Because the simple act of caring for the world is itself a victory. We must take a stand—not because it will necessarily lead to anything, but because it is the right thing to do. We never know what can or can't be done; only what must be done. I dedicate myself to an impossible cause.

JOANNA MACY

*"Be of service not knowing whether you're
a hospice worker or a midwife."*

Dedicating yourself to an impossible cause is no easy trick. Reckoning with the climate crisis, with its intersecting layers of everything from race to earth chemistry to bad timing, can be overwhelming. It's easy to become demobilized by denial or despair. With everything that's on our existential plates, it's understandable why so many of us wall ourselves off from the truth, and consequently from each other.

While Gopal acknowledged that "catastrophe is inevitable," he followed it right up with, "but change is also inevitable; transition is inevitable." And he laid out some strategies to guide us forward. But how do we proceed at the heart level? How do we break down the walls inside our psyches, reconnect with our neighbors, and take those first steps towards a just transition together?

Cue my next encounter[1]—in Berkeley, California, one neighborhood over from Gopal's homestead—with eco-Buddhist, whole-systems theorist, and author of *Active Hope*, Joanna Macy.

Born in 1929, with a lifetime of environmental advocacy and empowerment work now behind her, Joanna Macy could be rightly called the grande dame of "spiritual activism." In the 70s she fought the nuclear industry and the threat of nuclear war, in the 80s toxic polluters, in the 90s and 00s corporate globalization, and now looming climate catastrophe. Sifting those experiences through her scholarly work in systems theory and the Buddhadharma, she developed a potent, heart-centered approach to social change which she has since outlined in a host of books, articles, interviews, and most notably, a set of practices she calls the Work That Reconnects.

The Work That Reconnects is both a general approach and a specific set of techniques to bring people together to confront the painful realities of our time and find a more life-affirming footing for taking them on. Participants don't "get over" their personal despair, but by feeling it and sharing it with others—usually in a multiday workshop format—they access their passion, and compassion, to work for change. By acknowledging the terrible costs of ecocide and injustice, they uncover their motivation to both resist and build a better world. Since pioneering this approach decades ago, she's trained thousands of people in it, including my kayaktivist friend Lois.

In her book *Active Hope*, Macy writes: "The greatest danger of our times is the deadening of our response."[2] And much of what follows are wisdom-tools for how to stay awake in a time such as ours. The central framework is a spiral with four ever-cycling stages: *Gratitude*, which grounds us in love for life and helps us be more present; *Honoring Our Pain*, in which we have the courage to share our private anguish and turn it outwards as compassion; *Seeing with New Eyes*, in which we feel the larger web of connections and our own power to change; and finally, *Going Forth*, where we draw upon the full impact of these transformations to become more resilient as we head off to undertake social change. The Work That Reconnects contains a large repertoire of practices for each of these moments in the spiral process.

It's a radical, soulful, and very accessible framework. And threaded through all of it is the critical distinction between passive and active hope. Passive hope is, well, passive. You're sitting there hoping, and if the odds look bad, there's no point in trying to do anything. Active hope, on the other hand, is *a practice*. It's about "becoming active participants in bringing about what we're hoping for."[3] And, crucially, since active hope "doesn't require our optimism, we can apply it even in areas where we feel hopeless." Which seems to be exactly the kind of hope we need these days.

I wanted to know more about this hopeless hope. How did it work—not in the abstract, but for her? Was she actually hopeless? If so, what life-strength allowed her to sustain *her* engagement? While *Active Hope* was brimming with positive, engaged energy, some of her more recent articles and interviews had a darker tone. When asked in one of them about several scientists who thought we may have already entered into runaway climate change, she said: "I suspect that they are right. Logically they are right: we don't have a snowball's chance in hell," before going on to make a strong case for "acting with passionate dedication to life" regardless. "How lucky we are to be alive now—that we can measure up in this way."[4]

To my delight, Joanna knew of me, too. Upon witnessing the Climate Ribbon (see page 61) she'd remarked upon the resonances it had with her own work. Also, someone had once given her a copy of my book *Daily Afflictions* for her birthday, and she was quite fond—her assistant told me via email—of at least one passage from it. So, I had an *in*. And a good thing, too, as she was not officially doing interviews anymore. I came to her South Berkeley home for early afternoon tea.

She was an even more spry and sparkling version of her author photo, and seemed as curious to meet me as I was to meet her. She reminded me of Maude from the cult classic *Harold and Maude*. (Harold: "Maude, you're so good with people." Maude: "Well, they're my species.") She had Maude's joy and spunk, and something else: a kindness, both in her eyes, and in her at times fragile voice. She heated water for tea, and laid out a spread of cucumbers and dates from her neighbor's garden. I was the one taking up her time, but here she was taking care of me: *Would you like honey? A knife to cut the cucumbers? Is your recorder-thing working?* And she continued to check in on me in this way throughout the interview. She radiated kindness and a sorrow-laced wisdom. We spoke of Lois for a few minutes, and then jumped in. Not surprisingly, she asked me about my work before I could ask her about hers.

Joanna: So tell me what you're working on now.
Andrew: It's a book about our climate predicament. Working title: *I Want a Better Catastrophe*.
Joanna: Yeah, that sounds about right. It certainly looks like the Western Antarctic Ice Sheet is going to melt and fall in. Then what?
Andrew: Right. Then what?

Joanna: My apocalyptic sense is very strong. I feel we have very little time. I rarely hear any people—except, say, Richard Heinberg—talk honestly about how fast our window is closing before all is lost. A lot of us really think it's too late. So, how do you keep on going?

Andrew: Exactly. Social movements run on hope. You want to be able to invite people into something positive and meaningful, into building a new and better world, not just how to cope with a world that's going to become unspeakably worse. Yet, if you're paying attention to what's happening to the planet, you know that unspeakably worse is a possibility, even a probability—if not already an inevitability, to hear some folks tell it. So, I feel like I'm constantly navigating between my private sense of doom, and keeping up a positive public face. It's a struggle, a predicament, a paradox—even a cheat—and part of what I'm trying to figure out by writing this book—and talking to you—is how to navigate it in good faith.

Joanna: It's hard to invite people into more honesty than they're accustomed to, more honesty than our culture generally tolerates. In America, there's this very strong need to be hopeful and comfortable. You get the message early on that if you want to get ahead, don't talk about the really dark stuff. If you want to have friends, don't tell them how bleak it is. But, inviting people into their honesty can be very liberating. And letting people know they're not alone in all this can be very empowering.

Andrew: You designed the Work That Reconnects to do just that. How did you get started with it?

Joanna: It was back in the 70s, when the nuclear threat—which never went away, mind you—felt much like the climate threat does today. I realized I had all this information—the apocalyptic consequences of a nuclear accident, as well as how radioactive reactors are even when they're not having an accident—I'd been pulling it together for a lawsuit—and I wanted to tell people about it, but nobody wanted to hear. Well, that was very odd, I thought. Is it that people don't care? I soon realized, no, they do care, in fact most folks already knew how bad it was. And it hurt them to know. I quickly realized that to harangue people about it is the last thing you should do. What you need to do is let them talk. You need to create ways for people to tell you how bad they already know it is.

Andrew: And also that there's paths forward—or at least ways to cope—that they may not have imagined.

Joanna: Yes. My scholarly work is in systems theory and the Buddhadharma, and some of my early experiments were grounded in meditation practice. One practice was focused on how people can tolerate pain, moral pain. In the practice of meditation you feel physical pain—my nose is itching, my knees are cramped, whatever—and you learn to just be with that pain. So, I wondered: If we can learn to be with that personal physical pain, can we also learn to be with the pain in the world?

I began inventing ways to give people a chance to speak what they already knew. Open sentences were one of the first techniques. People worked in pairs, asking each other to fill in the sentence: "When I think about my world, what I'm scared of happening is —." Or: "As I look at my world, what breaks my heart is —." It was a very simple technique, but just speaking these sentences aloud to one another was very powerful for people. You're immediately in the thick of the conversation, in the reality of it all, in a way that would never happen on its own. If you asked folks to just sit down and talk about what they're worried about, it wouldn't happen. But that open sentence opens everything up.

When I did my first public workshop, I had written an article called "How to Deal with Despair." I'd written the article on the heels of a year-and-a-half-long dark night of the soul. I'd come out of it, and I thought I'd learned a lot about despair and how to deal with it. The idea was to write the article and get on with the rest of my life.

Andrew: The best-laid plans…

Joanna: Yes, exactly. I was going to put it all behind me. Well, it was published in *New Age Journal*, and hundreds of letters came in response. These letters were boxed and sent to me in Sri Lanka, where I was doing a year of fieldwork and Buddhist community organizing. Here's the most interesting part: none of the letters complained that I hadn't told them "what to do." Because I hadn't. All I had done was share a few processes—pretty embryonic at the time—for how to deal with despair and what it can do for you to walk through that gate. No one complained that I hadn't told them how to stop nuclear war, or how to stop clearcutting, or how to fix any of the other things wrong with the world. They just said thank you; that's all. It was all *thank you*. Each one in a different way said thank you for showing me I'm not crazy. Thank you for showing me I'm not alone. They were very moving to read.

On the heels of the article, I was asked to do a workshop. I said, "Okay, I'll do one." I entitled it "From Despair to Empowerment." Which turned out to be the first and last time I ever used those words—"from" and "to"—because I realized you don't leave it behind you. From then on, it was always "despair *and* empowerment" because the question of how to deal with your despair is ongoing.

You see, this was 1980, and Ronald Reagan was running for president and although the nuclear arms race with the USSR had already begun everything seemed so "normal." I was astonished that anybody would even come. But they did. A broad group of folks, from businesspeople to hippies, and I was terrified because I thought, "I really don't know how to do this." Anyway, even before it was over, people were asking me to do another one here and another one there.

Andrew: You were learning as you went.

Joanna: Yes, and what I was discovering in those early years, was how essential it was to do this in a group. People see that they're not the only one, and coming out of the workshop, people were charged up to work together, to get out there and change things. That was what I expected, and it was worth everything. But then something else happened. Something changed in the way they spoke. It was a shift in identity. I saw them start to move beyond their self-interest to speak as the Earth; it was as if the Earth was working through us. This I had not expected, and it blew me away. It was a liberation from the separate ego. And, once that happens, then you're ready to do so much more. Your caring is so big.

Andrew: And how has the work evolved since?

Joanna: The work now is very centered in gratitude. We don't go right into our pain for the world. We've stopped using the word despair.

Andrew: And why is that?

Joanna: There are so many other more immediate feelings. There's anger. There's fear, dread, and overwhelm. There's deep, deep sorrow. In speaking the grief you realize that it's not the kind of grief or rage or fear that can be equated with what you encounter in your own life or even in your family. It's categorically different from "the slings and arrows of outrageous fortune." You're grieving for the whole show. You're grieving for our entire world, Jesus Christ!

Once you allow that in, it brings you into a realization that you are vastly more than a separate person, that you're a doorway or window into a much vaster identity. And you can count on that identity to help you; it can work through you. You can undertake things. You could say, oh, run for city council, or stop a pipeline. You realize there's something ready to move through you. It's like grace. In my Christian childhood, we used to think that grace was something from God. But grace comes from whatever you're acting on behalf of. If you're acting on behalf of the Earth, the Earth is right there at your back, ready to come right on through you and for you. It's been a very beautiful journey.

Andrew: Your journey from Christian grace to eco-Buddhism?

Joanna: Well, that, yes, but everything. When we started working together on *Active Hope*, my British co-author, Chris Johnstone—who is not a Buddhist—told me I needed to keep two Buddhist terms, *bodhichitta*, and *Bodhisattva*. "But," he said, "you need to explain it in one simple phrase." "Well," I told him, "here's three: The Bodhisattva is someone who has a boundless heart; someone who realizes there's no private salvation; someone who acts on behalf of all beings."

Andrew: And *bodhichitta*?

Joanna: The motivation to act for the sake of the whole. And more and more I see people stepping up into both of these roles.

Andrew: Tell me what you're seeing.

Joanna: There's something incredible happening. So many people acting on behalf of the whole. They're from all walks of life. Whether it's people organizing farmworkers here in California, or people opening their homes to refugees in Greece and elsewhere, they're all over. Something is happening, and I'm beginning to see it as an evolutionary juncture. We are evolutionally showing a capacity to act on behalf of the Earth. That's certainly been true of noble people in the past, but not in such numbers.

Andrew: Do you think of yourself as a kind of midwife—one of many, of course—of the Great Turning?

Joanna: After that first workshop, a seasoned activist turned to me and said, "Why are you doing this? Why?" I started to say something like, "So people can be more effective agents of social change." But what came right from my solar plexus was: "So that when things come apart, we will not turn on each other." Now I think if that were all I had to offer, it would be enough.

If we're going to go out, let's do it well. Maybe if we know how to do it well, then it's all worth it. We have such an incredible history, so many noble souls, so much creativity, so much shared endurance, so many fighters for justice. If we go out well, it could be beautiful, and you'd be proud.

When you wake up, when you truly wake up to the moment we're in—and especially when you wake up together—you don't say, "Oh, I wish it would last forever." You don't care about that, because all of Time has come together for you. You don't say, "Actually, I wanted all this to go on for another hundred years." Because you are in the present moment, in all its fullness.

Andrew: So, you do the work both to help people become stronger, more resilient fighters for justice, and also good caretakers of each other in the here and now *even if* we're going down.

Joanna: I'm thrilled when I see people do beautiful, innovative, brave, creative stuff to talk back to the big jerks in power. I love that. And we absolutely also need to learn how not to turn on each other if things fall apart.

Andrew: You're saying: since we don't know how things are going to turn out, we must prepare for both possible futures? We must teach ourselves *now* how to serve the whole whichever way things go?

Joanna: Exactly.

Andrew: But some of us don't sit well with that kind of uncertainty. We want to know—does our species have a future worth living or not?—just to be able to orient ourselves. But we actually don't know, you're saying, and you've designed the Work That Reconnects to straddle that uncertainty.

Joanna: Maybe it's some moral cowardice on my part, but I don't think I actually can tell which way it's going to go. We can't tell because the natural world—and the social world—exhibit emergent properties. We may be much closer to an outbreak of sanity than we realize.

Andrew: In your work, you talk about uncertainty as a source of awakening and creativity.

Joanna: Yes, I believe we can be wonderfully strengthened by uncertainty.

Andrew: So, what about the people who are so certain that they just say, "Look, our time is up. We're done for. Deal with it."?

Joanna: That's what I've heard from Guy McPherson. For one thing, it's boring. There we go again. Who are we to say that, for heaven's sakes?! It doesn't help. It's boring. You're boring to be around; you're boring to be with yourself. If you're just sitting there, so certain that it's all over, then at some level you're glad about it. *See, I told you so. We're all going down.*

In the Buddhist path, there's great importance placed on the don't-know mind. The central teaching of the lord Buddha is the radical inter-existence of all things. Everything plays back and forth from all sides; everything is contingent upon everything else. So, to claim that you know the outcome is hilarious and pathetic.

Andrew: You can mic-drop pretty hard for an octogenarian.

Joanna: Ha.

Andrew: Now, this is all so much more challenging for people with kids, isn't it?

Joanna: Yes, absolutely. I have three children—in spite of growing up elsewhere, they've all settled here in California—and we all just had a big birthday celebration the night before last.

Andrew: Your daughter lives downstairs, you said.

Joanna: Yes, and my son Jack lives on our same block, six streets down. He works in the Department of the Environment for San Francisco County and City. We don't bother talking about how this might be the End, he knows how dire it is. And he's got these two beautiful, brilliant daughters, 15 and 18. It's heartbreaking and also beautiful to see him being such a devoted father while knowing so much. He's constantly working at the edges. He spearheaded the city to go zero waste. Now there's cities around the globe doing it. Right now, he's doing a lot of work around carbon sequestration in the soil. There's no way we can reduce our emissions quickly enough, he says, so the only way we're gonna make it through is by massively scaling up some form of carbon sequestration. He's the only one of the children I've co-led workshops with.

Andrew: How did that come about?

Joanna: When he was at College at Tufts, I did my first workshop in Boston. My daughter, then in high school, was visiting him, so they were both in town. She said, "Hey, mom. We're here and you're here. You're doing a workshop. We're going to come." I said, "You don't need to." I realized I didn't want them to. I didn't want to hear their despair. Or have them hear mine. Instead, I said, "Have you seen the aquarium?" But I couldn't stop them. They invited themselves. Everyone else in the workshop—probably 18 or 20 of us, in a dusty room up over Copley Square—were just blown away that I was bringing these two teenage kids.

Then I got on with things, and basically forgot they were there. In the middle of the day we did a despair ritual. People were softened up. It was very raw. I was brave back then. They were in different places in the grief circle. My daughter went to anger. And my son went to grief. He wept about species loss; he just sobbed. I was struck that they both went to what was least socially acceptable for their gender.

Andrew: That your kids—and people in general—are willing to take risks like that, is that what gives *you* hope?

Joanna: Oh, so much gives me hope. I'm a happy girl. But what I love about the work I do is that it makes us see that hope is less important. We in the West, particularly us Americans, enshrine hope. We're constantly taking our pulse as to whether we're hopeful or hopeless, whether we're optimistic. I think that's a waste of time! In the Buddhadharma, the term hope doesn't even exist. The point is to be present. In the West, the figure of hope overshadows so much else, it overmanages the present moment and the pulse of life. What gives me hope? What maybe gives me hope is that I'm glad to be alive. I love the world; I love life.

Andrew: What do you say to those who ask: *If there's no hope, if we have no chance, why should I spend my life fighting for an impossible cause, a hopeless cause?*

Joanna: Because it's too late for you to do otherwise. You're already worrying about how much trouble we're in, so there would always be in your mind, in your heart-mind, the realization that you're avoiding something, and so you're not going to be fully present. You're going to be at odds with yourself. Instead, I found that just about the most fun in life is to work with people on something that matters, even when you lose.

At this point, with all that's going on, if somebody isn't concerned, I'm not going to bother with them. There are enough people who are concerned. We don't need to have 100% of people on board. It'd just be so tedious to try to convince someone who's become such an escape artist to themselves, who's put a bag over her or his head. And, yes, there's a lot of them, but why should I decide I'm their moral rescuer? I'd much rather find people—and there are so many people—who are still alive enough to care.

Andrew: Choose your battles, and choose who to serve.

Joanna: There's a certain necessary triage.

Andrew: I read an interview with you in *Eco Buddhism* webzine, from 2014, in which you quote a Korean Buddhist monk who says, "Sunsets are beautiful, too." Does that ring a bell?

Joanna: Yes, it certainly does.

Andrew: So, what do you say to the people who have decided that their truest purpose now is watching, witnessing, the sunset? That the game is over—for our species, or at least our civilization—and our truest purpose right now is to witness the going-down, to experience the beauty of nature before it's all gone?

Joanna: It doesn't bother me. It's OK to make that choice. But there's so much more we could do. We could use this last time better. We could start by treating people better. It's embarrassing to go out as a species when we're treating each other and the world so poorly. I would like to go out with my hands washed and my face washed, and not just be a creep.

Andrew: Not just fall back to a smaller circle of concern.

Joanna: Yes, we need to keep the whole in mind. We need to have more pride in ourselves. If this is the end, I want to be my best self.

Andrew: Is that a moral commandment? Or more of an aesthetic commitment?

Joanna: Who knows. If it's the End, I want to do it well. Who in hell anywhere is ever again going to hear a line of Shakespeare, a phrase of Mozart, the sound of Bach? It's been a great run. So let's go out with pride, instead of just going, "I'll grab what I can."

Andrew: You're saying: if it's over, let's at least exit the stage with dignity, with nobility, with head held high —

Joanna: And have somebody doing some wonderful somersaults along the way! We've been able to do incredible stuff as earthlings, and with our friends, our older friends, the elephants, the owls, the mountains. For Christ's sake, let's treat them decently at the end! Let's pull it together!

We should have funerals for the mountaintops, you know. You need funerals for things, too. And, if you're going to a funeral, you wouldn't go with food smeared all over your vest. You'd have some beautiful music. You'd want it to be like: *Boy, they couldn't save their planet, but they did have a certain something.*

Andrew: One of the attitudes that surfaced in the conversations I've held is: "I'm going to drown with my boots on!" There's a dignity in it—similar, maybe, to the kind of dignity you're talking about here—and there's also a commitment to fight to the end because who knows what's possible. Or, as you say in the *Eco Buddhism* interview: "It looks bleak. Big deal, it looks bleak."

Joanna: Right. No whining. This moment is all that we have. Let's love one another, and do all that we can for one another.

Andrew: Well, I know you must go, and this seems like a strong place to end.

Joanna: Yes, but let's end with your *Daily Afflictions*. Can you read the piece about the agony of being connected to everything?

Andrew: Ah, woah, OK. That'd be an absurd, incredible honor.

Joanna: Read it. That'd be a beautiful close to our time.

And I read her the passage. Here it is in full:

The Agony of Being Connected to Everything in the Universe

What is to give light must endure burning.
— Viktor Frankl

Many of us have set out on the path of enlightenment. We long for a release of self in some kind of mystical union with all things, but the moment of epiphany when we finally see the whole pattern and sense our place in the cosmic web can be a crushing experience from which we never fully recover.

Compassion hurts. When you feel connected to everything, you also feel responsible for everything, and you cannot turn away. Your destiny is bound with the destinies of others. You must either learn to carry the universe or be crushed by it. You must grow strong enough to love the world, yet empty enough to sit down at the same table with its worst horrors.

To seek enlightenment is to seek annihilation, rebirth, and the taking up of burdens. You must come prepared to touch and be touched by each and every thing in heaven and hell.

I am One with the Universe, and it hurts.

After I read the closing line, we repeated it out loud together in unison. Joanna smiled. She knew better than most how hard it was to stay open and connected, especially in a time such as ours. She'd dedicated her life to waking people up to that adventure. I was glad to offer her a little something in return.

I ate the last two slices of cucumber, and we moved the tea cups and saucers to the sink. We hugged, took a selfie to send to Lois, and said our goodbyes. I was soon out on the sidewalk, in front of her home. The mid-afternoon sunlight was breaking around some cloud cover, I took off a layer, and decided to walk Berkeley's well-trimmed, traffic-calmed streets. There was much to ponder.

As with so many of these meetings, I hadn't known quite what to expect. I called them "Meetings with Doomers and Hopers," but few had fit neatly into either category, Joanna maybe least of all. "My apocalyptic sense is very strong," she'd said early on. Then, halfway through: "We may be much closer to an outbreak of sanity than we realize." By the end, however, we were discussing the existential etiquette for attending a funeral for our own species. And I wondered, what, really, is in store for us?

We might be facing Hospice Earth. Or we might be on the cusp of a Great Turning. There is certainly no shortage of terrible evidence to support the former, and in our interview Joanna wondered aloud whether it was "moral cowardice" that has prevented her from accepting that that's how it's going to go down. Ultimately, she asserted that we don't know—and according to her read of systems theory, we can't definitively know—and so we must prepare for both. To properly follow Joanna's path, it seems one must attempt to "be of service not knowing whether you're a hospice worker or a midwife." We must prepare ourselves for the End (of all things), while still fighting for a new beginning—all the while both loving life and honoring what we're losing.

It is not simple to prepare simultaneously for such radically different roles. It requires some skill in the art of living and loving, the ability to turn uncertainty into an ally, as well as some care in what stories we choose to tell ourselves. To see Joanna straddling it all with such grace and humor and tea and cucumbers, gave me, well, hope—or at least something that felt like hope. Not the optimistic kind of hope; more like Joanna's "life is a beautiful gift so let's honor it by doing all we can" kind of hope.

"Everybody," says Joanna Macy, "is going around as if it's normal to be preparing for the apocalypse, as if it were tolerable to not be fully alive." In our era of mass denial and self-deceit, it can be terrifying to look around and speak the simple truth of what we see. How do we begin to do this? Consider the legend of the Fisher King, as Joanna told it to me:

> Parsifal, a Knight of the Round Table, is on a quest. He finds himself in a wasteland (the first reference to "wasteland" in all of Western literature). Nothing can grow here. Eventually, he comes to the castle of the Fisher King. He finds the ruler wounded in the groin; he too has lost the powers of regeneration.
>
> There's a prophecy that the curse that is blighting the land will be lifted if a knight comes to the castle and simply asks, *What is the matter here?*
>
> Parsifal is brought before the King. Being a knight, he knows the proper etiquette; he knows you don't ask questions. So, he receives the hospitality of the ruler and the court, and although he sees that something is very wrong, he says nothing.
>
> And lo!—the castle vanishes before his eyes.
>
> Continuing on his quest, he eventually leaves the wasteland, and runs across Kundry the witch, who excoriates him for being such a lousy knight because he had neither the courage nor compassion to ask the king what was the matter. Parsifal is distraught; he falls into a terrible depression.
>
> Seeing that Parsifal is determined to return to the wasteland and redeem himself, Kundry finally consents to give him directions.
>
> Parsifal makes his way back to the wasteland, which is now more wasted than ever. He makes it to the castle of the Fisher King, who is sicker than ever. Without pausing, he goes straight into the hall of the king. There's an event going on. It's all very formal and forced; no one is talking about the sickness that has seized the land. Parsifal kneels by the royal hammock on which the king rests, and says, "My lord, what aileth thee?"

Light comes to the king's face, color to his cheeks, strength into his legs. As he stands, he and his courtiers look out the window where green shoots are already sprouting throughout what was the wasteland. The spell is broken.

Like the Fisher King, we too live in a kingdom in denial. The castle of our civilization also sits amidst a wasteland—a wasteland we ourselves have spawned. But by asking an honest and compassionate question—"What aileth thee?" (in effect insisting that, "No, it's not normal to be preparing for the apocalypse! No, it's not OK to only be half alive!")—Parsifal breaks the spell. And we can, too.

Like Parsifal (or Greta Thunberg, for that matter), we must learn to have the bad manners to speak the truth of our dark time, as well as the good heart to "see them through it." (See page 59.) To set off on our quest, we don't need to have it all figured out. We simply need to recognize that things are broken, and be willing to do what we can to repair the damage.

The first step is simple: with honesty and compassion, ask the question on all our minds. Then listen. And let that heartbroken human connection transform you, heal you, guide you.

Do we need hope for this task? No, says Joanna. In the Buddhadharma, the term doesn't even exist. It's an empty concept. The point is to be present. The pulse of life and connection is enough. And yet, deep in our hearts so many of us are still asking: Is there hope?

MEETINGS WITH REMARKABLE HOPERS AND DOOMERS

DR. JAMEY HECHT

"Witness the whole human story through tragic eyes."

On the question of hope, Gopal had no time for wishful thinking or uncertain ditherings; for him, hope rolled you hard into action or was worthless. He was certain we were going to survive. Joanna, meanwhile, practiced an "active hope," even if she wasn't sure there actually was any. In the wake of these interviews, any kind of hope would have sufficed, because my next meeting[1] was in Los Angeles with someone who had none.

My train rolled south out of Oakland down California's coast, through layers of beauty and destruction, both seen and unseen. On my right, the Pacific Ocean, blue and vast. Yet beneath its sparkling surface, a process of carbon-driven acidification was slowly choking the life out of it. And just a few hundred miles beyond its postcard horizon swilled a gyre of plastic garbage four times bigger than California itself.[2] To my left, the rolling hills and fog-bound forests of mid-Coast California belied another story: only 3% of the original old-growth forests that stood in 1849 (when white people arrived in force) were still standing, while drought and forest fire and the uncertainties of climate change threatened the rest.[3]

As I rolled into Los Angeles, traffic-clogged cloverleafs, paved-over arroyos, and policed enclaves of rich and poor rose up all around me. This built environment felt like an immovable monument to the tragedy of the commons and our ecocidal future. "The world began in Eden," Phil Ochs tells us, "and ended in Los Angeles." In his song of that same name, he sings:

> So this is where the Renaissance has led to
> And we will be the only ones to know
> So take a drive and breathe the air of ashes…
> Welcome to Los Angeles, City of Tomorrow.[4]

On the other hand, I could neither see nor smell the infamous smog that choked LA when Ochs had sung those lines. Thanks to tougher air quality standards, pollution is down 85% since the 1970s, and the number of ozone advisories has dropped from a high of 184 in 1976 to near-zero.[5]

Although the worst of the pollution that remains continues to have an outsized impact on Black and Brown and poor communities,[6] with forward-looking environmental policies including mandatory rooftop solar on all new building construction and an ambitious effort to generate 50% of its electricity from renewables by 2025 and 100% by 2045,[7] the state has become a global climate leader.

For the person I was here to meet, however, these late-game efforts and signs of progress did not outweigh the larger trends. He was convinced the world was ending, not just here in Los Angeles but everywhere. And he might not be wrong. Right or wrong, he had a unique—and uniquely tragic—perspective on how to live with that awareness, which I was keen to hear about in person.

By training a psychoanalyst and historian of literature, Jamey Hecht was also one of the most soulful "collapse bloggers" on the Internet. We had a mutual friend in common who put us in touch. And now here I was, in defiance of local custom walking the 30 or so minutes from my Mid City couch-surf to his apartment.

Alongside his psychological insights, what most struck me in Jamey's online essays was their compassion: "Whoever feels," he writes in "5 Reasons Why Some People Insist on Discussing Collapse, and Even Extinction," "the obvious emotional reality that elephants are non-human persons (they have self-awareness, love their children, mourn their dead, live by matriarchy, form deep social bonds, weep when sad, play joyfully, communicate, and so on) cannot bear the unbearable knowledge that these people are now being rapidly murdered out of existence."[8] Part of it was his heartbroken honesty. Here he is in the same essay laying out the case for why some of us are driven to speak the truth about our situation, even at substantial emotional cost:

> American civilization is an abusive parent who provides more material goods than most, but lies about just how violently he acquired those goods.... In such a family, some kids will prefer to keep the stuff and repress their own guilt and terror. This is not just so they can keep the presents! They do it because if they don't, their Dad's illusory goodness will disappear...and they will be flooded with a painful ambivalence that they are not equipped to process or contain.... The abusive parent has an addiction: oil [which] doesn't just drive the bully in charge, it also powers the profligate lifestyle that is all the kids have ever known. While some kids will need to stay with the abuser's program, other kids will find a way to speak the truth.... Speaking that truth will both risk the wrath of the abusive father, and alienate the kids who are still trying to love him. But in a regime of endless lies and unacknowledged open secrets, speaking the truth can feel so important as to drive us to risk ostracism and punishment. We have to do it.[9]

The oil addiction metaphor was familiar, but rarely had I seen it worked through with such psychological precision. What I had most wanted to discuss with him, however, was another essay, "Collapse Awareness and the Tragic Consciousness," in which Jamey wonders aloud how to make meaning in a doomed world. How, he asks, do we bear the "unbearable knowledge" that "we are staring at a near future of catastrophic warming, no matter what we do?"[10]

His argument is multilayered and relentless. First, he acknowledges the difficulties and emotional costs involved. "It is traumatic to realize this," he says, and so "one simply does not discuss it."[11] To do so, he argues, delivers us over to terrible feelings of helplessness (I can't stop climate change), humiliation (Exxon is more powerful than all of us combined), and anomie (what matters in a doomed world?). But for those of us who can't look away, he sounds his central theme: "The phenomenon of collapse is so frightening that the trauma of realizing it has to be mastered in a way that derives meaning."[12]

He then explores the tricky interplay of hope and meaning, admitting that "until just a few years ago" he believed that it was still possible to transition our civilization "towards a harm-reducing culture." And even though "most people continue to believe it possible still," he argues that "possible or not, it is vanishingly improbable—not as a lottery win or a bet at a roulette wheel is unlikely, where the problem-space happens to include a large number of equally unlikely outcomes, but as victory is unlikely in a war between equal armies after one side is decimated while the other is unscathed."[13] As he brought the allegory home, my blood ran cold: "Maybe the last ten green-shirted soldiers will somehow slaughter their remaining thousand black-shirted opponents—it is philosophically 'possible'—but everything speaks against its occurrence."

He then lays out the options:

1. "Recalibrate one's hopes, scaling them down so that the smallest of victories will count as a great 'yes' from the universe"; or

2. Go back to the data in a search either for hope or for "that dark certainty which makes despair into a solid resting place"; or

3. Use your remaining uncertainty to trigger the equivalent of a "restore" function in an electronic device, that deletes the painful knowledge and restores the comfortable illusions we once held; or

4. "Rest one's case within the limits of human knowledge," i.e., acknowledge that nobody really knows exactly when things are going to become utterly unlivable.[14]

If one finds these options unsatisfying, he offers another alternative: "the tragic consciousness."

It's a phrase, Jamey explains, that "literary scholars and critics have invented to describe the paradoxical effect of the tragic drama, where the observer experiences a strangely elevated mood after watching a sympathetic figure get destroyed by the gods, by society, by the entailments of his or her own mistakes. The material is miserable, and yet it elates us. The effect has something to do with what Aristotle called catharsis, where the story purges us of pity (which we feel for the figure on the stage, since he is doomed where we are safe) and terror (which we feel for ourselves, since we identify with him on the basis of a shared humanity and a shared (i.e., mortal) predicament. But catharsis is only a part of it."

He writes, "The tragic consciousness seems to require that we become witness to the whole story. It is this narrative completeness that grounds a story's moral complexity," and "answers the question of how a reasonable person could possibly come to this."[15] Which, he notes is "the same question psychotherapists are asked to consider when they get a client who challenges their sense of decency," and also similar to what is demanded of us in the practice of nonviolence, where "we are asked to consider how our opponents got to where they are—how they acquired their racism, or greed, or cruelty, etc.—in order to love the human beings beneath the history."

Once we know the full story—whether through tragedy, therapy, or nonviolent practice—"the ugliness has a meaning." Once we can sufficiently set aside our feelings of anxiety and blame and shame and see the full story, Jamey suggests, we can "contemplate our species' emergence, rise, and crashing decline," as a story "weirdly graced with an aesthetic and narrative completeness." "When it is too late for prudence or virtue," Jamey argues, "wisdom loses its ethical character and becomes a mostly aesthetic phenomenon." When we focus on the paths not taken, "we experience the disaster as a waste, a stupid mistake, a crime. It is all of those things, but when we push the counterfactual away and focus on what did happen, the picture changes. It becomes tragic." And from there the essay rolls tragedy and reality together in a conclusion that deserves to be quoted at length:

> Choose a tragic hero, and you will find that his or her hubris was avoidable—but only in a different world, or with a different inner character.... Tragic heroes do what they do for manifold reasons, the heart of which is human nature: we are the animal that does this. So it is with our destruction of the planet we loved.
>
> All animals in an isolated environment (like a vat, an island, or the Earth) do as we did, when they consume the available resources in a finite system until they overshoot the system's carrying capacity and begin to die off. If we are unusual in that we became aware of what we were doing...we are also unusual (though again, not alone) in our tendency to ignore warnings when our identity is involved. We did not lower our energy consumption because it would have been a return to weakness, childhood, helplessness, all the things industrial civilization fears and hates the most. Just as in Greek drama or Shakespearean tragedy, this fear-of-the-wrong-thing determines our fate and defines us in the universe.
>
> For me, these days, and perhaps for you, coping is a two-handed job: one hand holds the despair which must somehow be held (contained, regulated, bounded); the other holds the tools with which we must make our attempts to adapt.[16]

Here was an incredibly thoughtful person, who after much research, contemplation and angsting, had become convinced we were on the cusp of imminent catastrophe and very possibly extinction. Whether I dismissed his approach as a premature surrender, or respected it as a terrible kind of knowing—and I did both in the weeks leading up to our meeting—I was curious what kind of man Jamey would turn out to be. What was it like for him to know deep in his heart that we are doomed? I didn't have to agree with all his conclusions or share his sense of powerlessness to suspect I had much to learn from his dogged sensitivity to our predicament.

He met me at the door and welcomed me in. His apartment (actually, his office, art studio, and home all rolled into one) was small and dark and full of books; it felt more like New York than LA. The canon of Western literature and psychoanalysis had their shelves; so too did the canon of Collapse. There was Derrick Jensen leaning on James Hansen. Sally Weintrobe's collection of "Psychoanalytic and Interdisciplinary Perspectives on Climate Change" stood uneasily in between *The End Game* and Craig Dilworth's *Too Smart for Our Own Good*. On yet another shelf, George Marshall was holding forth on Why Our Brains Are Wired to Ignore Climate Change as Guy McPherson wondered whether The American Dream had Become a Nightmare and Dale Jamieson wearily told us Why the Struggle Against Climate Change Failed, and What It Means for Our Future. The impossible news went on for four full shelves, a shrine to Jamey's obsession and dedication. He may have been an amateur collapse theorist, but he could probably put more than a few professionals to shame.

He offered me a beer, and grabbed one himself. And we started talking. The most noticeable thing about Jamey was his voice. It had an intellectual, almost British lilt, that at first encounter could easily come off as pretentious and airy, but once I got used to it, I realized, no: the lilt was speculative, reflective, as if he was holding each of his ideas up to the light for his own critical inspection. He was intense and curious; his intellect fluid and wide ranging. It was obvious that he loved the life of the mind. "When people ask me what I think about things," he said, almost as a way of introduction, "I tend to name-drop books in an effort to fan out the subject, and increase its available surface area...." After some of this "fanning out," and some personal backstory in both directions, we zeroed in.

Andrew: In one of your essays, you describe yourself as a "bleeding-heart doomer." What do you mean exactly, and what were the stages that led you there?

Jamey: The bleeding heart part, which I appear to have said at some point, has to do with the politics of empathy, and the doomer part has to do with how impressed I am by the evidence that the habitat for our species on this planet is profoundly compromised and is in a cascading process of contraction and likely collapse.

A-a-and that ain't good. Not for us, and not for all the other species.

Now, given that all things begin and end in eternity, there had to be a generation on the scene when things became unworkable. Whether that was going to happen hundreds of thousands of years in the future, or right now and over the next coming decades, it had to happen eventually.

Given the evidence—and I find much of it, from Guy McPherson to Lester Brown and the Worldwatch Institute, convincing—that we are among the handful of generations present at the unraveling, we can, I think, reflect that somebody had to draw that straw, and we happen to be among their number. Why this perspective is soothing to me, I don't know, but I find it really reassuring. An end had to come at some point, and when it did, somebody had to be there to witness it. Turns out, it is us, and our children, and perhaps their children, but not too much further out than that. It stings like a wasp.

Andrew: I may not share your conclusions, but it still feels too real—if that makes any sense.

Jamey: It *is* too real. It's not amenable enough to the frames we try to put around it to handle it. It tends to dissolve whatever perspective or emotions we try to bring to it in order to tolerate it. It's like trying to hold radioactive material. You need one of those special glove-box setups, to even handle the stuff.

Andrew: If Tesla can make a battery to handle half of Australia's electric grid, you'd think they could make a glove-box to better hold our hearts, no?

Jamey: The goal, it seems to me, is to live in such a way that your one shot at happiness is not utterly destroyed. How am I to live in the truth on a doomed planet and not lose my shot at happiness? The answer has to be some kind of compartmentalization in the mind—you've got to find a part of yourself which can tolerate these truths, so another part of yourself can be spared that struggle.

Of course, we all know that compartmentalization comes at a cost. If the compartmentalization is deep enough, it reaches all the way down as a vertical split into the unconscious, and then you really pay a tax for it. However, it seems to me that a more conscious and deliberate compartmentalization, a more cultivated and informed compartmentalization, might be the road forward.

In this way, part of your mind is in the business of dealing with these facts and the rest of you is set up for a very different and sometimes contradictory goal, which is to not miss your shot at happiness. You live in the truth insofar as you occasionally touch base with the facts, but you live for the good in that you cultivate happiness largely through relationships with other people. Truth isn't everything.

Andrew: This deliberate compartmentalization is a life-hack for dealing with our "impossible situation." A way for us to still live both in the truth and for the good, in spite of what we know.

Jamey: Yes. Remember, Plato defined philosophy as having three main branches: the true, the beautiful, and the good—where truth was dealt with by epistemology, beauty by aesthetics, and goodness by ethics.

Andrew: Channeling Nietzsche: "We compartmentalize so that we do not perish from the truth."

Jamey: More or less.

Andrew: But in your estimation, the truth is going to kill us anyway.

Jamey: I wish I could say something more satisfying than that, but it's the best I can do.

Andrew: You've read widely and thought deeply about social and environmental collapse, and our prospects as a species going forward, and you're quite convinced we only have a few generations left. I don't want to debate your conclusions. My interest here is not to definitively determine which future we're going to get but, rather, to explore how to be human across a range of our possible futures.

Jamey: I respect that. And just for the record, I've read and listened to Guy McPherson's critics, and I was praying that they would convince me that Guy was wrong.

Andrew: We all want to live in a world where he is wrong.

Jamey: Yeah, including Guy. He himself devoutly wishes to be proven wrong. The thing is, a fellow like Richard Heinberg, who is profoundly aware of the energy profile of our civilization, or Joseph Tainter, the preeminent scholar of the collapse of complex societies, or Dmitry Orlov—

and I could name several more—while they may not say it as baldly as Guy, they know that what's already baked into the cake is severe enough that our current set of living arrangements will not endure, and that we probably face conditions in which our population will be, at the very least, reduced to something much, much closer to our natural carrying capacity on this planet in the absence of cheap and abundant fossil fuels. It makes sense to be nervous.

Andrew: Speaking of being "nervous," you've described anxiety as an understandable, if not always useful, emotion given our circumstances. Care to elaborate?

Jamey: Freud wrote about several different kinds of anxiety, and one of them he called "signal anxiety," which is an apprehension about some threat in the real world. The function of signal anxiety is to alert you to the presence of those dangers so you can do something about them, before the impending harm happens, or more to our point, before it has become inevitable that such harm will happen.

Andrew: So a very useful kind of anxiety.

Jamey: Some of the anxiety we feel about climate change is signal anxiety because it motivates us to do stuff in the hope that it will change things. But we find ourselves in a predicament, which by definition is not amenable to a solution.

Andrew: Unlike a problem, which might be.

Jamey: Right. Now, if you're in a predicament, signal anxiety is no longer of any use. It's outlived its usefulness. Since our mission in the world is to live in such a way that we can be both happy and ethical, it makes sense then to distinguish between that portion of the anxiety which can lead to an improved outcome, and that which can't.

Andrew: The trick, of course, is that people draw that boundary line—between what anxiety is useful, and what isn't—in very different places. It depends so much on whether you think we have a chance to turn things around, or at least survive—or not. Many would argue—including me—that there's still much that can be done.

Jamey: A big part of what Guy McPherson spends his time doing is spelling out how his message of near-term human extinction is not quietist, that he's encouraging people to live what he calls "a life of excellence" which includes striving to oppose the forces that face us with destruction even in the knowledge that we will not succeed.

Andrew: And you?

Jamey: In Jewish tradition, there is this very old notion that you're not responsible for fixing the world, but you are responsible for trying. When I was younger, I used to say, that it's almost impossible to fix the world, but the fact that it is just barely possible, even infinitesimally possible, demands that we try.

Some of my dearest heroes are people who gave their lives for the good and, indeed, the true and the beautiful, but chiefly the good. I am not among their number. I wish to survive. Not only because I have a child, and sisters, and parents, and friends, and all that, but because I love life and want to be here even though so much of life hurts really bad.

Nor do I want to live a life of despair. Even though the facts appear to warrant it, I decline to accept a despair which would deliver me over to bitterness and agony. Sure, there is an inherent bitterness and agony in confronting these facts for the first time. And, then, as you begin to live with them, more bitterness and agony. Ultimately, however, because conditions in this part of the world at this historical moment—especially for beneficiaries of white privilege—have not yet declined to a point of being unbearable, I still have detectable options available for the pursuit of feelings of wellbeing. And I avail myself of those options and those feelings even though I know that the incredibly complex, high-energy systems upon which we all depend for the continuity of our being are likely to come apart in the next few decades, or perhaps even sooner.

Kierkegaard says this age is both comic and tragic: tragic because it is perishing—

Andrew: And comic because it continues, right?

Jamey: Right.

Andrew: I read that in one of your essays and literally laughed out loud.

Jamey: You can't lie to yourself, but you can shift the emphasis onto the parts of reality that you can tolerate, and it behooves you to do that. If you want, you can die in the teeth of the truth, but it might be better to acknowledge that the truth is over there, and I see it, and I won't journey into its belly and be digested. That's different from pretending that it doesn't exist.

Andrew: I assume you're familiar with the notion of "therapeutic distance."

Jamey: No, actually. Tell me about it.

Andrew: It's the idea that a therapist should hold the middle distance:

close enough to have an empathic regard and rapport with the patient, but far enough away to not get sucked in.

Jamey: Exactly. Much of what I try and impart to my clients as a psychotherapist is to neither repress your feelings nor let them take you over.

Andrew: Not so different from how you suggest dealing with "collapse awareness."

Jamey: Yes. What I want to do is take a deep breath and then sink low enough in the pool that I can feel the bottom and push off from it. I want to know what's down there. I want to know how bad it truly is. I want to avoid a scenario in which my fear and anxiety prevent me from looking at the truth, but I also want to avoid a situation where I am hypnotized by the truth the way that a rabbit is hypnotized by a snake. I don't want to run from it, and I don't want to be eaten by it. I want to see it, and then I want to get the fuck back to going about my business.

Andrew: A healthier approach than, say, wallowing in the bad news, or hitting other people over the head with how bad it is.

Jamey: There's a very particular mode that you and I know a lot about, which is: *Oh fuck, I know the world is coming to an end far quicker than is generally known or acknowledged, and I am alone in the unbearable state of mind which that awareness imposes on me.* I want to help people get out of that spot. I can't do it by telling them: *Hey, you know what? Biochar*[17] *is a viable solution.* (Even though I think biochar is a viable solution, because I also think inertia and capitalism will make it almost impossible for biochar to be deployed on the scale and at the speed to sequester enough carbon for us to survive.)

Andrew: So, how *do* you get folks out of that spot?

Jamey: Well, one way is to "normalize our predicament" by saying *I, too, am aware of this, and yet I show up for interpersonal relationships, and community, and ice cream, and religion, and art.*

Andrew: I met with Joanna Macy earlier this week—I'm sure you know her work. You two are extremely different people, but she says much the same thing. She wouldn't use the word "normalize" I imagine, but both the healing and empowerment parts of her practice are based on bringing people together to, basically, look into each other's eyes and say: *I, too, am aware of all this. And here I am showing up for you, and for this moment.* It's not so different from what you're recommending.

Jamey: I know her work. And, yes, in that way, not so different.

Andrew: So, you act in a Shakespeare company, sing, sculpt, and write poetry. You're a psychoanalyst, a journalist-activist, a professor of literature. You speak how many languages? Anyway, you wear several more hats than the rest of us mere mortals even know how to put on. Do these overlapping lenses, or antennae, give you any special insights into our situation, into our predicament?

Jamey: In a flow state, you can cope with damn near anything. Artistic practice is one way to get into a flow state. As are religion, sexuality and relationship, communion with wild nature, cultivation of plants, meditation, maybe a certain class of relatively safe drug experiences…the list is long but very finite.

Andrew: Extreme sports?

Jamey: Maybe. It's not my thing, but, yeah, when you get into a flow state, you can tolerate damn near anything. "The game plays itself," as William James said.[18] Part of it is the pragmatic business of feeling good because you're doing something that you're good at, and your unconscious takes over, and you get to observe yourself in this flow state. But I think it's more profound than that. These experiences of creativity, sexuality, and religion are *veridical*, which is to say I think that they are not merely expressive but revelatory.[19] It's as obvious to my left brain that there is no God as it is obvious to my right brain that there is.

Andrew: That's quite a paradox, isn't it?

Jamey: To my left brain, however, it's not paradox. It's a contradiction in terms. It's absurdity.

Andrew: And can this left-brain/right-brain split help us navigate the moment we're living in?

Jamey: The right brain has access to truths which transcend our predicament. The left brain just doesn't. It chokes to death on the facts, and it can't get beyond them. It is stuck on an account of love which stops at dopamine, an account of music which stops at the score. Whereas, if you invite the left brain and the right into a room and say *I am going to expose you to Beethoven's First Symphony*, and you provide them each with a copy of the score, and then you play the music over the speakers, the left brain will go into the corner with the score, utterly ignore what comes over the speakers, and say *I've got the document! It's all here in black and white! I have mastered it.* Meanwhile, the right brain will be utterly uninterested in the score but be swayed like a branch in a flooded river by the passion of the music.

Who is right? Well, of course, each of them is in possession of a crucial part of the truth, but ultimately, it's the right brain that knows what matters most. The right brain knows *Oh, we destroyed the biosphere. It is tragic. We are here to witness it.* The left brain only knows that it's a totally regrettable disaster. There's nothing but bitterness, regret, and shame. The right brain can see this as a tragic predicament and say *Oh, there were too many of us. Because of the commercial exploitation of fossil fuels, we became like the unreflecting animals whom we held in contempt. This is hubris pursued by nemesis. The gods are in charge after all. Oh, well…. Soon we will lie down in the dust with our ancestors.*

Andrew: You're channeling Sophocles here. And Shakespeare.

Jamey: Well, those are my folks. And that's whom you need if you're going to cope with this shit.

Andrew: In your essay, you talk a lot about the difficulty—yet the necessity—of making meaning in this moment. You suggest that the "tragic consciousness" is the only way to give our awareness of collapse—the reality that feels like too much reality—its proper due.

Jamey: As we mature, we discover that the body will die and that everyone we love, including ourself, is ultimately perishing. The tradition in the humanities, particularly the poetic tradition, asserts a value of another kind: the acknowledgement and the perpetuation not of the human living body, but of the human image which compensates for the loss of organic, biological life. Says Shakespeare in "Sonnet 18":

> So long as men can breathe or eyes can see;
> So long lives this, and this gives life to thee.

Andrew: The "this" being the poem itself, and by extension human culture, yes?

Jamey: Yes. And the trouble with climate change is it says *Fuck you!* to all that. It says: you will no longer be able to exchange the cultural continuity of the human image for the loss of organic life of the human body, because the human image is no longer culturally continuous. It will be brought up short.

Andrew: By the collapse of civilization. By our own extinction?

Jamey: Yes. And here enters the salience of that wonderful book, *The World Without Us* by Alan Weisman, which speculates about just how the seemingly "permanent" infrastructure we've created is likely to decay in the absence of our maintenance of it. That book adds a beautiful piece to our puzzle, because once you realize how fucked we are, the next thing you often wonder is whether the biosphere will be able to flow over and around what we have inflicted upon it. If the book is any guide—and I think it is—it will take an incredibly long time, and it will be an incomplete process, but the biosphere will eventually reassert itself.

Andrew: And this gives you a kind of solace?

Jamey: I take a great deal of heart from that reflection. I really do.

Andrew: More than a few of the people I've spoken with, who, as you do, assume it's all over except for the dying, take heart from that reflection. I call it the "We're fucked, but this Rock is going to be fine (um, 100 million years from now)" attitude. And I get it. I can see why it's a reassuring attitude *if* you think it's already all over. In fact, it reassures me on my worst days. *If*, however, we still have a chance to turn things around, it can be a profoundly demobilizing point of view. So much hangs on where we draw that line.

Jamey: Extinction happens. This won't be the first mass extinction by a long shot.

Andrew: Sure. The Permian mass extinction took out 97% of the species that existed on the Earth at the time.

Jamey: There's a book about it called *When Life Nearly Died.* And yet I don't feel the least obligation to work through, in my spirit, the mind-bending loss which that extinction entailed, nor any of the other five mass extinction events. The sixth one, however, is happening while I am on the scene. I am participating in it by occasionally flying back to New York maybe once a year to see my family. I'm implicated in it. I'm vulnerable to it. It affects me in a way that the previous five extinction events don't. But from the perspective of eternity, all six are the same!

Andrew: But we don't live from the perspective of eternity. We live from the perspective of humanity.

Jamey: That's why the right brain is so important—especially now—because it makes tangential contact with eternity. (A tangent makes contact with a circle at only one point, but it makes contact.)

Andrew: There's a Zen expression: "My heart is on fire; but my eyes are as cold as ashes." It's an invitation into non-dualism, into holding both parts of a paradox, into operating in two contradictory modes at the same time.

Jamey: The right brain's jewel is the left brain's trash.

Andrew: The awareness that we're heading towards imminent collapse puts us in a predicament. It puts us in an impossible situation. The way to live with it, and in it, is to embrace the whole paradox.

Jamey: The right brain has this capacity for paradox; the left brain regards it as mere absurdity, as a mere contradiction in terms.

Andrew: The moment we're in is asking our right brain to seriously step up its game.

Jamey: It most certainly is.

Andrew: How? What does Jamey the psychotherapist have to say?

Jamey: I became a psychotherapist in part because of my own experience with depression and mood swings.

Andrew: Doctor, heal thyself.

Jamey: Exactly. Ralph Waldo Emerson says, "Our moods do not believe in each other." When you've coped with depression, you know how enveloping it can be, and how seductively it can convince you that the darkness which it discloses is the sum total of reality. Then it passes.

Andrew: Yep. Been there enough to know that.

Jamey: I figured.

Andrew: Go on.

Jamey: Depression has a way of teaching you that the darkness has its place. It is not the whole story. You can't afford to leave it out; and you also can't afford to give it everything. It must simply have its place— a place at the table, but not the head of the table. A place in the car, but not the driver's seat. You can't get rid of it, but you can generally prevent it from steering you into a level of difficulty you can't tolerate.

Andrew: Okay, so who *should* be driving the car? You're saying despair should be sitting in the back seat, but who's driving?

Jamey: Your most compassionate, nurturing, protective, adult self—the part of you that can tolerate what you know.

Andrew: And protect the part of you that can't?

Jamey: Yes. There's a film—which I regard as a great film—called *Melancholia*. A planetoid roughly the size of the Earth collides with our world obliterating the entire human race. In the days leading up to that impact, a man lives out a hypermasculine script of buying propane tanks and all kinds of emergency survivalist gear while his wife instead pitches a tent and sits in it with their child, awaiting the end without telling the kid what is happening. She's holding herself (holding the truth that she has to tolerate) while holding the kid whom she loves.

The man kills himself prior to the impact. The woman awaits it in the wheat field with her child. The person who is driving the psyche is that most adult, most nurturing, protective self who can handle the reality of the situation and say, *Come on, I got you, buddy. Come with me, we're going to be OK.* But the world is ending. *Yes, I know. Everything ends. Come with me.*

I wish I could say something more nutritious, but that's all I've got. Bowie is dead, why shouldn't we die someday? We have to, anyhow. It's all right. Mortality is a given; it's not some sort of shameful error.

Andrew: You're convinced we're definitely going down, and fast. You've shut the door. Joanna Macy still holds that door open. She thinks we might make it through, though we very certainly may not. I pressed her on it. Was she just hedging her bets? No. And she's designed her work to prepare us for both possibilities. I sum up her existential invitation as follows: "Be of service, not knowing whether you're a hospice worker or a midwife." I think I'm in her camp: Keep fighting for the best catastrophe we can get, while being a tragic soulful witness to this terrible moment, however it turns out.

Jamey: The man who was the pianist on the Titanic played the piano all the way down. That was his destiny, or dharma, as some would say. The potency of the metaphor lies in part in the inertia of the ship headed toward the iceberg—that you can't turn an ocean liner around just by recognizing that it's headed in the wrong direction.

Andrew: If you knew soon enough—

Jamey: If you knew soon enough and you had the attention of the people whose labor you required, you might be able to turn the ship around, yes. But lacking either the time or people's attention, your knowledge would be that of Cassandra. There's a line in the Oedipus Tyrannus which I translated, where Tiresias the prophet says "Damn, damn. How terrible it is to understand where understanding is useless."

Andrew: How terrible it is to have signal anxiety when the signal is useless?

Jamey: Right. If understanding were useful, our anxiety would be signal anxiety, alerting us to a problem that we could then solve. But—while signal anxiety must be respected—in a predicament, as I said before, it is not useful.

Andrew: So…what *do* you counsel in a predicament?

Jamey: To the degree that they are available, seek out the sources of human wellbeing: loving relationship, spiritual life, the arts, civic participation, and nature. That's most of what's there. Put yourself at the top of the list of the people who deserve your compassion. See to your own wellbeing as you would cultivate the wellbeing of a younger person who depended upon you. Were you your own child, you would not be cruel and exacting, nor always judging and measuring. You would be kind, and decent, and gentle.

To conclude: The best way for people to cope with a broken world is to cultivate their most nurturing and protective adult self, and turn it to face the traumatized child self who cannot tolerate the facts.

Andrew: *Melancholia*, except—

Jamey: Yes. Instead of the mother taking care of the kid, you take *yourself* by the hand—joining two intrapsychic parts of yourself—and say: *Come with me. I will be with you no matter how nasty Hurricane Katrina or whatever else becomes.* You do the best you can to protect that vulnerable part of yourself just as you're protecting the other people who depend on you. So while you're still here, you can have some kind of access to joy, or at least coping. That's your job!

Your job, your dharma in the world—even at the end of the world—is to become a non-miserable person, with less bitterness and scorn for yourself, and with greater capacities for sadness (not depression) and joy. Put more plainly, at the cost of some nuances: Your job at the end of the world is to become a happier person.

Your job at the end of the world is to become a happier person.

Your job at the end of the world is to become a happier person. If I'd come looking for paradoxes, well, here was a doozy. Hell, at the end of the world, it's hard enough to simply be a person, never mind a happy one.

I appreciated Jamey's willingness to make what good he could of the darkness he saw, but I couldn't quite tell where his willingness ended and his willfulness began. In spite of his commandment to be happy, was he himself happy, sitting there in his sunless office-slash-studio-slash-apartment, filled with books and artifacts and *memento mori*? Was it really all over except for the dying to come? He had a first-class intellect and he'd read his way through most of the Collapse literature, but, ultimately, who was he to say?

Luke, his bookshelves whispered to me throughout our meeting, *join me on the Dark Side.* No, Jamey. No. Thanks for the beers, and for inviting me into your home—and into your lucid, beautiful, half-defeated soul—but no, no. Or, at least, not yet. For better or worse, there's still some signal in *my* signal anxiety. And even if we green-clad soldiers are hopelessly outnumbered, even if those thousand black-clad warriors of Capital and Inertia are closing in, even then, we must fight, we must resist, we must do all that we can. Right? All the way back up the West Coast on Amtrak's Starlight Express, I was hounded by that question-cry of right? Right? Right?!

And I was hounded by wonder, too. Wonder at the inventiveness of the human spirit, for starters. Look at the last three people I'd met with. They were all hashing through the evidence. They were all reckoning quite genuinely with our predicament. And yet look at the vastly different places they had landed and premises they'd chosen. For Gopal, our survival was "axiomatic." We were wired for cooperation. We'd survived far worse. We would make it through. It was going to be ugly, but one way or another we were going to make it through. The critical question for Gopal was *how*—because the more justice we could bring about, and the more equitably we could share the inevitable suffering, the less terrible would be our fate.

For Joanna, our ultimate fate remained unknown—decidedly so. She'd very consciously chosen this uncertainty. Why? Was it her faith in a universe rich in emergent powers? Her Buddhist heart-mind practice of not knowing? Or a wanton decision to live—in spite of it all—in the most beautiful, life-loving way she could imagine? In any case, she was living her life and doing her work in service to our two possible fates, reinvention and extinction both.

And, then there was Jamey. Almost as strongly as Gopal believed we would survive, Jamey believed we were done for. We all need to tell our-selves a story, but "we're done for" is a grave choice indeed. How much of this choice was on the merits of the evidence, and how much was, well, a choice? A good fit with his character, training, and sensibilities? There's rarely a shortage of evidence once you've decided what you're looking for.

So, maybe it's not really about the evidence at all. At the end of his "5 Reasons…" essay, Jamey discusses the futility he feels trying to reduce his carbon footprint:

> When I reach for the feelings of well-being that would come from an experience of personal agency and instead feel totally power-less, I turn to the only thing left on the shelf: my awareness. I can't fix reality, but at least I can keep acknowledging it.[20]

Jamey feels powerless, and so he focuses on his awareness. Which is more than most of us are willing to do. And by doing it with such lucidity and compassion, he's able to bring up from the bottom of the pool a treasure trove of insights that can help us all through our "collapse awareness." But what does he offer the millions of us, myself included, who haven't given up on fixing reality? Who are still getting a signal from our signal anxiety, and acting on it. If your sense of power (or, in this case, powerlessness) hangs on trying to police your individual carbon footprint in the face of the vast destructive power of multinational capital (an effort Jamey describes as "99.9% symbolic"), well, then, yes, you're going to turn away from trying to fix reality and just get better at acknowledging it. But if your sense of power is rooted in being part of a community that is doing something about the problem, then you're way more likely to keep on trying to fix reality.

I've been part of several social movements, from nuclear arms reduc-tion in the 80s to battling sweatshops in 90s to the fight for affordable healthcare in the 00s. We didn't win all our battles, far from it, but we made a difference, and the beauty of that experience—and the sense of power that came with it—lifts me up still. Anyone who sticks with their activism into middle age learns how to keep the faith. (Yes, you lose the faith, but you also learn how to find it again. And lose it again, and find it again.) You learn a certain eyes-on-the-prize kind of discipline. You find that doing something at the scale of the problem (even if it's more of a "predicament" than a "problem") still helps you face the problem, er, predicament.

All the time I'm asking myself whether, in the face of our overwhelming ecological crisis, I'd prefer to have a sense of agency, even if it eventually turns out to be fictitious. And I think I would. (If my plane is going down, rather than a helpless and terrified passenger, I'd want to be the pilot focused on furiously trying to regain control. Even if it still went down. Maybe *especially* if it still went down.) And so, whether warranted by "the facts" or not, I'm still trying to fix reality. And I'm not alone. Millions of us still believe we can achieve a tipping-point level of awareness and action that will force governments and corporations in the direction of viable solutions. We still have faith—and we still feel we have a reason to have faith—in humanity. Then again, turning the truism back upon myself, there's rarely a shortage of evidence once you've decided what you're looking for.

Jamey Hecht re-read our interview in March 2022 and asked me to add the following to reflect a shift in his perspective since we'd spoken:

> Again and again, I've been shaken by emerging bad news about the biosphere. Again and again, I've been surprised and relieved when the worst predictions proved untrue, or, at least, premature. From 2002 to 2006, my friend Dr. Guy McPherson (if I understood him) seems to have thought we would be in far deeper shit by, say, 2020 than was the case. This does not warrant optimism, but it has shifted my perspective. And as Andrew poignantly suggests, our personal circumstances can influence our worldview no less than can the data we confront. My own life is better now. My heart is more open. I read Joanna Macy more receptively today than I did even a few years ago. She did (and does) a better job with the human side of climate change than I; I'm grateful for this.

Guy's biggest concern seemed to be abrupt, rapid, catastrophic climate change triggered by the "methane time-bomb." It's the thawing Arctic permafrost, and the frozen methane crystals on the coastal ocean floor. Warming from CO_2 has been releasing more and more of that methane, a superabundant and highly potent greenhouse gas that's been implicated in some of Earth's previous mass extinction events. This may still very well end agriculture and wipe us out. But in the first years of this century, "Near-Term Human Extinction" (NTHE) seemed to me totally inevitable. Because (and only because) that seemed to be the task at hand, I developed the position I took in this interview. I was asking myself: Suppose the worst is true. What then?

The climate crisis is an existential Rorschach Test.

We tend to think of our civilizational crisis as a reckoning with reality, a moment of truth that will bring out our best or worst selves, a stress test that will show us what we're really made of. And well it may be. But maybe what we're really made of is simply who we've always been? Folks say, that as we grow old, we "become more who we are." Having seen two aging parents into the grave, I'd have to agree. What if facing the likely collapse of civilization, and the possible extinction of humanity, has a similar effect? Yes, it could be a reckoning that inspires us towards fundamental social change and personal transformation; however, it could also just make us more who we are.

What is it about Jamey Hecht that brings him to his twilight philosophy? Jamey spoke admiringly of one Theodore Ronald Brailey, the pianist on the Titanic, and how in the crux of that unexpected moment in the middle of the North Atlantic in 1912, he courageously stepped up into his destiny and dharma and kept on playing as the ship went down. Well, how much was it Jamey's own dharma to see this man as a role model for our time?

For Guy McPherson (a strong influence on Jamey), the allegory of the Titanic also looms large. There are no lifeboats on Guy's Titanic. What is it about a Guy McPherson that drives him to be a prophet of doom? If we take Guy's story about himself at face value, we'd have to believe that his relentless rationality and duty-bound dedication to the facts led him to the very harshest of conclusions, and once there, he could hardly do other than become the town crier of doom. In that story, the facts were there, it just took someone singularly dispassionate to be their advocate. But there's another read: every serious debate *needs* its outliers, needs someone to anchor the far wing, someone to champion the worst case scenario, simply in order to cover the whole spectrum, and have the full conversation, and allow the rest of us to think and feel and imagine and contemplate all the possibilities, including the worst. There's yet another read, almost a dramatic read: this was Guy's destiny, his fate. His whole life long, his stubborn character has been grooming itself to play this part.

Climate scientists, sifting through the data, see the shape and direction of things to come, and they are terrified. Some, like Guy, pronounce a death sentence on the species; others sound one timbre or another of existential alarm. The data, especially as it trickles out to the rest of us, functions a bit like a mirror, like tea leaves telling us how things might go, or even as a cosmic Rorschach Test (those suggestive ink blot drawings psychologists place in front of their patients to induce free association), where our fears and hopes swirl in and out of focus, showing us what we want and who we are.

If we want there to be a sliver of hope, then we find it, and not just because we need there to be a sliver of hope, but also because there's just enough grey area in the data to allow it. Likewise, if some part of our deepest self, transfigured by the crisis, is most at home in a world that is doomed, then we are more likely to read the data that way.

There's enough room in the data, enough variables, enough un-knowns, to permit many different interpretations. But Jamey and Guy have made a strong choice for the worst one. At some level, I have to wonder whether they actually *want* to live at the end of the world. There's something pure about it that might appeal to a philosophically inclined personality. I myself feel that pull sometimes.

At times the crisis seems a theater scrim upon which we project our truest self. But maybe it's not our truest self; maybe it's our most dra-matic, or terrified, or heroic, or dissembling self? Maybe it's all our many selves? The crisis is a scrim upon which we do a full shadow puppet show of all the competing personae within us. Yet that description is too pas-sive, because this process is neither neutral nor just in one direction. It's a fluid and highly combustible encounter between facts and character and story—between "reality" (or at least that portion of reality we're willing to see) and who *we* wish to be. As we search for a place to stand, a story to live in, a world we can abide, we fit the facts (each of us according to our own ethics and algorithms) to suit those needs. The facts of climate chaos matter, but exactly what the crisis calls forth from within us, what burden it demands we shoulder, what opportunity we think it gifts us, are largely up to us.

For Gopal Dayeneni the crisis is the field on which the age-old battle between extractive capital and life-giving community will be fought. For Joanna Macy it is an awakening, an invitation to "serve people's exquisite beingness in the here and now." For Guy McPherson it is a stoic accep-tance of the near-term extinction of our species. For Jamey Hecht it is a question of how to know the worst and still be happy.

And what about me? In the final analysis, isn't this whole investigation just a Rorschach Test by me on myself? Isn't how I frame the facts and how I curate all the various threads and perspectives I come across just my own way of making sense of our situation so that it aligns with my own fundamental sensibility? So that *I* have a story *I* can live in? Things are grave, but not terminal; bleak, but still hopeful in a dark, difficult, cosmic-ironic way. Of course they are: because that's me; that's my own existential sweet spot.

And so, I take Jamey and Guy and these other prophets of total doom seriously because I feel they have something to teach us (and because, well, you can never have enough nihilist street cred). But I don't actually want us all to go down, and so I weigh their near-term human extinction predictions as too extreme. Yes, I want there to be a path forward. But not an easy one. I don't want us to be saved in a way that lets us off the hook, that short-circuits our necessary reckoning. I don't want some whiz-kid inventor to step in at the last moment with some newfangled tech solution to save the day. (Of course I do, but I also don't.) No, I want our path forward to be difficult and do its necessary work upon our souls. I want it to redeem us and transform us.

There's something Puritanic or karmic—or, yes, tragic—in me that wants us to have to deal with our own mess; that wants the repressed to return with a vengeance and deliver her cosmic justice, and teach us what we need to learn so that we will live differently on this Earth. And so, that is the story I tell. That is the way I "curate" the facts. *Oh, it's bad, folks. It's so bad that only a few of us truth-seeking souls are willing to see it and tell it like it is. And, of course, I'm one of them, hello! Oh, but it's not soooo bad, it's not extinction-level bad. There's still a path through. But it's a difficult path, and we must have the vision and stamina and "death eyes" to walk that path. So, come, let us go. Bring your resilience and your kindness and your gallows humor, and be ready to be reborn, because this will change everything.*

WHAT IS STILL
WORTH DOING?

*We don't have a right to ask whether we're going to succeed
or not. The only question we have a right to ask is
what's the right thing to do? What does this earth require of us
if we want to continue to live on it?*

— Wendell Berry

Jamey Hecht was convinced humanity wouldn't last more than three or four generations. Gopal Dayeneni was equally convinced we would survive—however, we'd have to survive one of the most catastrophic centuries in human history. Either way, it's a hard sell.

Usually, if you're volunteering for a cause and trying to rally support, you'll knock on your neighbors' door with a message of hope and possibility. Something along the lines of: "If we act now, neighbor, we can...." But how do you adopt that hopeful, can-do approach in good faith if you think we're mostly doomed? How are you going to talk to your neighbor then?

Do you say: "Wanna sign this petition to wreck the planet in 100 years instead of 50?" Or: "Hey, neighbor, I'm here today seeking beauty and dignity in failing to stop the inevitable. Care to pitch in?" Or: "My group is learning how to die as a civilization, wanna join?"

If you truly believe everything is going to hell, what "cause" could you possibly be signing people up for? "We're doomed! Follow me!" is tough to rally around. At the end of the world, what, actually, is still worth doing? (And, how do you get people to do it with you?)

What would Paul Kingsnorth do?

In 2014, the *New York Times* ran a long profile on Paul Kingsnorth entitled, "It's the End of the World as We Know It…and He Feels Fine."[1] At the time, it was deeply confounding. My comrades and I were gearing up for the largest climate march in history. We were going to bring half a million people into the streets of New York, change the conversation, get a global treaty, and, well, save the world. And here was this guy—this eloquent, heartbroken guy—saying, *Um, nope. Impossible. That's a fool's errand.*

If he'd been a PR flack for Big Oil or a garden-variety cynic, we would have paid him no mind, but he had the bonafides. He was a long-time environmental activist. He'd hunkered down in tree-sits to block road expansion through his beloved English countryside. He'd been in the streets of Genoa with 300,000 others battling police and trying to shut down the G-20 and stop the ravages of corporate globalization. He'd run environmental defense campaigns. He'd been deputy editor at *The Ecologist*, Britain's leading environmental journal.

He'd paid his dues, he'd fought the good fight, and he'd finally had the courage—is that we should call it?—to give up hope and speak the terrible truth of our predicament. And, boy, could he speak. His words were poetic and haunting, full of the kind of sadness that can only come from a deep love of life and the natural world. Was it lyrical realism? Or sad-sack defeatism? We could not decide. We just added his doubts to our own pile, and soldiered on.

Our climate heroes were Bill McKibben, Chico Mendes,[2] Naomi Klein, Vandana Shiva,[3] Rebecca Solnit. Dogged, hopeful warriors who told hard truths but kept the faith. Paul's message was different. "All of these changes are coming," he said in the *New York Times* profile that summer, "things that you value are going to be lost…but you still have to live with it, and there's still beauty, and there's still meaning, and there are still things you can do to make the world less bad."[4]

After years of trying to "save the planet," he'd come to the painful realization that things were only going to unravel further. Rather than continuing to peddle a false hope, he sought out others who'd come to the same heartbreaking conclusions, and in 2009, after a long series of conversations,[5] he and Dougald Hine penned *Uncivilization: The Dark Mountain Manifesto*.

You pick up a manifesto expecting something exciting, a call to arms full of fire and passion and hope and vision and the heroic clash of armies and ideas. You expect some bluster. *Uncivilization* was blusterless. It was not a call to arms; it was an elegy. An elegy that spoke in grave poetic particulars about loss. "We are already responsible for denuding the world of much of its richness, beauty, color, magnificence and magic," it read, "and we show no sign of slowing down."[6]

The Manifesto was a requiem not just for the Earth, but for civilization, for the certainties and stories we'd always counted on—the idea of Progress; the idea that humanity is the center and pinnacle of Nature; that Civilization is, in spite of all its flaws, on balance a good thing. But the Manifesto wasn't just a requiem for the *dominant* stories of civilization, but also for many of the alternative stories—"progress can be tamed to serve humanity," "people are ultimately more powerful than the systems they create," "there's always hope," and others—that I'd staked my life and identity on. For those of us still trying to fix things, the Manifesto was a wrecking ball to our hopes, a dirge for our dreams.

But Paul wasn't starting a doomsday cult. He wasn't saying, *We're doomed! Follow me!* He was saying, *We're doomed-ish. If you think so too, let's meet up. Maybe together we can figure out how to be properly alive in this moment and what good we might still do.*

In his 2012 essay Dark Ecology, seeing no way to fundamentally halt the destructive momentum of ecocide, Paul counsels us "to be honest about where you are in history's great cycle, and what you have the power to do and what you don't."[7]

After dismissing a host of things he considers a "waste of time," including political reform, eco-terrorism, romanticizing hunter-gathering, new tech, and most of what passes for environmental advocacy, he lists five things that he thinks are NOT a waste of time: withdrawing, preserving non-human life, getting your hands dirty, insisting nature has intrinsic value, and building refuges. What would Paul Kingsnorth do? These five things. Let's take each in turn.

Withdraw. "Withdraw not with cynicism, but with a questing mind," Kingsnorth counsels.[8] And he himself has taken this advice, withdrawing in 2013 to a farmhouse in Ireland with his family. "Withdraw to examine your worldview," he advises, because "all real change starts with withdrawal."[9]

Preserve non-human life. Paul likens our civilization's ecocidal assault on the Earth to a colonial conquest. Channeling that analogy, he writes: "The human empire is the greatest threat to what remains of life on Earth." He asks: "What can you do—really do, at a practical level—about this?"[10] His suggestions include buying some land and rewilding it, letting your garden run free, and putting your body in the path of a bulldozer.

Get your hands dirty. "Get away from your laptop and throw away your cell phone," he advises. "Get[11] out there and do physical work in clean air surrounded by things you cannot control." It's the only way to "learn what's real and what's not."[12]

Insist nature has intrinsic value. Shrugging off whatever "deep-ecology" or other labels you might get tagged with, Paul calls upon us to wonder at the sacredness of life. "Marvel at what the hell this thing called *life* could possibly be. Value it for what it is," he says, "and have nothing but pity or contempt for people who tell you that its only value is what they can extract from it."[13]

Build refuges. Given that ecocide and social collapse will kill off so much of what we most value, Paul wonders, what we can preserve and how? Imagining the librarian of a monastery in the Dark Ages, "guarding the old books as empires rise and fall outside," Paul asks: Can we "create places or networks that act as refuges from the unfolding storm?.... What power do you have to preserve what is of value—creatures, skills, things, places?"[14]

"None of it is going to save the world," he admits, "but then there is no saving the world, and the ones who say there is are the ones you need to save it from."[15] For Paul, the crisis is as big as civilization itself, and we're headed for a big fall. In accordance with these views, he makes his list. For other folks, the crisis is smaller, or fixable, or centered in a single realm—politics, say, or food or carbon. They, likewise, make their lists.

Towards the end of the essay, Paul notes: "It will be apparent by now that…I've been talking to myself." He's sounding himself out, making sure that he believes that these five things are indeed still worthwhile, and that he can still do them "with some joy and determination."

What about the rest of us?

It's never too late to fail to save the world.

We cannot halt civilization's ongoing assault on nature, Paul Kingsnorth tells us. While I admire Paul's honesty and eloquence, he's also telling me that nearly everything I'm doing, and nearly everything everyone of us in the climate justice movement is doing, is not worth doing.

Things are only going to unravel further, he tells me. And I agree.

We need to be honest about where we are in history's great cycle, he tells me. And I agree.

There is no saving the world, he tells me. And even here I agree.

But here's where we disagree: I believe we must still try—and, yes, fail—to save the world.

Why? Because, in history's great cycle, we're not just on the cusp of a great unraveling, we're also in the "decisive decade" of the climate emergency.[16] Yes, we're going to fail to save the world, but exactly how badly we fail at it in this next crucial handful of years will matter across that entire great cycle of history, and even beyond, over geologic time.

Noticeably absent from Paul's list of things still worth doing: passing a Green New Deal; engineering a fast and just transition to renewable energy; or, in fact, pretty much any notion of climate justice. Yet all of these things are still worth doing, even if we fail at them.

Because even though we're in a predicament we don't yet know how to solve, we also have a problem we absolutely must solve. And even if that problem is an "unsolvable" super-wicked problem, we have to at least *try* to solve it. Because even in failing to solve it, we are doing something essential: blunting its worst impacts and buying us all time. Time that is crucial for the Paul Kingsnorths of the world to build their refuges, find a still-livable place to withdraw to, and defend the non-human life they hold so dear.

There is no single Solution that will prevent climate catastrophe, but there are many small and medium-sized solutions[17] that can slow it down, mitigate the damage, and help us adapt. Which means we need a way to evaluate them.

Let me propose that we ask three questions of any proffered solution:
1. Is it a real solution? (Or fake greenwashing bullshit?)

2. Is it a just and democratic solution? (Or the same old inequality and exploitation, only now powered by renewables?)

3. Is it an on-time solution? (No matter how real or just it may be, is it simply arriving too late to make a difference?)

One of the groups I've worked with, the NY Renews coalition (nyrenews .org), is a great case study in the contest between real and false solutions. In 2019, they were the driving force behind the Climate Leadership and Community Protection Act (CLCPA), a landmark law requiring a zero emissions mandate by the year 2050, encompassing electricity, transportation, buildings, industry, and agriculture across all of New York state. Unlike "net zero" or "carbon neutral" approaches that include significant flexibility for polluters to avoid directly reducing emissions, the law explicitly establishes a "hard" mandate, as well as critical commitments on equity and environmental justice for the most impacted communities.[18]

However, in wanting New Yorkers to have a direct say in their own clean energy future, the law left many of the specific policy choices for achieving the emissions targets in the hands of a public decision-making structure and planning process bringing together government, advocacy, and private-sector stakeholders over a two-year period. And now—unfortunately, but maybe predictably—the fossil fuel industry is rushing into that process with a host of false solutions including biofuels, biomass, biomethane (aka "renewable" natural gas), green hydrogen, and waste-to-energy.

In 2021, NY Renews commissioned a 50-page report, "False Solutions. Gas and Trash: How the fossil fuel industry is holding back a just transition,"[19] detailing how these industry-promoted solutions are actually causing "more greenhouse gas emissions than they reduce," "more local pollution" and "adverse public health impacts" especially for environmental justice communities; as well as diverting land use from food to energy, depleting the earth's ability to recycle carbon, and contributing to water pollution and severe water stress.[20]

Green hydrogen, for example, is rarely "green", since 95% of it is produced using fossil fuels. (Using dirty energy to produce clean energy is generally a poor way to reduce emissions.) Meanwhile, biomass—basically, woody matter burned directly for energy—while touted as carbon neutral, is anything but, since recycling carbon from the atmosphere by regrowing trees takes decades while wood burning for energy is adding copious emissions today, all while worsening deforestation.[21]

In our interview, Gopal Dayaneni described capitalism's fundamental response to the climate crisis as a "festival of false solutions" (see page 137). For him, a real solution not only has to be real (i.e., not fake greenwashing bullshit like the examples above), but also address the root causes of the crisis. For Gopal, "Clean energy is not a solution to the climate crisis. The reorganization of the very purpose of the economy is the solution to the climate crisis." While "energy needs to be renewable," he says, "as much, if not more, it needs to be democratized, decentralized, distributed, and decolonized." Justice is not an add-on; it is central. "Climate Justice is the only way to win," says Gopal, "because it is from the injustice that the crisis has emerged."

Guided by this perspective, NY Renews is trying to implement a Just Transition in the fourth-largest state in the country. Every step of the way, they've been met by the self-interested maneuverings and dissembling of a fossil-fuel industry that rightly sees such a transition as a threat to their very existence. Fossil fuels will eventually lose. But in the process, they might slow things down so much that we will all lose.

Which brings us to our final diagnostic: Is the solution on time? Bill McKibben describes climate change as a "timed test."[22] "If we don't win very quickly on climate change, then we will never win," he says. "Winning slowly is the same as losing — that's the core truth about global warming."[23] It's exactly this crucial time-bound aspect of our climate challenge that led us to launch the Climate Clock, and to use it to try to rally the world to #ActInTime.

The time-bound nature of the crisis was brought home to me recently, as I watched a friend step through "Navigating Our Climate Predicament" flowchart (following page 98). When he came to the "What will it take?" section, he began to circle through the options. What will it take? Fewer showers? *Ha! Yes, that'll help. But only a tiny bit.* Stronger policy? *Definitely, yup.* Better stories? *Sure, definitely need those.* Serious tech? *Yep. But we need so much more.* And on he went around the circle from degrowth and resilience to resistance and revolution. *Yep but. Yep and. Yep. Yep.* Until he hit "Everything & Everyone."

Ah, he said. *Yes: everything and everyone!* And in that moment his attitude was so upbeat. He had a hopeful, roll-up-your-sleeves, whatever-it-takes kind of energy. But then he followed the flowchart path to the question "Is there still time?" and his enthusiasm cratered. He wasn't quite sure what to think, or where on the flowchart to go next.

I call this the "It's gonna take everyone to change everything, um, yesterday" effect.

Our future feels haunted. We sense a dark threshold lurking some number of breaths, some number of footsteps in front of us. A tipping point after which the feedback loop of climate impacts causes runaway warming of the planet and no amount of subsequent reining in of the fossil fuel industry or lifestyle changes can ever get it back under control.

This threshold looms large in our minds as a point of no return, a "game over" moment for a habitable planet. There's a range of opinion as to exactly what that threshold is, when we might cross it, whether we can claw our way back (via, among other things, large-scale drawdown of atmospheric carbon), and once crossed, how we might try to survive it. Scientists have flagged 1.5°C of warming as a "code red" for humanity.[24] We're currently on track for 2.7°C or worse. Some argue we can still change course; others, that it's already "too late."[25]

But too late for what? Because here we are, it is now, we're still alive, and we're still our brothers' and sisters' keepers. Yes, we can't save the world. Yes, we're too late. But there are better and worse ways to be too late to fail to save the world. In fact, there are so many different ways to be too late to fail to save the world, that there's got to be one that's right for you:

Market-based solutions. If you don't mind solutions that often make the problem worse, and you don't care too much about justice or democracy, and you trust in the "invisible hand" to convince the fossil fuel industry to protect the planet and the well-being of future generations (neither of which show up on their balance sheets or vote at their shareholder meetings), well, then you can throw your lot in with purely market-based solutions.

And if you notice yourself justifying this hands-off approach with blind faith in some new-fangled tech that's still on the drawing board, along with the boundless capacity of humans to innovate their way out of any problem, then you'll want to get together with Bill Gates, Steven Pinker, and Oded Galor,[26] and snort some *Techno-Utopianism*.

What really matters, however, is, "not the new things we do, but the old things we stop doing," observes George Monbiot. "Renewable power, for instance, is useful in preventing climate chaos only to the extent that it displaces fossil fuels."[27] And the "invisible hand" of the market is simply unable to do that at the speed required to avoid ecological breakdown; what's needed is a radical political intervention in the market that interrupts business-as-usual and shutters Big Oil & Gas.

"All of the Above." If you're open to mixing some government policy in with your market-based solutions, but have a hard time making decisions, you can go with an Obama-style "All of the Above" approach: just do it all—nukes, "clean" coal, solar; real and fake solutions both—because, well, anything else is too politically costly even if it wrecks the planet. Far better, however, to take a *Some of the Above* approach, because *some* clean tech—wind and solar, for example—are much cleaner and cheaper than others, and scale faster, too.[28]

"No Regrets" strategies. If, say, you're in a purple state trying to appeal to an ideologically mixed constituency (that might even include some climate deniers who nonetheless prefer drinkable water over nondrinkable water), you might adopt a "No Regrets" approach. These are efforts (like home energy efficiency,[29] anti-pollution measures, wilderness protection, specific infrastructure improvements) that "generate net social or economic benefits irrespective of whether or not climate change occurs."[30] By building "resilience to future climate shocks while also delivering near-term benefits," no-regrets actions are worth doing no matter what climate scenario plays out.

Credit: Courtesy of Joel Pett

Drawdown. If your singular focus is stabilizing the planet's climate and moving us off of a fossil-fueled economy to a 100% renewable one, then the Drawdown Project's solutions toolbox is your bag. Initiated in 2017 by ecologist Paul Hawken, Drawdown[31] is a ranked matrix of the 100 most globally impactful solutions to the climate crisis, from better refrigerant management (a surprising #1) to educating girls (#6) to farmland restoration (#23) to wave and tidal electricity generation (#29) to smart thermostats (#57). When combined with (as yet untested and unscaled) carbon sequestration strategies like direct air capture and intensive silvo-pasture, they claim to have the potential to not just reduce emissions, but actually "draw down" enough excess carbon out of the atmosphere to create a stable, livable climate over the long haul.

Transition Towns. If you believe, along with founder Rob Hopkins, that, "If we wait for governments, it'll be too little, too late. If we act as individuals, it'll be too little. But if we act as communities, it just might be enough, just in time,"[32] then you'll probably want to throw your lot in with Transition Towns. It's a global movement of communities plotting out their own transitions to a circular economy, while simultaneously preparing for a future of extreme weather disasters, compromised ecological conditions, and economic shocks.

Just Transition. If you agree with Gopal Dayaneni that there is no meaningful way to tackle the climate crisis without centering justice, and that the very purpose of the economy must shift from profits for the few to a guarantee of "the rights of Mother Earth, and the right of people to have access to the resources required to create productive, dignified, and ecologically sustainable livelihoods," then a Just Transition is going to be your bag.

We've seen how NY Renews is trying to move such a vision forward in New York State. In Canada, a broad red-green coalition has gotten behind the LEAP manifesto,[35] a roadmap for transitioning Canada off of fossil fuels to an economy "based on caring for the Earth and one another." Meanwhile, across the USA, the *Green New Deal* offers a framework that, in spite of getting scaled back in DC, has the potential to overcome the usual jobs vs. environment vs. justice splits (see page 158), and bring on a fast and just transition to a 100% renewable-powered future.

Meanwhile, to apply the principles of just transition on a global scale would require the rich countries to own up to some of the climate debts—estimated to reach a staggering $57 trillion by 2035[36]—they owe to the poorer countries for the "negative externalities" (i.e. social and economic costs) inflicted on the Global South by the historic emissions of the Global North.

Degrowth. If you're a "less is more" type who's made a sober assessment that we've already way overshot the limited carrying capacity of the planet, then degrowth might be the solution pathway for you.

While few mainstream economists are willing to acknowledge it,[37] given our limited carbon budget, and the general impossibility of infinite growth on a finite planet, there's a growing awareness that a habitable planet will require some degree of planned economic contraction, or degrowth[38] amongst the world's richest nations.

Proponents have mapped out a strategy of "prosperous descent"[39] focused on partially "powering-down" the world's richest economies and helping local communities and ecosystems become more resilient in the face of the many shocks—both ecological and economic—that are unavoidably on their way. One model is the human-centered *Lean Economy*[40] proposed by late British economist David Fleming whom we met in the "How do you want to decline?" chapter (see page 106). Another formulation is *The Great Simplification,*[41] popularized by ecologist and energy analyst Nate Higgins.

Degrowth has been defined as "an equitable downscaling of production and consumption that increases human well-being and enhances ecological conditions."[42] Not exactly capitalism's strong suit. And, while no national economies have yet re-aligned themselves on this path, prefigurative local examples can be found[43] bubbling up from grassroots movements based on permaculture, voluntary simplicity, local food initiatives, local currencies, worker cooperatives, and eco-villages. "You might actually enjoy it,"[44] says Samuel Alexander, research fellow at the Sustainable Society Institute.

Deep Adaptation. If you think social collapse is right around the corner, you might want to prepare yourself for Deep Adaptation.[45] Professor of sustainability leadership Jem Bendell is convinced we're on track for "near-term social collapse," but instead of raising the white flag at the end of the world, he's developed a "map for navigating climate tragedy" that combines resilience, relinquishment, and eventual restoration. Given the growing signs of collapse all around us, his framework is now blossoming into a global social movement.[46]

Finally, if none of the above work for you, strap yourself in for the *Desperate Last Resort of Large-Scale Geoengineering* (or as it's referred to in politer society, "Climate Restoration"[47]), in which nervous men in lab coats attempt to manipulate our climate back to stability by, say, injecting sulfate aerosols into the stratosphere from tethered balloons in order to reflect solar radiation back to space —aptly described in a recent video explainer as, "A horrible idea we might have to do."[48] Or brace yourself for the *"Benevolent Eco-Dictator"* option, where a radical, ruthless, charismatic ecologist is swept into power, expropriates the fossil fuel industry, and forces through the extreme measures required to preserve a livable planet.

Some of these pathways have been forced on us because we failed to take better ones decades ago.[49] Some are obviously at cross-purposes with one another. Some should only be taken as a very last resort. No one, however, can say they don't have options.

Still the question lingers: Are we too late for any of it to matter? But what do you mean *matter*? Matter enough to save humanity from itself? We don't know. We just don't know. But to matter at all? Yes, definitely. Every one of these partial solutions that can at least check off the first two boxes—is it real? is it just?—matters infinitely. Because every unwanted pipeline that gets stopped by an Indigenous community doesn't just mean millions of tons of additional carbon that will not be further poisoning all of our skies. It also means that one more aquifer has been protected, one more historical wrong righted, and one more community has become more resilient and more able to live on land that is sacred to them. ("No Regrets" strategies for the win!)

No matter what happens to the big global mess we have got ourselves into, with coal slurry still poisoning the hollers of West Virginia and millions dying every year from preventable respiratory illnesses tied to climate change,[50] there's every reason to quickly transition off of coal, oil, and gas—and make our renewable-powered future as just, democratic, and resilient as we can.

If we can do one thing, solve one problem, right one wrong, then, of course, we should do it, solve it, set it right. Even if we don't yet know how to "save the world"—or don't even think that's possible anymore—there are still a host of things we can do to reduce or slow down the death and destruction, secure some justice—and certainly some dignity and kindness—in the here and now. All of which might also give us a fighting chance down the road.

It's never too late to fail to save the world.

Why the fuck am I recycling?

To recycle or not to recycle, that is the question. Or certainly a question that, with all this talk of fake and real solutions, a lot of us are asking. Not necessarily out loud, but in that secret doubting place inside us every time we drag our paper and plastic to the curb. Why, *really*, am I doing this? How, in the face of the vast planet-destroying inertia of industrial civilization, is my little drop in the recycling bucket doing any good at all? And come to think of it, what really happens to all this stuff once it leaves my curb?[51]

These doubts dogged Jamey Hecht, and they dog me across all that I do to reduce my carbon footprint. Besides recycling, I also compost. I rarely use the A/C except on the worst summer days. I live in a tiny apartment in a multi-unit building in a big city, so I get good efficiencies on winter heating. I don't have kids. I bring my own shopping bags to the grocery store, and I try to re-use the plastic bags (and water bottles) I inevitably end up with anyway. Knowing that the livestock industry is a huge contributor to deforestation and global warming, I don't eat red meat or poultry, except for turkey once a year on Thanksgiving. My microwave is a second-hand donation from a friend, and it's worked fine for the last 15 years. I don't own a car. To get around the city I mostly walk, bicycle, or take the subway. And I've gotten my flights down to a relatively modest two per year. I've got my footprint slimmed down so far I'm like a feather on the Earth. A resource-intensive, middle-class, American feather.

And why do I do all this? I have a friend who takes ten flights a month for work. Is reducing my flights to two per year—or even to zero—going to make any difference? Americans put 100 billion (that's "billion" with a "B") plastic bags into landfills every year. How are the handful I reuse going to change that? And yet I feel I must do these small things.

Only a global ban on single-use plastics is going to stop the plastic pollution killing our oceans.[52] Only a hard cap on fossil fuels, the whole-sale adoption of renewables, along with a few other large-scale systemic interventions, is going to reduce carbon emissions at the speed and scale needed. Anything I do to slim down my footprint is just pissing into the wind. But piss I must. Why? It makes me feel better. It gives me a sense of agency. It models good behavior. But more than all this, I think it's actually a kind of prayer.

Every returnable bottle I take back to the grocery store, every soup can I rinse out and put in the recycling bin, is a secular offering. A sacred vote of confidence in the future. A moral act that aligns my habits with my values, and keeps me on a principled path. It's an act that maybe matters less for its direct material impact, than for the work it does upon me, the recycler.

But what really happens up the chain, after we ritually rinse off our filmy milk jugs and fold our newspapers and place them in the bin, and then carry our offerings—always on Friday morning, the mandated day of deliverance—to the curb to meet their remaker? *Forgive me Earth, for I have sinned* we whisper into our recycling box confessionals. The catechism of the three Rs is gospel. We say it like a rosary *Reduce-Reuse-Recycle* over our paper and plastic offerings. Our perfectly sorted box will be first to the curb, and the eco-pieties—or tragic eco-waywardness—of our neighbors will be duly noted. Will our plastic and paper waft up to heaven and be heard? The ways of the recycling authority are mysterious indeed.

Most of us realize there's some degree of self-trickery and social performance going on here. We know that fierce political action at the scale of the problem is what's really needed, and we're only recycling to keep up appearances, to stay on the right side of some vaguely felt karmic accounting.

But, the three Rs *can* actually make a difference. The Jump, (takethejump.org), an eco-responsibility movement that launched in 2022 under the slogan, "Less Stuff; More Joy," aims to achieve 25%[53] of the carbon emissions reductions needed to stay under 1.5°C by encouraging citizens of wealthy nations to make six shifts in their individual consumer behavior:

> End clutter—Keep products for at least seven years.
> Travel fresh—Try not to own a personal vehicle.
> Eat green—Follow a plant-based diet.[54]
> Dress retro—Try not to purchase more than three new clothing items per year.

Holiday local—Keep flights to one every three years.

Change the system—Take at least one action (anything from peaceful protest to switching to a green energy supplier) to nudge the system.[55]

Except for flying about six times too frequently, and buying a few too many pairs of underwear, I'm happy to say that I check most of these boxes. I'm also happy The Jump recognizes that not everyone is in the same position to pursue the program. They're founded on the notion of "equal access but different responsibilities," and suggest that "trying is enough, just start."[56]

Yes, Reduce-Reuse-Recycle is only a fraction of what we need to do, and recycling is only a fraction of that fraction, but with the future of the planet at stake, fractions of fractions count. So, let's get cracking on the big system-wide changes we need, but let's also recycle, let's hold on to our appliances as long as we can, let's fly less, let's try to go vegan (or at least "consume no animal products before dinner," as author Jonathan Safran Foer modestly suggests).[57] Let's slim down our footprint wherever possible.

Doing so is not only good spiritual practice, it's smart politics. Because when our opponents can't trash our ideas, they try to trash our character—often, by cynically turning our own eco-morality against us. You want to be able to say to those people, "Yes, I recycle, thank you very much. Now let's talk about eliminating that $423 billion dollars in fossil fuel subsidies..."[58]

I've still got some work to do on my three Rs, but when I carry my big bag of clinking aluminum and clunking plastic to the curb, when I bicycle uptown instead of grabbing a Lyft, when I dump my compost into the weekly city collection bins on top of a big pile of mulchy garden and kitchen detritus from hundreds of my neighbors, I feel slightly more aligned with my values. I'm not walking my whole talk, but I am walking a few steps of it. It brings me into better integrity with myself and my values, and models that integrity to the larger world.

The father of a good friend of mine was dying of cancer. He took up a macrobiotic diet and meditation, not because he thought it would make a difference, but because he wanted to demonstrate skin in the game. It was a "votive act" to show he wanted a cure; a therapeutic gesture, unlikely to make a difference to anything but his own peace of mind. He wanted "the consolation of thoroughness"—the idea that at least he tried with all his might.

Our situation is not so different from his. As we face our planetary crisis, we don't know whether anything we're doing will make enough of a difference, but we owe ourselves, our planet, our children, and all future generations this same consolation of thoroughness. Maybe this is why the fuck we recycle.

We have met the enemy and he is us.
No, them! But also us. But mostly them.

*The earth is not dying, it is being killed, and those
who are killing it have names and addresses.*

— Utah Phillips

"We have met the enemy" went the famous 1960s-era *Pogo* cartoon, "and he is us." *Pogo* creator Walt Kelly was referring to the American experience in Vietnam. Yes it was a war, but the enemy we thought we were fighting wasn't Communism, North Vietnam, or the Viet Cong—it was ourselves. *Our* folly and paranoia; the illogic and momentum of *our* military-industrial complex; the unintended consequences of *our* empire-building grandiosity; the return of *our* repressed.

In Vietnam, we met the enemy and he was us. In our climate-changing 21st century, we meet the enemy everywhere and we (still) can't decide whether he is us or them. Most of us agree that global warming is a man-made catastrophe, but what man made it? Who is this enemy? Is it Us or Them? And if it's Us, how do you declare war on yourself? And if it's Them, is it all of Them or certain specific bad actors and vested interests? Or maybe it's some out-of-kilter system we've created that we can no longer properly control? Wait, is that an Us or a Them?

It's a tricky question to ask about the most complex problem humanity has ever faced. How we answer it tells us where we think the problem lies, what we think our responsibility is, and how we might go about tackling it.

Stripped down, it's an age-old question:[59] Does the fault lie inside ourselves or in the world around us? Team Us (quite reasonably) asks: How can we possibly change the world if we can't change ourselves? Team Them (also quite reasonably) replies: A corrupt system will always produce corrupt behavior; we must change the world first.

Team Us: We can't build a healthy society out of broken people. The most effective way to work towards a better world is to rectify our own patterns of domination, selfishness, and denial—and then be a model to others. We must heal our own split psyches. If we don't do this, then with whatever power we attain, we will simply revisit those same patterns on others and on Mother Earth. Change starts with oneself or not at all.

Team Them: Personal change does not equal political change. No matter how much we recycle (or how many meditation retreats we go on), polluters will keep poisoning poor communities unless they are stopped by concerted political action. As individuals, we're all somewhat flawed, but that's no reason not to work together to save the world. Let's start now!

Do you have a home-team favorite in this eternal debate? When it comes to our climate crisis, many of us do. "Humanity, as a whole, is the meteor," Paul Kingsnorth, star player on Team Us, argues. "It doesn't mean we're all equally culpable, but as a result of our appetites, and our attitudes, and the stories we tell ourselves, we are wrecking the planet."[60]

His teammate, social philosopher Charles Eisenstein, author of 2018's *Climate: A New Story*, contends that when we frame our climate predicament as a war and assign an enemy, we're "disguising a problem that we don't know how to solve as a problem that we do." He puts it bluntly: "Fighting the enemy is futile when you inhabit a system that has the endless generation of enemies built into it. That is a recipe for endless war."[61] Instead, says Eisenstein, we need to face the fact "that 'they' includes every one of us that participates in this civilization."[62]

Sure, responds Team Them, but some of us "participate" more than others. *A lot more.* Historically the developed world is responsible for 79%[63] of all greenhouse gas emissions. The average American still has over 150 times[64] the carbon footprint of the average Ugandan. The super-wealthy jet set around the world from mansion to mansion as if there was no tomorrow, making sure that for the rest of us there won't be. We're not living in the *Anthropocene*, we're living in the *Asshole-ocene*; humans aren't destroying the planet, assholes are. And the biggest assholes of all are the assholes in power who are permitting, even encouraging the destruction.

Team Us hits back: "That's the trouble with the standard left-ish social-change paradigm," says Paul Kingsnorth,[65] "it always divides society into the bad guys in power versus the rest of the oppressed masses who want to change things. But when you look at the ecological crisis we're in, that's not really the situation. We're all part of the problem." "We've created this enormous machine," he says, "that none of us is actually responsible for. We were born into it. However hard we try, we can't step outside it." *If the enemy is an asshole*, Team Us is saying, *that asshole is us.*

Sure, says Conceivable Future (conceivablefuture.org) co-founder Josephine Ferorelli (see page 328), but if we had better options, we could make better choices. "The reason that it costs so much carbon to be an American," she argues,[66] "is because the system is poorly designed, not because everyone is an asshole." The fault lies, Team Them is saying, with the folks in power. They're the ones spewing junk science, murdering enviro journalists, and putting profits before the planet; they're the ones making it extremely difficult to redesign the system.

But we ourselves don't yet know how to redesign the system, says Team Us. And by pretending we do, we often make it worse. There's no simple technical fix here. Humanity is in a longstanding spiritual crisis, a "crisis of Being,"[67] to use Eisenstein's phrase. We must re-learn how to be; we must reawaken to our interconnectedness with a Universe that is alive in ways we we've been trained not to see or feel.

"Look," says Patrick Reinsborough,[68] former Organizing Director of the global climate group 350.org and a top spokesperson for Team Them. "People are good about responding to threats when they're aware of them, but our choices have been concealed from us. We've been kept in the dark by a 35-year-long misinformation campaign." Indeed we have. Exxon knew[69] the implications of climate change since the 1960s. They not only kept that information from the public but later spent millions deliberately undermining the science and blocking policy efforts aimed at addressing the problem.[70] And Big Oil and their DC enablers are still at it, doubling down on fracked gas, tar sands, and a score of catastrophic "carbon bomb" extraction projects in spite of massive public outcry and dire warnings from scientists.[71]

"Okay," I pressed him, "so, if it's mostly Them, how much? Can you put a number on it?"

"It's 98% Them."

"Which still leaves 2% that is Us?"

"Yes, and people can detect bullshit and hypocrisy a mile away, so it's important to acknowledge our own complicity in the system, even if it's relatively tiny."

Patrick has his ratio. Everyone else has theirs. Wherever you land, it's fair to say it's not 100% in either direction. In some forever-to-be-debated proportion, we have met the enemy and he is Us *and* Them. We need both teams.

In fact, *which* team you choose probably matters less than *why* you're choosing that team and *how* you're on it. Because you can join a team—either team—to cop out, or throw down.

Don't join Team Us in order to say, "hey, the problem is all of Us, so I only have to be responsible for my own little bit"; join because you're ready to reinvent yourself and all the human systems around you. And don't join Team Them because you're looking for someone else to blame; join it because you're thinking "the problem is Them, and it's up to me to stop Them."

Gooooooo teeeeam(s)!

No need to choose between mitigation, adaptation, and suffering; just get good at all three (especially suffering).

On the question of whether it's "too late" to prevent catastrophe, physicist and former Obama White House science advisor John Holdren tells us, "We basically have three choices: Mitigation, adaptation, and suffering."[72] Which is like telling someone to please choose between: (1) Unpleasant and painful things you can do to reduce how unpleasant and painful things are gonna get; (2) Unpleasant and painful things you can do to get used to how unpleasant and painful things have already gotten; and (3) Suffering.

If this sounds like a poor set of options, welcome to the 21st century.[73] Nonetheless, we grit our teeth and make our choice. And then, given how high the stakes are, we tend to insist that our choice is the correct one:

> "Guys, we have to keep trying to do unpleasant and painful things to *prevent* a catastrophe (even if we don't think we can anymore) because anything less will let government and corporations off the hook!"

> "No way, man, that train left the station long ago. But we can still do every unpleasant and painful thing possible—from radical emissions reduction to carbon sequestration—to *mitigate* the impacts."

> "Are you kidding? Those are useless techno-fixes. The only thing left to do now is every unpleasant and painful thing we can think of to *adapt* to our new abnormal, totally reinventing how we live so we can survive in resilient communities."

Meanwhile, others take a page from the Jewish grandmother in the light bulb joke: "I'll just sit here in the dark and *suffer*."

All this righteous jockeying can leave the rest of us confused about which set of unpleasant and painful things we should get on board with. But hold on. According to Holdren, "We're going to do some of each. The question is what the mix is going to be. The more mitigation we do, the less adaptation will be required, and the less suffering there will be." Which makes perfect sense: The mitigators need the adapters because we're well past the point of being able to completely mitigate our way out of this crisis. The adapters need the mitigators because fighting to the death over the last acorn north of the Arctic Circle is not the kind of environment we want to have to adapt to.[74] As for suffering, we're all going to need to get better at that.

So, let's spend less time arguing over which unpleasant and painful thing is the one correct unpleasant and painful thing, and just get on with the whole range of unpleasant and painful things we need to get on with. Plus suffering.

We need to do the impossible, because what's merely possible is gonna get us all killed.

It always seems impossible until it's done.

— Nelson Mandela

We can debate whether the problem is Us or Them. We can endlessly parse false and real solutions, and fight over what—given prevailing political realities—we're willing to consider "good enough." We can argue whether to focus on mitigation, adaptation, or suffering. We can quarrel over whether our salvation lies in individual actions like recycling, or in collective action that brings about systemic change. And these are all important debates. But you might wonder: why debate any of it if, ultimately, our task is impossible? But here's the thing: given what's at stake, we can't afford to fail. It can't be impossible. And given how quickly time is running out, it needs to not be impossible as soon as possible. Even if it's impossible.

"Don't let the perfect be the enemy of the good," the political pragmatists tell you. And sometimes they're correct. If, say, we're talking about a multi-decades effort to reform an unjust and inefficient healthcare system, you could see how getting something imperfect like Obamacare through Congress—because it's the best deal you can get at the moment—and making it better as you go, might be the right strategy.

But what if "the good" will lead to runaway climate change, the collapse of civilization, and the destruction of everything you love? Then maybe "the good" is simply not good enough. In fact, given the relentless logic of climate change, "the perfect vs. the good" doesn't even begin to capture our dilemma; it's more like "the bare-minimum-needed-to-not-go-extinct vs. going-extinct."

So, if you're debating possible climate strategies with someone, and they say to you, *don't let the perfect be the enemy of the good*, what they're really saying is don't let what we absolutely must do right now to save ourselves be the enemy of the best deal we can get right now which will kill us all.

It would seem a no brainer, then, that we should get on with what is absolutely necessary and be right quick about it. I mean, if that's what's absolutely necessary, what's the point of doing anything else? And that's exactly when the other half of our dilemma kicks in. Because what's absolutely necessary is, well, impossible.

We know what is absolutely necessary to stay under 2°C (immediate moratorium on all new fossil-fuel extraction, controlled degrowth of the world's richest economies, WWII-level emergency mobilization to make a fast and just transition to a post-carbon economy, etc.) and it's politically impossible.

Meanwhile, the most ambitious edge of what seems actually achievable ("net-zero by 2050," etc.), is utterly insufficient. In fact, it could very possibly get us all killed. So, what's our move? Do we focus on what we know we need to do, even though there's no chance of getting it done? Or do we focus on what we actually can get done, even though it won't ultimately save us? The 21st century can be a real bitch sometimes.

Your choice will likely depend on who you are.

If, like me, you're a tragic optimist, you will set your sights on the goal that is necessary yet impossible, and give it your all, hoping that the impossible somehow becomes possible before it's too late. (After all, there's nothing more inspiring than a smart, dedicated, reality-based person acting as if the impossible were possible to actually make it so.)

On the other hand, if, and also like me, you're a can-do pessimist, you will set your sights on the most ambitious goal you think you can pull off even if you know it's insufficient to the task, trusting that in the unlikely event (remember, you're a pessimist) of achieving it, you might just create the conditions for an even more ambitious goal that *is* up to the task.

But what if—and also also like me—you're a compassionate nihilist? You recognize the cosmic futility of both these approaches, but you also recognize their profound and heroic humanity—what then? Well, you could offer back rubs to any of the stressed-out people engaged in these heroic efforts. Back rubs and donations and volunteer time and whatever talent you have to offer (including writing a book about the grand dilemmas we face). Contrary to conventional wisdom, you don't actually have to believe in anything to start giving a shit.

adrienne maree brown

"How do we fall as if we were holding a child on our chest?"

We have to do the impossible, and we're 30 years late getting started. But it's not too late—in fact, it's *never* too late—to do something right, to do something good, whether that's building a refuge, implementing a Green New Deal, recycling (if only for "the consolation of thoroughness"), or telling better stories.

In search of better—and more visionary—stories, I headed to Detroit. After Jamey Hecht's doomslinging, I needed to talk to someone who'd learned how to hold the gaze of our looming dystopia while cultivating the seedlings of utopia in its unforgiving ground.

Cue my seventh and next-to-last meeting, with adrienne maree brown.

adrienne wears many hats: doula, community organizer, pleasure activist, emergent strategist, "organizational healer," apocalypse theorist, and evangelist of visionary fiction, to name just seven.

In the oos, we'd crossed paths at several direct-action training camps during her tenure as Executive Director of the Ruckus Society, a national network of trainers in Greenpeace-style blockades and media stunts, and since then, at a "new economy" conference or two.

She works primarily as a facilitator, but because of the faith, time, and creativity it takes to do this kind of work in a deep way, she calls it "organizational healing" rather than the usual "strategic planning."

"I live in the post-apocalyptic shape-shifting city known as Detroit," she says in a profile piece in *Yes Magazine*.[1]

And that's certainly how the city looked as my bus rolled in, passing, first, a litany of shuttered and burned-out homes; then, the abandoned hulk of the Michigan Central Station, arguably the world's most famous monument to urban decay; until, off in the distance, the glittering sore thumb of the Renaissance Center, the almost equally famous, awkward, corporate-driven attempt at urban revitalization that left the city's neighborhoods behind.

The decline of Detroit's auto industry, white flight after the '67 riots, and decades of mismanagement hollowed out the city's tax base. Recently placed in receivership, the city had more than once cut off gas and water to thousands of mostly African-American residents until those communities organized to get it turned back on again. Amidst the decay, community gardens and grassroots experiments in self-reliance were thriving. Meanwhile cheap real estate had encouraged an influx of new, younger, "hipper" (aka whiter) residents, bringing with it the usual problems of gentrification.

Disembarking downtown, the city felt eerily quiet, near-empty eight-lane streets echoed back to an earlier time when cars were king. I spied a solitary share-scooter abandoned on the sidewalk.

Most of adrienne's work supports communities around the country that, like Detroit, are "directly impacted by the changing climate and our racialized economic system."[2] Amidst such conditions, she often witnesses groups with a severely limited capacity to imagine. "I've found myself on the edge of hopelessness," she says, "slowly devastated by the ways we treat each other when we can't see a way forward."[3]

And so, she has become a champion of imagination in the broadest sense, and of visionary science fiction in particular. Much of this she owes to Octavia Butler. *Parable of the Sower*, and the "change is constant" theosophy Butler developed in its pages, have become special touchstones for adrienne, a well of inspiration and guidance she goes back to again and again.

She likens social change work to "science fictional behavior," and thinks of visionary fiction as a "medicine of possibility" for her and the communities she works with. Armed with this understanding, and as part of a broader mission to "decolonize the imagination," she has encouraged a wide range of community organizers to experiment with visionary fiction. In 2015, she and her colleague Walidah Imarisha gathered together 20 such stories in a collection titled—in a nod to Butler—*Octavia's Brood: Science Fiction Stories from Social Justice Movements.*

The story that adrienne herself contributed to the anthology, "the river," is instructive. Told through the eyes of a Black water-woman, with a mix of sci-fi horror and magical realism, adrienne imagines Detroit's river as a supernatural ally. Churning with memory and history, and rebelling against the industrial toxins dumped in it over decades, it gurgles into being: an eerie, impossible wave grabs the state-appointed mayor, as well as a smattering of newly arrived white hipsters (drawn by "the opportunity available amid the ruins of other people's lives"), and pulls them down into its murky depths. In the ensuing exodus, the city is returned into the hands of those people "too deeply rooted to move anywhere quickly."

Beyond the hard-edged socioeconomic fable, this is a story of Nature becoming an ally, guide, and leader—an approach which is present throughout adrienne's work. That theme winds through her 2017 book *Emergent Strategy: Shaping Change, Changing Worlds.*

Here, Nature's patterns become organizing models: underground networks of mycelium exemplify interconnectedness; the flocking behavior of starlings illustrates decentralized leadership; ferns suggest a fractal-like way to scale up from small solutions to whole systems; while the toughness and explosive seed pods of the dandelion weed evoke resilience and regeneration.

adrienne weaves these and many other insights into an "emergent strategy" framework, that is one part radical self-help, one part materialist spirituality, and one part a set of methods that communities can use to plan for the big changes that are coming.

Many of these changes, adrienne acknowledges, are going to be unpleasant, unfortunate, even apocalyptic. Does she give up hope and check out? No. She wants to know: "What do we need to imagine to prepare for it?" "How do we articulate a compelling economic vision to sustain us through the unimaginable, to unite us as things fall apart?" "How do we experience our beauty and humanity in every condition," no matter how poorly things turn out?[4]

These are similar to the questions that set me off on my own search, and the tensions they capture are very present when we speak. The Green New Deal, for example, might be for adrienne a compelling utopian project and story, but she emphasizes that it can't be imposed on Detroit—or anywhere—from above. It's got to be "nourished," "grown," and "cared for" from the local level on up. I notice how flavored her language is here with nature metaphors. (And throughout our interview, I wonder at the invisible shifts these metaphors are making to my own deep thought patterns.)

At some point she says: "Look, if we care about our next *what*, we've got to change our *how* now." Which seems to sum up her whole approach, as well as, possibly, the central challenge of our time.

I'd hoped to dig further into these whats and hows with her while we hung out together in Detroit, but thanks to an old cell number, a lost email, and star-crossed schedules, we instead caught up with each other by phone some weeks later.[5]

Andrew: Besides being the bestselling author of *Emergent Strategy* and *Pleasure Activism*, and a nationally in-demand trainer and facilitator, you also co-host a podcast called *How to Survive the End of the World*.
adrienne: I do, with my sister.
Andrew: Your motto is, "Learning from the Apocalypse with Grace, Rigor, and Curiosity." So…what have you learned from the apocalypse lately?
adrienne: The latest lesson is that we're not going to have a soft, easy transition. That might sound a bit, *oh, duh*, at this point, but I still notice so many people around me becoming more fragile in the face of crisis rather than less fragile.
Andrew: How so?
adrienne: Too often, people—including me—show up expecting every single thing to be already pre-thought and taken care of for them, and every single need to be met. But then I check myself: *Wait, that's not how I imagine showing up in the apocalypse.* I imagine showing up in the apocalypse more like how you have to show up to a DIY action training camp where you better have brought your sleeping bag and bug spray, otherwise, you're going to be sleeping on the dirt and getting bitten. We have to take stronger responsibility. We need to remember that whatever access we have is access we have to create together.[6]

So, that feels like one of the biggest lessons to me: It's not going to be a soft, easy glide and then, *oh, things get a little harder*. The weather is not going to get just a tiny bit warmer. Climate impacts are not going to be subtle anymore. In the communities that I care about, the changes are not subtle at all. Everything is actually super drastic and super dramatic, and so much is happening at once. Drastic, dramatic shit is happening in the Amazon, and here in Detroit, too.

Which is why I tell people, and train people, to get in a better relationship with change. Because the changes will be drastic, and they won't be changes we like.

Right now, many of us are still in the privileged phase of the crisis; we're living through "the golden age of global warming." You know: *gosh, it's balmy in November*, and the like. We've been the frog in the boiling pot for a long time, but now we're really starting to cook, now it's actually happening to communities that we love and care about.

I used to think our central challenge was: How do we turn and face the fact that our relationship with the Earth is broken? But the more I've come to understand how it's all connected—how, say, climate drives forced migration, and how capitalism drives both, etc.—my focus has shifted. Now, the question I'm really interested in is: How do we survive the end of capitalism? That's been a big shift for me. I don't think capitalism works for us as a species; I think it's falling and is going to fail.

Andrew: A tough ideological pill to swallow for a lot of people, us included. Because at a certain level, capitalism is everything we know. It's the water we swim in. As has been said, *It's easier to imagine the end of the world than the end of capitalism.*

adrienne: Exactly. It's like "The Nothing" in *The NeverEnding Story*. The kid in the movie—a very formative childhood movie for me   is reading a book which turns out to be about not just our world but also the realm of dreams and the fantastical, and there's this "Nothing" that's spreading over everything. If it swallows everyone's dreams, then the real world, too, will cease to exist.

I think of capitalism that way. It seems to be selling us so many bright shiny things, but what it's really selling us is a nothingness—an emptiness, a hunger inside, a sense of not having. We need to figure out how to orient ourselves to capitalism. We need to understand what it actually is, and what it's doing to us, and to our children, and our dreams.

I think of myself as post-capitalist rather than anti-capitalist. Capitalism is here. What I want to be is beyond it, right? It has served a function. Technical progress, material wealth, etc. But to survive, we must get beyond it. I want to get to that place where it is a layer of our past. Like when you go to the Grand Canyon and you're looking at all the exposed layers of geologic time, including the one all the way back when the ocean covered everything. Capitalism is that ocean now, but I want to get to a place where I can look back at it and it's just this thin layer of crappy plastic. Then, on top of that, you've got the dirt again.

Andrew: The full title of your show is "Learning from the Apocalypse with Grace, Rigor, and Curiosity." How do you enlist such beautiful qualities of the human spirit to learn from such a terrible thing?

adrienne: For one, I'm no longer convinced that the apocalypse is this terrible thing. That's also been a big lesson for me. It comes from living in relationship with so many people who are suffering under current conditions. My father grew up in poverty, and a portion of my family grew up in poverty. I come from that. Other parts of my family have gone through evictions and foreclosures and all kinds of hardships. I have been moved into movement because of the AIDS crisis and police murders and all the rest. We manage to pull some beauty out of it, but there's so much about this existence that we're already in that's actually a shit show.

How do we partner with the potential ending of it in such a way that we can clear out the things that are not aligned with our long-term survival as a species? How do we partner with the apocalypse in a way that allows the best things to come forth? Our society is so unfair and unjust and imbalanced, maybe the best thing that can happen is for it to fall. But how do we fall gracefully? If I have a child on my chest, how do I fall in such a way that I protect that child, right?

Andrew: That's probably the most beautiful way I've heard anyone describe Collapse.

adrienne: Most of us do have children on our chest. We have babies that we love, and we're, like, OK, capitalism must die but not my kid. I don't want my nibblings [a portmanteau of "nieces/nephews" + "siblings" adrienne learned from Chicago healer-organizer Tanuja Jagernauth] to be going through *The Road*. That's our epic tension: these systems have to end. But how do we turn and face the fact that they're ending, and figure out how to *partner* with that ending?

It wasn't so long ago that I was still in the "we've got to stop the bad; we've got to save the world" attitude. But the folks at Movement Generation [which includes previous interviewee Gopal Dayaneni] got me to turn and face the apocalypse. Their week-long training was the first time that someone broke it down for me in a way I could understand: nothing that we do now is going to stop so many of the terrible things that are coming at us. We are unavoidably heading into an era of super-storms, super-droughts, and world-wide fights over resources. That's coming. We can avoid some of it, but not all of it. I feel like that was such a blessing to my life to just fucking hear that and be like: all right, cool, now what?

Andrew: That was a key turning point in your understanding. And how did it change what you did in the world?

adrienne: I was like, OK, all this is unavoidably going to happen, so, then let me try to shift *how* it happens. I want to shift how we are orienting to change and how we are orienting towards each other. This led me to *Emergent Strategy*. I want to speed up how quickly humans can fall in love with each other, how quickly they can make an assessment of whether someone is trustworthy or not. I want to speed up the process by which people can identify the right role that they can play in community. I want to redistribute how we value the different kinds of roles that are in community. I want to devalue the amount of power we give to celebrity and famous people, and increase the value of nurses and emergency workers and mediators and people who are like, *hey, I know martial arts.*

Andrew: What are some of these other less tangible skills people may not realize are important?

adrienne: We need healers. We need people who know how to teach. We need people who know how to build community. And I'm, like, oh, community organizers are actually highly prepared to be leaders during the apocalyptic time that is coming. *If,* that is, we can get out of our "resist and demand" mode and get ourselves in right relationship with the changes that are coming. Because the mode of a lot of community organizing—resist the bad ideas of the people in power and demand some change from the people in power—is not going to cut it in the future.

When I'm operating in that mode, I might be making a demand of power, but in doing so, I'm still keeping the power dynamics in place. And the people who have power have shown us over and over again that they are not interested in redistributing that power. So, we'll just have to take it, carve out our own power, build our own community autonomy.

Most of my work now is as a mediator, and facilitator. I'm an "organizational healer." I know how to help a group move through to consensus quickly. And I want to train an army of facilitators, who can unite people as they're facing these crises. We need to be ready to do that. And the only way we'll be ready is if we have good practice under our belts of being with each other in better ways. So, I'm focusing my attention on how we are with each other and how quickly we can form community with each other, how quickly we can fall in love with each other.

I've watched all the apocalypse films. The good leader is the one who knows how to get us together and get us moving. That's the one who also ends up getting to shape the direction. I want loving community to shape the direction, not might-makes-right power.

Andrew: You're saying our apocalypse "bug-out bag" shouldn't just have water filters and a gun in it. You're drawing attention to the overlooked "soft" skills—community organizing, consensus building, say—that are going to be of great value as our situation worsens.

adrienne: A lot of people who are developing hard skills right now are doing it from a place of deep fear. They're terrified of the future, terrified of other people. They're thinking, *how do I hoard and lock down and just survive no matter what?* I'm like, who gives a fuck about surviving no matter what? If it's going to be a total horrific shit show, who wants to be here for that? Not me.

I'm interested in thriving. I'm interested in a beautiful life. I'm interested in raising children that are happy. I'm interested in having lots of orgasms. I'm interested in feeling *it's good to be alive.* Yes, we need to learn how to survive, and, yes we need to be cultivating the spaces that will help us survive. But I'm not interested in living in a bunker underground indefinitely. There's nothing that I've experienced that makes me think that would be a good life for me.

Andrew: What are some of these other soft skills, intangible ones people may not realize are important?

adrienne: Well, knowing how to be joyous with what you've got, for one. It doesn't take as much as we think to live a relatively good or relatively beautiful life. I don't make a ton of money. I've never had a ton of money. In fact, I have lived in a ton of debt. But my lifestyle is still relatively easy, relatively comfortable. I live sometimes paycheck to paycheck, but I recognize the privilege I still have relative to most of the world.

When I travel to other places where the majority of people have significantly less than I do, I find over and over again that they're making so much more of it. They're making more of the experience of being alive, of the emotional and spiritual connective tissue of life. The conclusion I draw is that there's something about having access to everything (that many of us in the US have) that leaves you with nothing. How then do we cultivate the skills to recognize what is enough?

I don't think of redistribution the way a lot of people think about it. I'm all for taxing billionaires and raising the minimum wage, but it's also important to ask: What does it look like to have enough? How do we get familiar and intimate with "enough," and how do we start to commit to that as a shared radical desire that doesn't feel like settling, or having less?

I rent. And I live in the same place that I've been for ten years. Recently someone was teasing me. They were like, *hey, you're a* New York Times *bestselling author now, so, are you going to buy a house?* But this already feels like an abundant space, and I don't need more than this room and this place. If I needed to raise a child in this space, I could figure it out. I have enough, and I've had enough my entire adult life. Just because I can get more doesn't mean I need more. That fundamental sense of "enough" is really important to me.

Andrew: In *Emergent Strategy* you talk about "moving at the speed of trust." Why is that so important? And how can it serve us when the apocalypse is moving at the speed of capitalism?

adrienne: Moving at the speed of capitalism is how we've gotten here, right? When you move at that pace, you forget so much that is crucial to a decent life, and you overlook so many factors that later become fissures in the foundations of things. In the US, we have fissures in the foundation of everything about this nation because it was built very quickly with so much injustice at the core. That injustice still impacts me every day.

I think the trick of trust is that it's only slow at first. You know this, Andrew. You've been on many actions, right? When you're starting out with an action team and you don't know each other, and some of you have never scaled a wall with a banner before, or what-have-you, then trust can be very slow. You're like, wait, hold on. We need to make sure we have agreements and a whole process in place just to make sure we're all clear with each other. But when you have a team that's been through the fire a few times, it's a different story. Now, you have your people. Now, you'll call 'em up, and you'll be like I don't have to tell you the whole deal, but I need a banner that says this, and a team that's ready by tomorrow noon, and OK, boom, and then it's fast, right?

Andrew: And to navigate the crises that are coming at us, we're going to need many teams, many community networks, that have that kind of trust, that can respond with versatility and speed and unity as the shit hits the fan, yes?

adrienne: Exactly.

Andrew: In *Emergent Strategy*, you paraphrase Grace Lee Boggs, saying that "Building community is to the collective what spiritual practice is to the individual." What does that spiritual practice look like to you, given the crucial importance of resilient communities, especially as conditions worsen?

adrienne: Before I became a professional facilitator (before I realized you could get paid for that kind of thing!), I thought of myself as an "organizational healer." I'd come into an organization where something's not feeling right, and find the part that's hurting, I could figure out the root cause, and help to alleviate the stress. I've done healing at an individual level, organizational level, alliance level and, in some places, at a movement level.

When you look at something that is broken, it's important to see the wholeness in it. In this moment, one of our key tasks is to try to get people clear on what is actually broken.

Let's not go around offering false solutions. Let's not go around saying, *oh, this is the problem, and here's a solution to it*, when it's not actually a solution.

Take me, for example. I've got chronic traumatic pain in my lower back from abuse. If someone were to suggest I just get massages, that would be a false solution. Yes, I can go get massages forever, and it'll give me some temporary relief. But, structurally, I will still have that trauma in there that has created a wrong shaping, something that's broken inside.

When I work as a somatic bodyworker, I try to see not just the person's body, but her mind, her story, her personality. I try to see all the parts, and try to understand from that whole system what needs to be liberated, what needs to be removed, what needs to be invited in, what needs to be strengthened.

I feel the same way about social movement work right now. There are no temporary solutions for the big issues we're dealing with. Because what we're actually dealing with is that capitalism is living inside of all of us. Competition, dominance, patriarchy—all the shit we're fighting against—is *in us*, and so it shows up in all of our organizing work.

So, instead of going off and critiquing someone else or making a demand on a power structure outside of ourselves, what I'm trying to do is bring people's attention inward and ask: How can we be honest with ourselves about the way capitalism is playing out in our lives and in our organizations and start to shift that?

Imagine if philanthropy was able to turn and face itself and recognize that it's hampering so much of our moves towards justice? It's still so concerned with keeping its capital and keeping its dominance over the communities it serves—and we're so caught up in this nonprofit industrial complex—that our movements can only go as far as rich people decide they can go. That's ridiculous, right? Ever wonder why it's so hard to do good alliance work? Because we're trained to pit ourselves against each other for the grants and resources we need to stay afloat.

The root of the problem, here, as it is pretty much everywhere, is that we're being pitted against each other just to survive, rather than encouraged to collaborate in order to generate abundance.

In both my healing work and my social movement work, I try to see as big a picture as I can, and then come up with as precise a solution as I can. You don't tell someone, *Here's the 30 billion things that need to be fixed.* You tell people, *Here's the one step. Here's the one practice.* And if you begin to practice it, it will unlock the other healing modalities, the other pathways.

I could do a billion different things to try to heal my lower back, but if I do one thing, say, I stop eating bread, that would immediately reduce the inflammation in my body, which would immediately begin to move me in a different direction. And *then* I won't need as many massages because I won't be inflamed all the time, right?

Andrew: Speaking as a healer, then, what's the next right thing we must do as a society?

adrienne: Grace Lee Boggs was my mentor and a big reason why I moved here to Detroit. She was always asking the question: "What time is it on the clock of the world?" For our species, right now, the time on the clock of the world is: We are already in the crisis, we're already in the boiling pot. We just need to know it. So, how do we turn and face where we actually are?

Turning and facing that truth immediately shifted my priorities.

So: let's not give any of your attention to the climate denialists. They're a small minority, and getting smaller, and they're in the pot too, so they'll eventually figure it out for themselves. Also, don't spend too much of your energy on electoral politics. Whoever gets elected President is still chief of an empire that's falling.

Where we need to focus our energy is getting into real relationships with people. People you can touch and be around. People who, when the power goes off, when the water stops flowing, when the local utility turns the heat off—all of which happens in Detroit regularly—are in networks and know how to connect with each other and help each other. Every time people show up for each other, those networks get stronger and deeper.

So, change where your attention is flowing. Stop giving it to things you can't actually touch or change. Give it to the things you can actually tangibly change in your lifetime. If everyone was doing that, we'd have plenty of energy and attention and time for the work that we need to do.

My conversation with adrienne shifted my perspective on a host of fronts, from healing to social movements to the apocalypse. It's been a slow burn absorbing it all.

We're in for a hard fall, but, paradoxically, adrienne notes, what we may need most are soft skills. The unravelling is going to intensify things. Many of our old models—including the old model of community organizing—will no longer suffice. Pressuring and petitioning those with institutional power will no longer deliver the goods. Instead, adrienne is saying, we need to take power into our own hands, and build what we need from there.

The "apocalypse" may not be such a bad thing after all, she said. Which initially startled me, but then I understood her deeper point: Due to poverty and violence and the ravages of capitalism, life is already such a "shit-show" for her and for many of the marginalized people she works with, that an apocalypse that threatens to "change everything" would in many ways be a welcome development—*if* it could be properly channeled.

For her, channeling the apocalypse meant, first and foremost, understanding that our crisis was fundamentally caused not by fossil fuels, but by capitalism. adrienne described herself as "post-capitalist," an identity that seemed as much an act of imagination as a commitment to wholeness and healing. For her, capitalism wasn't just an economic system with ultimately ecocidal consequences, but a spiritual disease selling us an emptiness, "a sense of not having."

Capitalism gets inside us, she said, and it gets into our movements and into our philanthropy, and deforms who we are and how we operate. As a healer—whether of bodies, psyches, or organizations—one of adrienne's roles is to help us see how capitalism is hurting us, and help us imagine our way beyond it. She wasn't just interested in how to have a post-capitalist economy, but how we could be post-capitalist people.

She acknowledged there were "no temporary solutions for the big issues we're dealing with." And like Gopal, who described capitalism's fundamental response to the climate crisis as a "festival of false solutions," adrienne was laser focused on distinguishing false solutions from real ones. If we can do that, she said, we'll get through this. In fact, we might end up better off.

For adrienne, as for so many—Tim DeChristopher and Roy Scranton come to mind here—seeing our circumstances plainly became a kind of liberation. Letting go of the hope of preventing catastrophe allowed adrienne to fundamentally reorient herself. She chose to get more deeply grounded in community, and to focus very strategically (and soulfully) on adaptation. How do we "partner" with this apocalypse?, she wants to know. How do we survive the end of capitalism? How do we fall as if we were holding a child on our chest? She, too, is looking for a better catastrophe. And she turns to visionary fiction to help us imagine our way there.

Dystopia: If the Zombie Apocalypse comes the Day After Tomorrow, will Max still be Mad?

A weird time in which we are alive.
We can travel anywhere we want, even to other planets.
And for what? To sit day after day,
declining in morale and hope.

— Philip K. Dick

In the outro to *Octavia's Brood*, adrienne says, "We hold so many worlds inside us. So many futures. It is our radical responsibility to share these worlds, to plant them in the soil of our society as seeds for the type of justice we want and need." She describes science fiction as "the perfect 'exploring ground,' as it gives us the opportunity to play with different outcomes and strategies."[7]

Kim Stanley Robinson's 2020 novel *Ministry of the Future,* which cascades through the next few decades of climate politics imagining mechanisms like "carbon quantitive easing" and climate freedom fighter attacks on fossil-fuel infrastructure, is a perfect example of this kind of strategic exploration.[8]

"If we want to bring new worlds into existence," says adrienne, "then we need to challenge the narratives that uphold current power dynamics and patterns." Borrowing a phrase from our comrade Terry Marshall, this clash of stories becomes an "imagination battle."[9] Somebody imagined a racialized economy, someone imagined white supremacy, Hollywood keeps imagining dystopia and apocalypse, and the moment we question these narratives, says adrienne, "the moment we begin to dream of justice, of liberation, of right relationship, we become imagination warriors." Our task, she says "is to co-dream visions more compelling than oppression, and more honest than supremacy. And then move from imagination all the way to new practice."[10]

In this task, we're not only up against 500 years of colonial thinking, but a large chunk of mainstream sci-fi. That's where so many of us rehearse for the end of the world. and it's a key terrain where this "imagination battle" will play out. Take, for example, the utopia vs. dystopia war that sci-fi has been waging in my own head pretty much from jump.

Like any nerdy American man-child growing up in the 1970s, I was a creature of sci-fi & fantasy. My first act of rebellion was staying up all night to read *Charlie and the Chocolate Factory*. I remember my tender five-and-a-half-year-old consciousness being bent 360° sideways by a screening of Kubrick's *2001* in Manhattan's cavernous Ziegfeld theater, so disturbed by the final few scenes that I emerged onto the sidewalks of midtown Manhattan expecting to be swallowed live by Jupiter-strong gravity. In middle and high school, I built more graph-paper Dungeons & Dragons mazes than I went out on dates, and I was so taken by Asimov's notion of psychohistory from his *Foundation* trilogy, that I proposed to put a scientific basis underneath it in several of my college application essays. (Dur. No wonder I didn't get into any of my top choices.)

As an adult, my first exposure to gender fluidity came from Ursula K. Le Guin's Hugo Award-winning *Left Hand of Darkness*, set in a world where people are not male or female, but "potentials" and every six weeks, depending on factors both psychological and circumstantial, "molt" into female or male or neutral. The works of Phillip K. Dick accelerated my philosophical turn in college and after. The protagonist in *Bladerunner* (and the PKD book, *Do Androids Dream of Electric Sheep*, it was taken from) goes by the name Deckard, which is a clever stand-in/namesake for Descartes, whose philosophical dictum, "I think, therefore I am" is turned by the film into a question, as it ponders whether synthetic humans are truly human (or possibly "more human than human").

Sci-fi suffuses our future-imaginary with heroes, visions, nightmares, and pop-philosophical dilemmas (red pill or blue pill, anyone?). By extrapolating the present into the near or far future—or even wholly other realities—it delves into not just hot-button social issues like automation, surveillance, and social collapse, but into fundamental aspects of being human—love, gender, memory, intelligence, identity, hope. It is an act of world making, and when it takes an apocalyptic or dystopian turn, as it so often does these days, it's an act of end of the world making, as well.

Because all these decades that sci-fi has been blowing my mind open to new worlds and possibilities, it's also been digging a deep post-apocalyptic neural groove into my brain. Whether by zombie or virus or alien invasion or meteor or robot-revolt-singularity or climate catastrophe, we are so often led to the end of the world in the movie house, that we assume it will eventually happen in reality. Increasingly, from *Mad Max* (where road bandits fight over the last drops of gasoline across a parched dystopia) to *Avatar* (where a space-mining company desperately colonizes other planets in search of rich deposits of "unobtanium") to *Interstellar* (where a blighted corn farmer heads through a tesseract in a black hole in order to find humanity a new home—I know, I know), the apocalyptic driver is some combination of resource exhaustion and ecological collapse. So much so, that dystopian sci-fi has spun off its own climate-specific sub-sub-genre known as "cli-fi."

Some of this apocalyptic canon is gut-wrenchingly desolate, delving Biblically deep into existential questions of human cruelty, hope, God, and meaning. Yet even those that merely transpose a Western gunslinger or wandering Samurai storyline onto a dystopian landscape often take up the big questions: *What kind of world are we going to get? How can we stay human in it? Are we more likely to survive via rugged individualism or cooperation? And, ultimately: Is there still hope?*

The worlds our fictions conjure are not just fantastical trash; they can be our most articulate and precise alerts to the likely manner of our future enslavement and doom, as well as possible paths we might take towards liberation and survival.

Our current dystopian imagination tends to conjure three kinds of worlds: The Empty World, The Teeming World, and The Workable World.

The Empty World is that familiar post-apocalyptic landscape where much of the physical plant of civilization (in a state somewhere between shambles and perfectly preserved) is in place, but people are mostly gone. This is the setting of *Zombieland, The Walking Dead, Mad Max, I Am Legend*, and so many other films we know and love and hate. These films are often just set ups for lone-gunslinger or gangland genre fare transposed into a dystopian landscape. But even these are asking some deep questions: *What are our ethical bonds to one another in a broken world? What code do we live by when there is no State?*

The Teeming World imagines a different kind of apocalypse. Not a nuclear war or virus or mysterious supernatural culling that wipes most people out, and leaves the rest to fend for themselves, but a combination of accelerating trends of global warming, ecological collapse, fossil-fuel depletion, overpopulation, species extinction, economic inequality, and social segregation, that lead to a world teeming with people whose systems of healthcare and governance and livelihood are profoundly degraded. This is the claustrophobic, urban-gangland landscape where the whole world has become a sci-fi Sao Paulo favela or AI-enhanced 1970s-era South Bronx. This is *Bladerunner*'s dark claustrophobic neo-noir LA of 103 million inhabitants, *Elysium*'s Gaza Strip Earth, *Children of Men*'s burnt-out polyglot England turned internment camp. Here the questions tend towards: *How do we survive? How do we live (and love, and hold on to the past, and orient to the future) in a collapsed or collapsing world?*

Finally there's The Workable World: a livable society set in a shattered future. The world, say, that John Michael Greer imagines in his *Retrotopia*,[11] set in 2065, in a North America where the USA has broken up into separate nations after multiple economic and ecological shocks led to civil war. Most of these new region-nations are still pursuing the dream of Progress, but one of them, the Lakeland Republic, located in the upper Midwest (capital, Toledo), has chosen a different path of equity and appropriate technology. Here people make their own shoes, travel in french-fry-grease-powered streetcars and have rediscovered the joys of community.

Or, *New York 2140*, where Kim Stanley Robinson imagines a New York City on the other side of climate catastrophe. Here, "The super-rich live uptown, in a forest of skyscrapers near the Cloisters. The poor live downtown, in Chelsea, which is half-submerged."[12]

Whether our post-carbon and post-catastrophe future is going to turn out like the flooded, elite-friendly futurism imagined by KSR, or the more modest, back-to-the-best-of-the-past *Retrotopia* imagined by JMG—or, for that matter, something closer to the apocalypse of *Mad Max*—we don't know. We're not there yet.

It's certainly helpful and instructive to have a gallery of Workable World futures to set our sights by, but back here in the early 21st century, we're not post-anything yet. We're still facing into the horrors of climate catastrophe and social collapse; we're still poisoning our future with carbon and erecting heartless Walls and refusing to act at the scale of the problem. And most days, things look bleak as hell. For so many of us, the question isn't so much *which* livable future are we going to get, but whether we're going to get any future at all. We're wondering: Is there any hope? And our dystopian imagination has much to say.

The question of hope is asked most brutally in Cormac McCarthy's *The Road*. The world of *The Road* is an Empty World: bleak, barren, unrecognizable; the Earth is effectively dead; the feeling is of Hell, utter solitude, despair, a true End Time. McCarthy's prose is spare, plain, and darkly prophetic; its relentless atmospherics and brutal truths (humans will do almost anything—from murder to torture to cannibalism—to survive) are studded with Christian allegory.

Like *Melancholia*, there is a family of three, a Trinity: the man, the boy (no one has a name), and a ghostly third: the wife and mother, who, seeing nothing to live for in this world, has taken her own life before the events of the novel begin. The reader can certainly appreciate her choice. As the man says to himself, "On this road there are no godspoke men. They are gone and I am left and they have taken with them the world." And in this world-less world, it's hard to find any basis for hope.

And yet, the man and the boy keep walking their hellish Road. The boy is basically the man's God—God as love, God as compassion, God as innate moral sense, God as something to live and die for, a way to "carry the fire" through Hell and across the generations. And even in this blasted desolate world, there are tiny moments of redemption. Water appears again and again as a marker, a brief reprieve from the suffering. The ocean they're trekking to; the grey, freezing waterfall pool they swim in; the one bath they take—each of these moments offers a brief interlude from the monotonous horror, an occasion—almost—for joy.

When I met with Tim DeChristopher, he described the classroom exegesis of *The Road* as the most important inquiry during his three years of divinity school. Every item, tool, scrap of food, or hasty shelter the man and his son stumbled upon they received like a gift, evoking in Tim an overwhelming feeling of gratitude. As I read the book (in one excruciating day amidst the beauty and teeming life of Pt. Reyes, California), I too couldn't help but feel a profound appreciation for so much that I often take for granted: the simple caw of a crow, the sun in the sky, sleeping in a bed, the simple fact that the town of Pt. Reyes Station had a fire department, and how I was able to chat with the townsfolk instead of being eaten by them. Maybe this is the greatest virtue of apocalyptic literature: not just to alert us to potential future catastrophes, but to help us appreciate all we have here in our blessed pre-apocalyptic.

In *Children of Men*—a Teeming World set in a chaotic, flooded, soft-fascist near future—for unexplained reasons no child has been born in the last 20 years. Suicide kits have been distributed to the citizenry. There is, in a very literal sense, no future, no hope. And then our protagonist, ex-rebel Clive Owen, is handed an infant. It's the only child in the world. There is a Greenpeace-esque ship waiting to receive and safeguard the child, and he must ferry her and her mother to safety, avoiding the authorities, as well as his old domestic-terrorist buddies who want the child for their own nefarious purposes. The visual spectacle of the film is mesmerizing: a portrait of a chaotic and claustrophobic world, exhausted, at the end of its ropes, and utterly hopeless. Until now, that is. Because anyone who sees the child or hears her cry, is suddenly shot through with an almost impossible, miraculous ray of hope.

In Christopher Nolan's *Interstellar*, we find another hero in another dying world. It's a dust-storm-blasted world. Humanity has responded to eco-crisis by giving up war and engineering to focus on farming to the point that schoolkids are taught that the Apollo missions were a hoax, and the protagonist's grandfather wonders aloud about how we ever thought six billion humans could "have it all." Yet blight keeps driving one crop after another extinct, until corn is all that's left and it's not long before it too will succumb. Humanity has only one shot: a last-ditch effort to find a new suitable home in another galaxy, reachable via a wormhole placed near Saturn by a mysterious helpful "They." The film's plot and pseudo-science and dialogue (um, "that's relativity, folks.") fall apart many times along the way but there's stuff worth mining in its odd narrative of hope. The driving force of our species' salvation? Maverick Matthew ("We used to look up at the sky and wonder at our place in the stars; now we just look down and worry about our place in the dirt") McConaughey, astronaut turned dirt farmer turned astronaut again, who blasts off from his family to find a new home for humanity.

How is all this going to work out? "We'll find a way like we always have," says one of the astronauts gamely. But it turns out the scientist who has sent these brave souls on their mission has been lying to them. The grand theory of gravity he's been working on all these decades that will get humans off the Earth (once McConaughey finds a viable place to go) is a crock. Our Mystic Keeper of the Inscrutable Theory merely wants to give folks the illusion of hope, droning on across the interstellar audio link with a monotonous repetition of Dylan Thomas's "Do not go gentle into that good night; Rage, rage against the dying of the light." But that is small consolation to McConaughey who, circling a black hole and aging far faster than his children back home, is being robbed of his life's treasure while on a fool's errand.

But, of course, when Hollywood, quantum mechanics, and ever-stubborn humanity itself team up, nothing is impossible, and into the Black Hole our hero goes, entering a tesseract that bends gravity and time so that he's able to send his daughter (now grown up and the smartest theoretical scientist in the land) the critical information she needs to complete the impossible theory and save humanity. And who does the mysterious helpful "They" turn out to be? Why, ourselves, of course! "They" is actually a time-stitched "Us," motivated by love and survival instinct, to reach in from the future (aka a black-hole-enabled alternate time-space dimension), and help ourselves out in the, um, present. If that's the most viable survival plan we've got, no wonder people are down in the mouth about our chances. It makes you wonder whether Dylan Thomas had the best advice after all.

Interstellar is part of a slew of dystopian films with pretty "hope-a-dope" narratives. If you take a step back from the CGI, you can usually find a cast of classic tropes and archetypal figures.

There's the Messenger/Savior. Think: John Connor from the *Terminator* series, or astronaut-farmer McConaughey himself. They bring warnings from the future, and maybe also a world-saving Fix. Whether that's the suspicious quantum theory from *Interstellar*, or some just-concocted antidote central to so many zombie-plague and pandemic scenarios, this Fix has the power to save humanity (though first, usually, the hero's girlfriend).

There's often an Ark involved. Whether a spaceship, a pod, a boat, or even a train, The Ark is humanity's last home and hope, the vessel holding the last remnants of the race and/or the seeds of our future. In *Battlestar Galactica* the entire fleet is effectively an Ark, containing the last 50,298 members of the human species, in search of a new (or is it old?) home in the stars. The entirety of *Snowpiercer* takes place on a steampunk train barreling through a frozen wasteland Earth with the last of humanity aboard. As a band of underclass rebels battle their way from the back of the train to the front through increasingly surreal layers of class society, the train serves as both an allegory for runaway industrial civilization and an Ark fraught with class struggle.

Across the blighted dystopian landscape, the Savior and his Posse set off in search of a Sanctuary. If not always a full-on Promised Land or

Eden, the Sanctuary is at least some kind of shelter, safe haven, sanctum, oasis, or hideout, where civilization and reciprocity and kindness—or simply the once normal functioning of nature—hang on.

In *Road Warrior*, this Sanctuary is little more than a ragged desert fortress barely holding out against marauding gangs. In *I Am Legend*, it's a fortified hippie-ish commune somewhere up North that Will Smith's anti-zombie serum must be ferried to. In *Elysium*, it's a high-tech off-world gated community: a sumptuous space-station for the wealthy elite orbiting a Gaza Strip Earth. The only hope in this world, at least for the teeming, toiling billions on Earth, is to break into this Elysium and make its rich tech available to all (particularly to the terminally ill son of Matt Damon's ex-girlfriend.) In *Waterworld*, the Promised Land is an island, a spot of dry land—mythical or real, we don't know—in a totally flooded world.

In so many of these stories, it's unclear whether this Promised Land is a real and achievable destination, or just a mirage—just a way to keep our hopes up, and keep us moving. In *Mad Max: Fury Road*, the desert oasis Furiosa is seeking turns out to be exactly this kind of mirage. If it ever was a real place, it is no longer; it might as well be a legend.

Our dystopian canon asks: Will The Messenger warn The Savior to lead The Posse to find The Path to carry The Fix from The Sanctuary to The Ark before The Flood? And, all too often answers: Tune in next week.

But next week is too late, because back here in reality reality we're still not listening to our Messengers (climate scientists), we don't have a magical Fix (desktop fusion and direct air carbon capture are decades, if not forever, off), and our Arks are literally flooding (the "impregnable" Global Seed Vault[13] on Spitsbergen Island, half-way to the North Pole from Norway, flooded[14] during 2017's unprecedentedly high temperatures). Oh, right, and: there is no Savior, and we must become The Posse we've been waiting for.

Our dystopian imagination can be very seductive. At its most apocalyptic, it promises to solve the ultimate problem; to wipe the slate clean; to bring all of humanity's troubles and frustrations and failures—millennia of not getting it right—to a dramatic finale. Sci-fi often serves up the apocalypse in its Platonic form: a flood that washes us all away; a virus that takes us all out in a matter of weeks; an "I told you so" eco-morality play whose terrible swift sword justly smites us all down for disrespecting the Earth upon which we depend.

A book like Alan Weisman's *The World Without Us*, less sci-fi than a non-fiction "thought experiment," invites us to imagine that pure day after: a perfect quiet. No jackhammers. No cars honking. No wives and husbands yelling at each other. Nature has the reins again. And there's something eerily satisfying about this. It feels like the solution to a problem. On those really bad days, the days I don't feel up to the tasks before us, the days I feel defeated by my own species, I take a strange solace here. The disappointed idealist in me looks out over the sprawling mess we've made of things and thinks: yes, let's just pull the plug and end it now.

The thing is, the apocalypse is already happening, and it's anything but clean, anything but quick, anything but a morality play. Those least guilty are being slaughtered. Millions of people die every year due to fossil fuel air pollution, many of them children.[15] The climate change-fueled carnage of the Syrian civil war has displaced over three million civilians. 1,000 Americans were killed by Hurricane Katrina, overwhelmingly black and brown, and from the poorest, most low-lying neighborhoods. Most were not killed by the rising waters as much as by the systematic inequality, decades of neglected infrastructure, as well as acts of outright racist violence.

Of the one hundred killed by Hurricane Sandy, two were children—two- and five-years-old—pulled out of their mother's arms by a raging sea as she huddled in her car at the edge of Staten Island. An hour earlier, amidst lashing winds and rain, she'd knocked desperately on a door in search of shelter, but the man had refused her entry. He happened to be white; she happened to be black, and now her children happened to be dead.

Climate Catastrophe isn't going to spare the innocent; Climate Catastrophe isn't going to put humanity out of its misery. Just the opposite. The apocalypse isn't going to happen in one cleansing swoop, or in some poetic poignant finale; it's going to happen in ugly stages; it's going to be collapse by a thousand (budget) cuts, and at each cut there will suffering and families torn apart and people at each others' throats, as well as extraordinary acts of sacrifice and cooperation and resilience. It won't make anything simple; it will be full of chaos and complexity and impossible decisions. It will be both heroic and banal. It will be human-all-too-human at every turn. And in many parts of the world it already is.

This is the apocalypse we are looking at. Unlike our fantasy apocalypse, it resolves nothing about the human predicament. It is not an escape from being human; it is a forced return. It's not a clean end to humanity; if we're lucky, it will be a shattering new beginning.

To help us get through the next century, we don't need more stories that seduce us into being consumers of our own doom. We don't need more stories that peddle false solutions and flimsy hopes. We need stories that help us face our uncertain future and imagine our way through.

One of the tools that can help us do this is visionary fiction, whose core elements adrienne describes as follows:

> …explores current social issues through the lens of sci-fi; is conscious of identity and intersecting identities; centers those who have been marginalized; is aware of power inequalities; is realistic and hard but hopeful; shows change from the bottom up rather than the top down; highlights that change is collective, and is not neutral—its purpose is social change and societal transformation.[16]

These qualities are exemplified by Octavia Butler—the author adrienne refers to as her "north star"—in her most popular book, the Nebula and Hugo Award-winning *Parable of the Sower*, which charts a quite different path through Collapse.[17]

Set in a dystopian LA, the world Butler conjures in *The Sower* is a broken and ecologically-compromised—but very believable—extrapolation of our own world. There is growing social chaos amidst the remnants of a failed state. Society is a patchwork of corporate enclaves and semi-functional suburban cul-de-sacs that have banded together and erected perimeter walls. The rare times when it rains, people can still grow things in gardens. Our mixed-race protagonist Lauren Olamina, far wiser and more courageous than you might expect from her 16 years, lives in one of these banded-together communities in the outskirts of what was once LA. She conceives/receives a prophecy, a Taoist-like, nature-centric, humanist religion in which "God is change"—a precept both ontological (reality is always in flux) and imperative (we must grow; we must change the world; in fact, to survive, humanity must spread its seed to other worlds). With the Earthseed prophecy as her guide, she makes plans to head north where climate change is less severe. As social chaos grows and the community comes under attack, Lauren grabs her bug-out bag and flees with a few comrades. It's a story of flight, community building and collective resilience. A motley but intrepid band begins anew. There is prophecy. There is a plan. There is hope.

As for the classic tropes of hope, they're there, too: A Messenger (Lauren) leads her Posse (a rag-tag-band of climate refugees) to a Promised Land (the verdant Acorn community they're all trekking north to), carrying a Fix (the EarthSeed prophecy). But here the Fix isn't a magic techno-fix, it's a creed of resilience and change. And the Messenger isn't the all-too-common violent White Savior dude, it's a 16-year-old mixed-race climate refugee with a gift for prophecy and community building. And the World they inhabit isn't some fantastical CGI doomscape, but, for better and for worse, a recognizable extrapolation of our own. The story isn't neat and pretty, but it feels like a Map, even a Strategy, we can follow.

And what about visionary fictions set in our own (non-extrapolated) present-day world? What maps or role models are on offer? Consider Halla, the 50-something Icelandic bike-riding eco-guerrilla from the 2018 film *Woman at War*, who, when not holding heartless industrialists to ransom, is trying to adopt a child. Or Reverend Ernst Toller, from Paul Schrader's 2017 indie *First Reformed*, who pathetically, tragically, heroically, struggles to jibe his religious faith with our ongoing assault on nature.

Or Hushpuppy, from Benh Zeitlin's 2012 genre-defying *Beasts of the Southern Wild*, a particularly stunning portrait of how to survive (and stay true to yourself) in the face of climate-fueled disaster. Is it a monster film? A coming of age film? A social portrait of defiant Louisiana bayou dwellers? A political parable of climate catastrophe and resilience? It's all of these, and at the center is our most unexpected heroine-protagonist yet. Six-years-old, motherless, and almost ferally wild, Hushpuppy lives in an uneasy standoff with her dad in a ramshackle pair of buildings along the bayou, part of an overlooked multiracial community called the Bathtub.

When the big storms come in, if the levees up North hold, the rich communities stay safe, but the Bathtub floods. In the little schoolhouse the teacher slash medicine woman talks of dinosaurs and evolution and climate change, and in Hushpuppy's mind she imagines fearsome beasts released from the Antarctic ice as it cracks apart, slowly making their way towards her. A perfect child's-eye metaphor for climate disruption that should strike fear into every adult's heart, too.

But fear is the last thing Hushpuppy or her dad or their neighbors have in mind. Instead they're all about resilience and survival and joy and living free. Her father prods and goads and challenges Hushpuppy to be strong, in both muscle and mind, and as the storm builds, most in the community refuse to evacuate. They are determined to ride out this storm in their own way. There's several subplots (an island-hopping search for mom at houses of ill repute, an escape from well-meaning but paternalistic healthcare workers, a nighttime mission to dynamite the levee), and many scenes of close-to-the-bone bayou living (festivals, fires and fights; whiskey-drinking, chicken-killing and crab-eating), but the overall takeaway is clear: a storm is coming; we didn't stir it up, but all of us together—and each of us in our own utterly singular way—are going to have to find a way to ride it out.

Utopia: Our Afro-Indigenous-Trans-Eco-Socialist Futurism can beat up your Capitalist Realism!

Our choice remains: Utopia or oblivion.

— Buckminster Fuller

For adrienne the critical transition isn't from fossil fuels to renewables, but from capitalism to a post-capitalism grounded in social justice and operating in sync with the rhythms of nature. Capitalism has "served a function," she said—technical progress, material wealth, etc.—but now to survive, we must get beyond it.

An already hard task made even harder because of the way capitalism—like "The Nothing" in *The NeverEnding Story* adrienne likened it to—has insinuated its way into everything, even our dreams.

To imagine our current capitalist ocean as a future "geologic layer of plastic" detritus with post-capitalist dirt on top of it, adrienne turns to visionary fiction and utopian sci-fi. If we can first imagine it, then maybe we can create it.

Almost immediately, however, this utopian impulse runs right up against "Capitalist Realism." According to British theorist Mark Fisher who coined the term in a 2009 book of the same name,[18] it's "the widespread sense that not only is capitalism the only viable political and economic system, but also that it is now impossible even to imagine a coherent alternative to it." This "sense" is not so much an ideological argument as a "pervasive atmosphere...a kind of invisible barrier constraining thought and action," says Fisher.

The mainstream media does this policing of the imagination, chastising, for example, Naomi Klein for daring to question capitalism unless she can produce a total blueprint of a working alternative. And we also do it to ourselves.[19] Through what Fisher calls an attitude of "reflexive impotence," where we recognize capitalism's flaws but, lacking any faith that we can overcome the larger system that produces them, we engage in a self-fulfilling prophecy of apathy, outrage, and ineffectual action.

Unlike neoliberalism (which sells itself as a kind of capitalist utopia, celebrating the unfettered winner-take-all market as the triumphant destiny of History), Capitalist Realism is inherently anti-utopian, holding that no matter capitalism's flaws or destructive impacts, it's simply the only way to go. Together they sweet talk us into surrendering to the status quo.

But with the appearance of the quasi-utopian Green New Deal, the rise of Bernie Sanders and the democratic-socialist wave he ushered in with his groundbreaking 2016 Presidential campaign, and the growing strength of an intersectional climate justice movement that increasingly identifies capitalism as the culprit and has its sights set on radical systemic change, we may be witnessing the beginning of the end of Capitalist Realism.

I had my own direct experience of this seismic shift in action in December 2018 alongside hundreds of young people from the Sunrise Movement as we swarmed Congress for a Green New Deal. Our mass "people-powered day of lobbying" was not only impassioned, incredibly well-organized, and strategically focused on persuading key members of Congress to sign on to the GND, but, more than all this, in some intangible way, it made me feel that the once impossible was truly possible. Finally, the utopia we needed had a name, a plan, a path forward, and a movement gathering behind it—and that movement was both inside the halls of power (AOC et al.), and laying siege to those same halls, too.

As 20-or-so of us from the New York City area occupied our local Congressman's office—one of countless mini-occupations that day—I fell into conversation with one of my sit-in mates: he was a designer and architect and his briefcase was full of the future—specifically, a creative brief for the Green New Deal, complete with sketches, design strategy, and a retro-futurist aesthetic. Whether he would ever hear back from the staffer at AOC's office he'd sent it to wasn't the point. The point was: from that month's headlines to that man's briefcase, the once-locked doors of Capitalist Realism were being blown off their hinges.

At a 2021 talk on "the Future as Dream-Time," just before his tragic death, visionary anthropologist David Graeber fielded a question from an attendee who wondered whether he was "optimistic about the next hundred years for humanity" given that no one else seemed to be. Graeber paused and said something I believe adrienne, Gopal, Joanna, and Tim would wholeheartedly agree with: "I'm quite optimistic about the death of capitalism…I'm actually more worried that the next thing might be even worse. So…how could we have a stupider moment to tell people *not* to try to think of what a better system would be like?" [20]

Which is exactly why adrienne's invitation to imagine utopia in the midst of dystopia is so important.

Utopia is not a blueprint, it's not a step-by-step plan, it's not even a strategy; it's an act of imagination and an act of will. Like the poet[21] said:

> Utopia lies at the horizon.
> When I draw nearer by two steps,
> it retreats two steps.
> If I proceed ten steps forward, it
> swiftly slips ten steps ahead.
> No matter how far I go, I can never reach it.
> What, then, is the purpose of utopia?
> It is to cause us to advance.[22]

"This is a great moment for the reinvention of utopianism," said Graeber in that same talk. "The lesson I think we've learned about utopianism is not that utopianism is bad, it's that when you just have one utopia it's bad.... What we need is lots and lots of utopias...the more the merrier and more liberating it is."[23] Which is exactly what our current explosion of visionary sci-fi and cli-fi futurisms seems to be giving us.

Mainstream sci-fi, minus a Lando Calrissian sidekick or two, might have you thinking everyone in the future is white. Afro-Futurism tells us otherwise, in music, visual art, graphic novels, fiction, and film. Via Janelle Monae's album *Dirty Computer*, Marvel's *Black Panther*, and the novels of Octavia Butler and successors like N. K. Jemisin, Afro-Futurism grabs the reins of the sci-fi genre to imagine futures in which Black lives not only exist but matter. Take Wakanda, an enlightened, technologically advanced, Afrocentric civilization hiding in the heart of Africa that also speaks at a mythic level to the griefs and dreams of the African-American experience.

From museum shows to comic books, the Afro-Futurist imagination conjures everything from reverse diasporas, to super-sensory and tech-enhanced modes of carrying on the long march towards justice and liberation. It's "thrilling," says adrienne. "People wanted to erase us and... we're writing ourselves back in. We're creating stories that are rooted in African heritage and that articulate an African future."

Indigenous people are also picking up the reins of sci-fi.[24] Rejecting the fetishization of Indigenous knowledge so prevalent in the wider culture, Indigenous Futurism is transplanting trees in space, queering the space-time continuum through NDN time,[25] and turning AI into the voice of ancestors—an ancient cultural DNA reawakened to (selectively) rewire contemporary culture. These stories, says Indigenous scholar Daniel Heath Justice, "guide us forward into an ever-uncertain future, just as they guide us back home."[26]

In our meeting, Gopal Dayaneni argued for what one might call "Trans-Futurism," suggesting that "the transgender journey is the literal embodiment" of the larger socioeconomic transition our whole civilization must make. "There is wisdom in the trans experience," he told me, "that our other movements need if we are to reimagine our way forward." He highlighted the "complexity, emergence, and mystery" common to both journeys.

And that's just the tip of the dystopian iceberg melting away to reveal the many utopias we can build at the edge of our flooded future. In addition to adrienne's *visionary fiction*[27] (radical speculation to advance justice and liberation), there's *hopepunk*[28] (being kind as an act of rebellion), *retro-utopia*[29] (using cosmopolitan appropriate-tech to go forward by going backwards), and *solarpunk*,[30] (gritty, optimistic, post-carbon sci-fi), just to name a few of the sci-fi subgenres helping us imagine viable, non-terrible futures.

Solarpunk pairs well with an understanding that the future will require resilience and radical adaptation. According to a Trinidadian video-essayist who goes by the Internet handle "Saint Andrew," it's "a futurism that focuses on what we should hope for rather than on what to avoid." It offers "a shining vision of a positive future, grounded in our existing world," that "looks beyond the limitations of capitalism and beyond the current rift between humanity and nature." It accepts "that climate change, the consequences of centuries of damage, aren't averted in the future. Yet it still manages to incorporate hope." It imagines, "a future where we've got a lot of work to do, but we're doing better. We're using technology for more uplifting ends, like seed bombing drones and solar ovens."[31]

Our future might be trending dystopian, solarpunk is telling us, but we can—and must—*also* make it utopian. And not just uni-topian, but multi-topian. One big no to Capitalist Realism; many beautiful yesses to our Afro-Indigenous-Trans-solarpunk-retro-visionary-Eco-Socialist Future(s)! We need them all, because we can't leave anyone behind, and because the best people to lead the future are likely *not* the ruling-class people who fucked up the present.

Channeling Paul Kingsnorth, who describes the dominant stories our society operates by as "malfunctioning software,"[32] I like to imagine all these rogue utopias and alternate futurisms as people-powered algorithms hacking and code-patching our culture's faulty software. But if we're really going to hack the future, we must make our multiverse of utopias visible, compelling, common sense and, most importantly: operational. We must carry our utopias into the real world. "Saint Andrew" would agree. He lauds solarpunk for being not just a sci-fi genre, but a growing social movement that "emphasizes real-world application," focused on "what we do here and now, from DIY projects to larger organization."

America got a glimpse of solarpunk in action when the Sunrise Movement kicked off their 2019 nationwide GND tour with a utopian video, "A Message From the Future With Alexandria Ocasio-Cortez."[33] Set some decades into an ecosocialist future, and narrated by the congresswoman from aboard a New York-to-DC bullet train as she reflects back across all that it took to bring about the GND, it's a seven-minute tutorial in how to do visionary-futurist storytelling.

"You can't be what you can't see," it begins, naming a core principle. "People said the GND was too big, too fast, not practical," AOC narrates from the future, "I think that's because they couldn't picture it yet."[34] Which is a beautiful meta-moment for the viewer, as that's exactly what the film is doing for *us*. Then, with help from Molly Crabtree's animated watercolors, the film paints that picture, from winning the White House in 2020, to Medicare for All, to retrained oil-field workers and urban AmeriCorps-Climate kids restoring wetlands in coastal Louisiana with the help of Indigenous wisdom, to the full suite of "solutions at the scale of the crises we face...without leaving anyone behind" that unfurl across the "decade of the Green New Deal." And to its credit, it's not all roses and Kumbaya. As AOC's narration continues:

> Those were years of massive change and not all of it was good.... When Hurricane Sheldon hit Southern Florida, parts of Miami went under water for the last time.... As we battled the fires, floods and droughts, we knew how lucky we were to have started acting when we did.... The first big step was just closing our eyes and imagining it.[35]

For Naomi Klein, who conceived the project,[36] "The question was: How do we tell the story of something that hasn't happened yet?" At the sold-out culmination of Sunrise's tour, she told the crowd,[37] "We have all been raised in a culture bombarded with messages that there is no alternative to the crappy reality we have today." And then added: "If we're going to win a Green New Deal, we're going to have to start telling different stories about who we are and about the kinds of futures that are within our grasp."

Naomi's partner Avi Lewis,[38] who co-wrote the script with AOC, noted that such an effort requires "a totally different creative muscle than I've ever exercised." Using it, he admits, can be terrifying. "Having a brief flicker of hope that we could do something" about our situation "seems to open the floodgates of our repressed grief." Yet as AOC (and Lewis) say in the closing line of the film, "We can be whatever we have the courage to see."

Yes, in this future, as in all our likely futures, Miami is underwater, and droughts and fires rage all around. (After all, we must eventually give climate realism and the tyranny of the possible their due.) And yet, when I watch this seven-minute gem, the impossible feels slightly more possible.

What is utopia for? "It is to cause us to advance." So, let's imagine the world we *really* want, and let's advance. Even if this advance is no longer up the sun-drenched hills we hoped it would be, but down the treacherous rapids of climate catastrophe, we still need to know where we want to go. Otherwise, we won't be able to shipwreck ourselves there.

adrienne turns to visionary fiction to help us imagine our way through collapse to a post-capitalist solarpunk-ish utopia. She also looks to Nature and its manifold patterns to discover how we might get in right relationship with each other and all that we're in for.

When we talk about the crisis we're in, we often describe it as a problem rooted in our separation from Nature—our soul-killing, and possibly civilization-destroying alienation from Nature. We're not sure how to fix it, but we sense that any chance at healing or a livable future lies in her direction: we must repair our relationship with her, relearn her ways, resync with her rhythms, and let her wildness reanimate our own.

So, What Would Nature Do?

My next and final meeting was with someone whose profession—and, arguably, whole life—is an answer to this question.

DR. ROBIN WALL KIMMERER

"How can I be a good ancestor?"

Dr. Robin Wall Kimmerer, botanist, ambassador of Indigenous thought, and author of *Braiding Sweetgrass*, is one of the most celebrated naturalists in the country. What Would Nature Do? Her basic answer: what people have been doing for most of our history; and: what most Indigenous people are still doing. Oh, and: let's ask the plants.

We met up in the Adirondack Mountains of upstate New York, a far cry from Detroit. We're at the Blue Mountain Center, an ecology-oriented writing center in the middle of the largest state park east of the Mississippi. We've both been here several times before, to write, reflect, hike, or to clear brush, patch holes, or pitch in in whatever way we can. The place is like a home away from home for both of us.

Over the course of our few days together, we read each other's work and discussed topics ranging from ecological compassion, to the restoration (and "re-story-ation") of our relationship to the natural world, to Richard Powers' novel *The Overstory*. "Some of the trees in that novel have as much presence as some of the human characters living alongside them," I gushed.

"He could have gone further," she countered. "He could have given them full personhood. He could have treated them not as extraordinary objects of the forest but as actual beings."

I tell her I've been following with fascination and awe Western science's late-to-the-game understanding of plant intelligence and the far-more-complex-than-we-ever-imagined way that forests communicate and cooperate. I also tell her that, while I love the great outdoors, I'm fundamentally a city kid, one of those alienated denizens of the fabricated world who suffer from "nature deficit disorder."[1]

"Ha," she said. "You should get out more." I took her advice, hiking the grounds of the center, keeping an eye out for plants that play a starring role in *Braiding Sweetgrass*. "There's an abundance of *Umbilicaria americana* lichen here," she tells me. "In fact, the chapter about lichens was actually written right here at the Blue Mountain Center."

Lichen, it turns out, are not only one of the most ancient forms of terrestrial life, they're actually two beings in one: a fungus and an alga, "joined in a symbiosis so close that their union becomes a wholly new organism." Using the motif of the "wedding basket," an exchange of wedding gifts (and promises) common to many Native American traditions, Robin draws a series of parallels between this lichen "marriage" and her parents' marriage, which has lasted for over 60 years.

In the algal wedding basket: the gift of photosynthesis, the extraordinary ability to turn light into sugar; in the fungal wedding basket: the ability to reach its delicate threads out along the rock to find and dissolve minerals. One cooks, the other hunts; it's "a reciprocal exchange of sugar and minerals." Likewise, in her parents' marriage, "the balance of giving and taking is dynamic, the roles of giver and receiver shifting moment to moment." Both unions are committed to an "us" that emerges from well-matched strengths and weaknesses, and extends into, and often benefits, the broader community or ecosystem. Core to both unions: mutual reciprocity. Basically, a gift economy.

Throughout the book, she weaves together many such examples, drawing lessons from both the natural world and her own family and cultural history to guide us towards a more reciprocal paradigm. Instead of a one-way extractive and instrumental attitude to the natural world, she's telling a different story of who we are (or could be) as people. In the prologue to *Sweetgrass*, she paints this choice in mythic form, staging an encounter between Eve and Skywoman.[2]

In the storytelling traditions of the original peoples throughout the Great Lakes region, Skywoman is the first woman. She falls through a hole in Skyworld, toward a dark, watery earth, clutching a bundle in her hand. The water animals see her fall. The geese fly up to catch her. A huge turtle offers her his back to set foot on. But she needs land to live, and there is none. Loon, otter, sturgeon and others try to dive to the very bottom of the water where it is said there is mud. None succeed, some don't return. Finally, muskrat, weakest swimmer of them all, volunteers to go. He dies in the effort, but floats back up to the surface with mud clutched in his paw. Skywoman spreads the mud on the back of turtle, does a powerful dance and a song of gratitude, and the mud turns to land. Turtle Island is born. She takes the bundle in her hand—fruits and seeds from the Tree of Life that she grabbed as she fell—and scatters them across the new land. She tends to them, and nourished by the light pouring through the hole in the sky through which she fell, they grow, and the earth turns from brown to green. And with all this abundance, many of the water animals come to live with her on Turtle Island. A gift is given; in turn, a gift is received.

Skywoman created a garden for the good of all. Meanwhile, on the other side of the world there was another woman in a garden with a tree. But when she ate the fruit, she was exiled, and now, "in order to eat, she was instructed to subdue the wilderness into which she was cast."

"Same earth, same species, different stories," Robin writes. It's a collision of creation myths and civilizations:

> One story leads to the generous embrace of the living world, the other to banishment. One woman is our ancestral gardener, a co-creator of the good green world that would be the home of her descendants. The other was an exile, just passing through an alien world on a rough road to her real home in heaven.[3]

Eventually, the offspring of Skywoman and the children of Eve meet and, as Robin tells it, "the land around us bears the scars of that meeting, the echoes of our stories."

For Robin, however, this echo is insufficient. Because the story of Skywoman is not a cute relic of the past, but full of vital instructions for the future: about reciprocity, gratitude, humility, and the wisdom of the natural world. Unlike the Western tradition, with its hierarchy of beings that places humans at the top and plants at the bottom, the Indigenous worldview sees humans as "the younger brothers of creation." We're the ones who've only just shown up; it's us who have the most to learn. "Plants," notes Robin, "know how to make food and medicine from light and water, and then they give it away." How about us? For Robin, plants are *our* teachers. The question is: can we listen? (And even if we eventually do learn to listen, will it by then be too late?)

At some point, amidst all our talk of plants, I innocently ask her *when* she became a botanist. She corrects my colonial mindset in the gentlest way imaginable: "I think I was born a botanist. I simply am. I cannot remember a time in my life when plants were not my family and my companions."

For our interview,[4] we met in one of the center's cabins, and I made a fire. It was a noticeably poor fire: slow to catch; smoke leaking into the room. Is that because after our easy back-and-forth there was now a tape recorder involved, and I was nervous? In any case, we sat down by this undignified fire, and began.

Andrew: Can you tell me and my tape recorder a tiny bit about yourself? How does your background shape the approach you bring to the climate crisis?

Robin: The little bio is that I am currently a professor of plant ecology and the Director of the Center for Native Peoples and the Environment, which is explicitly designed to bring together scientific tools with the wisdom of traditional people.

Andrew: That's unusual in an academic setting.

Robin: Yes it is. The academy tends to be pretty hostile to such things. But I don't think of traditional ways of understanding as a challenge to science, per se. I believe Western science and "traditional knowledge" are both science. Science is a set of tools. We can use those tools for all sorts of things. We can use the tools of Western science without buying the whole scientific worldview. We can use scientific tools in an Indigenous worldview. That's really what I'm all about.

Andrew: You also mentioned a complicated ancestry. An Irish carpenter in your family tree. A mixture of Native and white ancestry.

Robin: I'm a member of the Citizen Potawatomi Nation. I'm of mixed heritage, like many of our people are. Our people were historically removed from our homelands in the Great Lakes, first to Kansas, and then to Oklahoma. Our people took three, four months to do that walk. The Cherokee talk about their Trail of Tears. We call ours the Trail of Death. In the context of Climate Change, I think about that part of my heritage a lot. We experienced tremendous climate change in a single season. Our people walked from the forests of the Great Lakes, to the prairie grasslands. If you look at climate maps today, you'll basically see that what is now the climate of Kansas and Oklahoma is moving to Wisconsin. Our people did that walk, and now—

Andrew: Nature is walking it back.

Robin: Nature is walking it back. If I've had an "aha" moment about climate change, it's that our people went through this radical climate shift— while also in the midst of tremendous cultural loss—and we survived.

Andrew: Oof. And this really wasn't all that long ago, was it?

Robin: Not at all. My grandfather was one of those children who was taken from his family at nine years old. His brother was only seven. My grandfather was brought to the Carlisle Indian School.[5] For me, growing up with those stories, and hearing what my dad had to say about it—and what my grandfather *didn't* say about it—I kept thinking: *Here was a school that was designed to make you forget who you were. So, couldn't there also be a school that would help you remember?* That's how I think about the Center for Native People and the Environment. Our mission is in many ways a response to my family's and my people's history of loss, which is a very formative story for me.

Andrew: You're a scientist, yet you emphasize the critical role of story, myth, and even prophecy in addressing the climate crisis. Why so?

Robin: As a member of the scientific community, I've been living with the knowledge of our climate threat for a long time. Originally—and naïvely—I assumed that once people knew what was happening, well, they wouldn't let it happen. But, for the most part, that's *not* how we've responded. The incrementalism of climate change—the way it happens molecule-by-molecule with no malice—is confounding. It can be hard for people to see and feel the danger, never mind act on it. I realized: we know it's happening, and yet we're letting it happen anyway. That realization struck moral terror into me.

Andrew: Not mortal terror, but *moral* terror. Say more.

Robin: If you know that what you're doing is wrecking the world and you're doing it anyway, that's a moral problem. This realization propelled me to not just do the science, but to become a storyteller, as well. I realized: information isn't going to save us, but stories might. And, in any case, it's too urgent, too important, not to try. Yes, I think we need a carbon tax. Yes, I think technology can play a role. Yes, we have to change our lightbulbs and our tech and our economic policy, but what we really have to change is how we conceive of ourselves.

Andrew: So, what is the central story that needs to be "re-storied" here?

Robin: The story I feel most compelled to try to disrupt is this story about who we are as people. We have this notion that we're just takers. (Hell, our government doesn't even call us citizens anymore. It calls us "consumers.") But we're not just takers. We've forgotten that we can be partners, too. That we can bring good things to the table. We have to change the story that we are inevitably bad players in a living world, otherwise, we just fall into despair and down the toilet, right? We end up in a place where there is nothing we can do. Because, well, it's just human nature. But it's *not* human nature. Yes, it's been human *behavior* for the last 500 years. But 500 years is an eye blink in the scope of human history. We've bought this story, and it's a destructive story. Maybe it's made life convenient and easy in an iPhone-ish way, but it hasn't made us happy.

Andrew: Unfortunately, it's the only story a lot of us can see. Our big lie is so big that it's invisible to us.

Robin: A big lie is invisible until we see another story. If we see another story, we can look at our own and say, well, wait a minute, what are the consequences that flow from living that story? This can be an extraordinary act of liberation.

I am no scholar of politics, policy, economics, or any such thing. (I'm a botanist.) And so, this may come across as quite simple minded, but it seems to me that we have a choice to look at the world as if it were a gift, or we can look at the world as if it is property. It's within our own power to make that choice.

Andrew: What role does language play in making that choice? In choosing another story?

Robin: In *Braiding Sweetgrass*, there's a chapter called "Learning the Grammar of Animacy."[6] It's about my own learning—late in life—of the Potawatomi language, which is so difficult because it is impossible in our language to speak of the living world as "it." For example, Andrew, you're sitting by the fire. So, I wouldn't say "it" is sitting by the fire.

Andrew: I'm not an "it."

Robin: Right. In English, we have a special grammar for each other. But in Anishinaabe, that human exceptionalism doesn't exist. All beings are treated as family, as relatives, as alive. That animacy is baked into the language, into the grammar. And so, knowing that the way we speak deeply shapes how we know the world, I wondered whether we could we come up with a way to "animate" the English language so that we no longer refer to our relatives as stuff? And so, I asked my language teacher, "Do we have a word for a being, an earthly being?" He said, "Yeah, we do. It's *bimaadiziaki*," which translates as *living being of the earth*.

Andrew: And your proposal?

Robin: To replace the word "it" when we're speaking of a living being with the word "ki." (The last syllable of *bimaadiziaki*.) Which slides—nicely, I think—along the pronoun sequence: he, she, ki, and it.

Andrew: But a table, say, would still be an "it"?

Robin: Sure. Tables are inanimate in Potawatomi, too.

Andrew: Okay.

Robin: The elegance really shows up when we go plural. And here we don't have to borrow from Potawatomi because we have our own perfectly good word in English, which is "kin."

Andrew: Just add an "n" to "ki."

Robin: Yes, and suddenly, we are talking about our relatives. We can say "it" for bulldozers and paperclips, but everybody else gets to be a person, and that takes away permission to exploit the world as if it was all stuff.

Andrew: It's a very beautiful idea. Hard to actually implement, I'd imagine, but worth it. We need to retrain ourselves around the values we believe in.

Robin: For me, it's a simple substitution. Today, for example, I was out walking, and there was a beautiful white pine. I sat down, and I could either lean my back against it, or I could sit with *ki*. Those are different things. One of them is intimate and personal and makes you feel like, oh, I'm sitting with my buddy, my kin.

Andrew: It's a way to give the world presence, agency, even selfhood.

Robin: Selfhood. Personhood. When I see birds. It's like, *Oh, kin are waking us up this morning. Kin are flying south for the winter.*

Andrew: I can see how it could re-map your brain, how it could help re-enchant the world.

Robin: Right. In fact, I did an experiment along those lines with freshman composition students. I gave them a talk about *ki* and *kin* and asked them to experiment with using it, and how it felt. Some of their responses were like, "…this is really stupid." But others are like, "Oh, my God, this makes me feel so happy. This makes me feel so connected."

Andrew: In the Hopi language, you don't "*have* a brother"—you "are brothered *by*."

Robin: Yes. What are nouns in English are verbs in my language, including land forms and different forms of water. The word for "bay," for example, is actually "to be a bay." *Wiikegama.*

There are many other words like that. To be a river. And technically speaking, even "to be a day of the week." Grammatically, we say "to be a Saturday," "to be a hill," "to be a mountain." That's one of my favorites. It isn't "a mountain," it's "to be a mountain."

Andrew: A language built this way makes everything come alive.

Robin: Everything is alive.

Andrew: Alive and wise, too. You describe plants as our teachers. They've learned how to survive for eons longer than we've had to—through periods of both bounty and scarcity. What can they teach us about how to survive the crisis we are facing?

Robin: Plants have so much to teach us. They could lead us out of this mess we've made, if we could only listen. Plants know. They are carbon specialists. Right? They're already converted to a 100% solar economy. Do they do it at great cost and suffering? No, they do it with beauty and generosity and flowers and berries. Plants are marvelous, marvelous teachers of reciprocity. Plants know what to do.

Look at the tallgrass prairie and the terrible price we have paid for the loss of that ecosystem. Prairies are amazing carbon sponges. Just as much carbon can be stored in the soil underneath tallgrass prairie as in a forest. But what is the rarest ecosystem in North America today?

Andrew: Let me guess: The tallgrass prairie?

Robin: Not too long ago, it was a third of the continent.

Andrew: And now?

Robin: Less than 4%.[7] We are terminating our teachers just when we need them most. As someone who engages with the living world as my family, it's not just morally painful, but spiritually painful. Incredibly painful.

Yes, we're learning from the tallgrass prairie. We're teaching ourselves new techniques, conservation tillage, low carbon farming and ranching, etc., but we're slow on the uptake. Meanwhile, the tallgrass prairie already knows how to remedy the climate crisis, but we've killed most of it off.

What will people do when they understand that the plants are sentient, esteemed teachers? What will it be like to suddenly realize that these beings who you've treated as objects, as the lowest of the low, are like holy people? What if we learn it, but it's too late? What do we do then? Imagining that moment, simply breaks my heart.

Andrew: There's so much about our moment to break our hearts. Grief is so laced into our times. What has the natural world taught you about how to grieve what's happening, while still seeing your way forward?

Robin: A core medicine for me is gratitude. It's very healing for me, especially when I face my plant relatives who are being assaulted at every turn. But it's also much more than a coping mechanism. When I teach and waken others to the notion that this world isn't property, but rather that this world is a gift, it can change outcomes, too. Because once we see things that way, we tend to also ask: if that is so, how then should we behave?

We have so many stories and cultural teachings warning us about the failure of gratitude. In every indigenous culture I know of, there are stories of what happened when people forgot to be grateful: the corn didn't grow back; the bison didn't come that year; the stream dried up. I take these stories very literally. They're telling us to notice what happens when we forget that the world is a gift. If we let the gifts that sustain us become invisible; if we take them for granted, if we don't reciprocate, if we don't steward them...then the buffalo don't come back. Gratitude has a spiritual purpose, but it also has a very practical purpose.

Gratitude also assuages some of the pain I feel amidst our extinction crisis. If the trees and ferns and mosses and corn know that I really love them, and I'm grateful for them, and I feed them, and I'm really, really sorry for what's happening, and I can truly say I'm doing my best here— that's medicine for me. It's not enough, but it's something.

We also have a lot of data showing us that people who practice gratitude consume less. Gratitude leads you to that sense of: *Oh! I already have everything I need.* I don't need an iPhone upgrade. I'm good with the simple version. I don't need the latest home entertainment system, I'm good with actually going over and knocking on my neighbor's door to talk to them. Gratitude cultivates a sense of abundance, which is the antidote to the false scarcity that capitalism implants in us.

So, the notion that gratitude is just an empty healing practice for your own self, I don't agree. I think it has agency. I think it has agency in the world. Living out gratitude can be a radical statement: *I have enough. I have enough.*

Andrew: *Dayenu*, us Jews might say. *It is sufficient.* Not quite the same, but you build off the traditions you inherit.

Robin: You absolutely do. Another cultural teaching I rely on is prophecy. We have a prophecy about this time called the Prophecy of the Seventh Fire, which is really a history of what happened to our people. It speaks about a moment—the moment we are in right now—when all of humanity faces a fork in the road. One path is dark and charred. The other is lush and verdant. But the prophecy tells us we can't just stride down the path we want. Because we don't yet know how. We have to turn around and pick up what our ancestors left behind for us, and bring them with us. Then we will know how to go forward.

Andrew: As you say elsewhere: "By our ceremonies, we remember how to remember."

Robin: Yes.

Andrew: So, what must we pick up and carry with us?

Robin: Biodiversity for one. In the prophecy, it is said that our plant and animal relatives will turn their faces away from us. They used to love us; now they don't. There once was mutual respect, but now only shame. They are almost lost, but we must not let them go. We must conserve that diversity, and bring them along with us.

We must also pick up our stories. We must bring along our world view—a world built on gift-gratitude and reciprocity. As a plant ecologist, I know that's how the world works. Natural law is based on reciprocity. And, human law can't stray far from natural law. If it does: disastrous consequences. This is a wisdom teaching that's been transmitted to us via story and myth. And yet, look at us and our economies based on perpetual growth: we have gone so deeply astray from natural law. How has this even become valid thinking?

Andrew: Capitalism demands it, and for a brief historical moment, cheap, abundant fossil fuels have made it seem as though endless growth is actually possible.

Robin: We must get human law back in alignment with natural law. This, too, we must pick up and carry with us. It will require ecological compassion. The understanding that all beings have personhood. The granting of rights to nature.

Andrew: If a corporation can have personhood, why not a river?

Robin: Exactly.

Andrew: As I'm sure you know, after a long fight, in 2017 the Whanganui

River in New Zealand was granted legal personhood,[8] and with it, quite a bit of legal standing to protect itself and its larger watershed.

Robin: Yes, that was one of the first examples. Legal personhood has been conveyed to any number of natural entities since then. It's a very positive development, and it's growing. But no matter how good our campaigners and legal scholars are, it won't really take root unless we cultivate a much deeper sense of ecological compassion.

Andrew: Which is why Richard Powers' *The Overstory* is such an important book.

Robin: Yes. It brings us to compassion for those tree beings. And we need that so desperately. Because I don't think the rights of nature will resonate with the public until we have deepened our ecological compassion. We need to understand this other being as a person; and *not* a human person. I get into all kinds of trouble in the scientific community where we are not allowed to anthropomorphize. I'm not anthropomorphizing. I'm not saying those trees are *like* people. I'm saying those trees are like trees, and they're persons.

Andrew: One of the takeaways from *The Overstory*: trees are intelligent, but we humans are simply moving too fast to notice. They smell, they communicate, they move (toward the light, toward water, toward nutrients), they warn each other of danger, they share, they even sacrifice themselves for one another.

Robin: They even hear.

Andrew: Hear?

Robin: Yes. One of my colleagues, Monica Gagliano, a cognitive biologist, has discovered that plants can hear.[9]

Andrew: How, exactly?

Robin: We already know that plants will turn on their own chemical defenses to pests. Sometimes they're warned by trees on the outer edge of the infestation. We also know that if a caterpillar chews on said experimental plant, that plant turns on its defenses. *I'm under attack.* That's biochemical. It's mechanical. It's oxidative.

Andrew: Right. This is the new science of trees. But where do the sound waves come in?

Robin: What Monica did was to record the sound of the caterpillar chewing, put the plants in sealed chambers so there was no possibility of chemical communication, and then play the sound of the caterpillar chewing. And lo! The plant turns on the defense.

Andrew: Wha?! And why is this not headline news?

Robin: There's a really good reason for that: we can't have a world where our food is sentient.

Andrew: That would cause a deep revulsion. And *then* maybe a revolution. (And along the way, a total surreal mental meltdown—just imagine if your pet dog of many years suddenly started talking to you.)

Robin: Our world would be completely changed by this knowledge, if we let it in.

Andrew: How could we let it in? It's a profound ethical problem. Even going totally vegan wouldn't help.

Robin: By the honorable harvest, by giving gratitude, by asking permission, by saying *I am going to consume you. But because you're my relative, I have to consume you in an honorable way, and with restraint. And I have to give you something back in return for the fact that you gave me your life.* This is the whole "honorable harvest" teaching.

It starts with how we are taught to pick things from the world—berries, firewood, etc.—but makes the case that these would be really great principles for an energy economy, too. We don't have to be rapacious. That is not who we necessarily are. We have built an economy that is relentless in its consumption and does not conform to natural law, but we don't have to do it that way.

When you say the world is just stuff and is not sentient and they're just objects, you don't have a moral dilemma about how you treat the world. You're not eating your relatives. You're eating property. If we conceive of the world as an object, we don't have to take moral responsibility for it.

Once you recognize the personhood of non-human animals, and our kinship with all living things, then you have to consume differently. Our [native/traditional] cultures are rich in teaching us how to do that. But can we scale it up?

Andrew: Can eight billion—ten billion—people share in an honorable harvest?

Robin: Let's remember: people in urban settings aren't going to have "traditional" systems of reciprocity. Part of the problem of referring to the wisdom of reciprocity as "traditional knowledge" is that it makes it seem like the only way we can go is backwards, back to what we used to do. This is not the case. Our people survived because they adapted to contemporary circumstances. Humanity, too, must adapt. The core teaching is reciprocity; we just need to figure out how to make it work, at scale, in our world now.

Andrew: So, do you have hope?

Robin: I honestly don't even know what hope means anymore. For me, I take guidance from Kathy Moore. The house is on fire. Your family is in it. What do you do? You put out the damn fire. You can hope all you want, but put out the fire. Don't stand outside debating about what kind of hose to use or whether you should use it or which fire department you should call. Just do what you have to do.

I can't call it hope, but when I think about the future—especially as I become a grandmother, and when I think not only of my human grandchildren but my frog grandchildren and my thrush grandchildren—putting out that fire is the best thing I can do. That's how I can be a good ancestor. It's not hope, but it's in the same neighborhood.

The opposite of hope, for me, is that we will have such a broken system that its natural capacity to respond will be impaired. And so, along with putting out the fire, I'm committed to creating conditions under which regeneration is possible. The metaphor I use is building good soil. How do we do that? Well, preserving biodiversity is key.

The question I ask myself, the question I invite my readers and the people I come across in my activist work to ask is: *What do you love too much to lose?* Once you have an answer, then, pick it up. Because the only way the thing you love is going to get through these narrows is if you pick it up and carry it.

We can't do everything. We can't respond to everything. I can sometimes make myself crazy by being reactive to this and reactive to that. Then I ask myself: *What do you love too much to lose?* And what are you going to do about it? How will you pick it up and carry it into the future?

Andrew: That is an extraordinarily clarifying—and motivating—question.

Robin: Yes it is. And it motivates people by love and commitment as opposed to motivating people by fear, which has not worked. Nor has information worked. Information and fear have proved to be dead ends, strangely enough. I don't quite get it, but they're not enough. But where they have failed, love and story might succeed. And, there, we might find common ground.

Andrew: *What do you love too much to lose?* What is so core to your being that you can't live without it? So essential that you will live and die for it? In the Climate Ribbon ritual (see page 302) we ask people a similar question. And once they've named that for themselves, we ask them to make a specific commitment. To act with others, and also to change their own behavior.

Robin: Kathleen Moore and I did a workshop together on this same theme. The metaphor that we used was a confluence of two great rivers. One river of *what do you love too much to lose* and the another river of *what am I going to do about it,* The place of power is in the confluence of those two rivers. We had a huge graphic of these two rivers coming together, and asked the participants to write their responses on both rivers.

I'm glad to report that people had so much love for the world—and so much fear at the threats against it—that pretty quickly there was hardly any room left to write on the river of *what do I love too much to lose.* But the responses on the *what am I going to do about it* river were simply not up to the urgency of the task. Most fell along the lines of *I'm going to buy organic vegetables.* Good for you. That's a start. But nowhere near enough. That was very disheartening.

Andrew: *Let's save the planet, said everyone, not knowing how.* The weaker response on that second river speaks to many things: the overwhelming size and complexity of the problem, the relative poverty of our "solutions imagination," our general unfamiliarity with acting collectively, as well as the way capitalism—and mainstream environmentalism—have trained us to only think in terms of individual, citizen-as-consumer-type responses, etc. Even as policies at the scale of the problem, policies like the Green New Deal, gain political traction, it's still hard to formulate *a what am I going to do about it* kind of statement...

Robin: Which is not to deny the power of that diffuse commitment and naming what you can because that is how culture shifts happen. And let me also share what one of my undergraduate students said to me. She's a promising, bright, committed young woman with whom I'm very close. When she was graduating and going off to do environmental work, I said to her that I was sorry that after all these years (it was just about the 45th anniversary of Earth Day) of my generation trying to fix things, that she had to still be fighting this fight.

Her response was so kind and heartfelt. (And I love it when the students are our teachers.) She said, "Dr. Kimmerer, don't you understand that this is the best possible time to be alive?" Which mystified me. How could an environmentalist say that? "Well," she said, "everything is hanging in the balance. What I do matters. I am so lucky to live in a time when my life matters and the choices that I make about where to put my energy matter." That gave me so much hope.

Andrew: And this wasn't just a fantasy story she was telling herself to feel better, right?

Robin: No. It's deep in her bones. She's living it out. She's an environmental activist. And she's not alone. Any optimism I have comes from being blessed to be in the company of idealistic, brilliant, energetic students who are not downcast, who know how bad it is, and just roll up their sleeves, and do what they need to do. It's hard work. But what else would you rather be doing?

Andrew: In closing, let me send a big thank you to this student of yours, and her peers, and to you for everything you're doing.

Robin: And you too. It's strengthening isn't it, to know that we're all in this together?

Andrew: It makes a huge difference. No matter what else happens.

Robin: It does. It does.

The fire kept going for a while after Dr. Kimmerer left the cabin. I sat there and pondered all that she'd shared. As with the previous interviews, I was reeling at the huge shift in consciousness she was offering.

Plants are not just alive, they are intelligent, they are beings, they are persons, even. And not just plants: mountains are beings, shorelines are beings, a rock that's become a home for lichen is a being. And once we see our proper place in this living universe, we realize we are not lord over it all, but kin and co-beings amongst it all.

At the political level, this paradigm shift, "takes away permission to exploit the world as if it was all stuff," says Robin. At the philosophical level, it literally changes what *is*. It changes what is alive.

To truly grok that the beings we've treated as objects, are actually more like holy people, requires a kind of spiritual awakening that is still unfolding for me. In the meantime, it shows up in daily rituals and little shifts in my awareness and attention. On my hikes now, I notice myself giving a nod to the big trees on my path, a little bow to the boulders, a tiny wave to the beetles, a gentle salute to the hawk soaring overhead. As a semi-vegetarian, it has me confused. Before my meeting with Robin, when I'd eat, say, a lox and cream cheese bagel, I'd say, "Thank you, Mr. Salmon." It had a quirky formality to it, but felt respectful. Now, what do I say? "Thank you, Mr. Salmon. Thank you, Mr. Red Onion,

Mr. Tomato, Mr. Caper, Mr. Poppy Seed, Ms. Cow (there's milk in that cream cheese), Ms. Hen (there's egg in that bagel batter)"? And how far do I go? There was some yeast involved, wasn't there? Nonetheless, it's helping me recognize with gratitude this wider circle of beings—including Mr. Deli Man, Mx. Delivery Person, and Ms. Farmer Lady—involved in my lunch. For all its awkwardness, my newfangled mealtime ritual is helping me to "remember how to remember," as Robin might say.

Her advice for how we get ourselves out of this mess is as revolutionary as it is charmingly obvious: Listen to the plants. "They are carbon specialists," she said. "They're already converted to a 100% solar economy." And they did it "with beauty and generosity and flowers and berries."

Her other advice: "Information isn't going to save us, but stories might." It's not what I expected to hear from a scientist, but after decades of science communication that's her conclusion. If at our core we think we're bad players in the natural world, we will simply continue to play badly, and eventually destroy our home. Instead, we must "re-story" who we are. The last 500 years of human *behavior* is not human *nature*. For the vast majority of our species' history we've been partners, not just "takers." We can decide now to once again look at the world as a gift, not as property. She believes this is a choice we still have the power to make; whether we'll do it in time she doesn't know.

When it came to the question of hope, she said, "I honestly don't even know what hope means anymore." In its place she takes courage from the history of her own people, who survived a radical climate shift while being subjected to tremendous cultural loss. Instead of playing games with hope, she simply notes that, "The house is on fire. Your family is in it. What do you do? You put out the damn fire." To help us focus on putting out that fire, and building good soil for regeneration, she shares stories, among them her people's Prophecy of the Seventh Fire, in which humanity has to re-learn the path forward; and the allegory of the confluence of two great rivers, in which the threat of loss propels action. These stories left me pondering: *What kind of ancestor do I want to be? What do I love too much to lose? What must I pick up and carry into the future?*

Across the days after our meeting, I realized Dr. Kimmerer's questions weren't just thought experiments, but heart experiments. Nor were they the first I'd uncovered along my journey. Such experiments can jolt—or charm—or slowly scaffold—us into new ways of thinking and being. I've gathered a handful of them here.

EXPERIMENTS ON THE VERGE

*And then the day came,
when the risk
to remain tight
in a bud
was more painful
than the risk
it took
to blossom.*

— Anaïs Nin

Our climate predicament is not just a technical or political challenge, it's a spiritual one.

For those of us already in the thick of its vicious impacts, how do we carry on? For those of us still waiting for it to arrive, how do we prepare for what we know is coming?

To help guide us, here are a handful of exercises, experiments, meditations, and a philosophical provocation or two, along with one deep conversation about whether to bring children into the world. Use them to work the muscles of your imagination, stretch the sinews of your heart, and tickle the darkest bones you still consider funny bones.

What do I love too much to lose?

We are tied together in the single garment of destiny.
— Dr. Martin Luther King

As I described earlier in the Five Stages of Climate Grief (see page 61), starting in 2014 I was part of a team of artists, activists, and faith leaders that devised a process called the Climate Ribbon to help people move through their climate grief. According to the official website (theclimate ribbon.org), the Climate Ribbon "is an arts ritual to grieve what each of us stands to lose to climate chaos, and affirm our solidarity as we unite to fight against it."

Here are simple instructions for you and your community to make and share your own climate ribbons:

Gather together in a circle with other members of your community.

Reflect deeply on the question, *What do you love and hope to never lose to climate chaos?* Be specific and personal. What's at stake for you?

Write your response on a ribbon.[1] Add your name, hometown, and age:

Share your ribbon by tying it onto a tree or passing it around your meeting circle.

the hope that my children can live healthy lives. Jon, Milwaukee

THE COASTLINE OF MANHATTAN - ANDY, LOWER EAST SIDE

The Sound of bees buzzing in the morning. Monica H, NYC

Next year's harvest.   Jacob L., Akron, OH

Miami, my city   Lynda Montoya, FL

Credit: Gan Golan, 2014

Examples of climate ribbon messages.

Credit: Josiah Werning

The Climate Ribbon Tree during COP21 in Paris in 2015, with thousands of ribbons from all over the world, each one symbolizing a hope and prayer, a desperate reminder to world leaders of all that was at stake.

Exchange: Consider the many ribbons in the circle (or tied to the tree). Find one that moves you deeply. Tie it to your wrist to signify your resolve to protect what this other person loves too much to lose. Whether friend or stranger, become the guardian of this person's story.

Witness: Read each other's ribbons aloud. After each ribbon-story is read, those assembled can together repeat, "We are with you," addressing the person whose name is on the ribbon.

As Robin and I realized during our interview, there is a strong connection between the climate ribbon ritual and the "confluence of two rivers" exercise she pioneered with Kathleen Dean Moore that follows here. So, take your ribbons with you and step into the river.

Step into the river.[2]

Gather together, as a class, a circle of friends, or a community group. Create a large graphic of two rivers flowing into one. When you're ready to begin, say these words together:

> We are all standing right now at the confluence of two great rivers. One is the river of "What do I love too much to lose?" The other is the river of "What am I going to do about it?"

Sit with these questions for a few minutes.

When you're ready, write your responses (you can have multiple responses to each question) on slips of paper or Post-it Notes, and add them to the big graphic.

(If you or your group made climate ribbons in the previous heart experiment, you can add them here.)

Stand back and observe what everyone has shared.

If you choose, you can acknowledge the commitment you just made by voicing it aloud and taking a step forward as if you were stepping into the water.

Discuss in pairs or as a whole group.

Everyone knows history moves in circles;
the surprise is how big the circles are.

— Greil Marcus

If we can no longer believe that History operates like this:

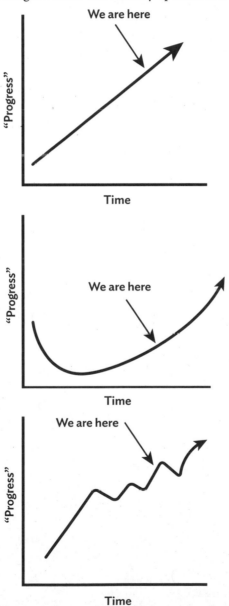

And we manage to avoid completely going extinct from this:

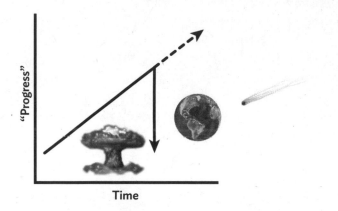

Or this:

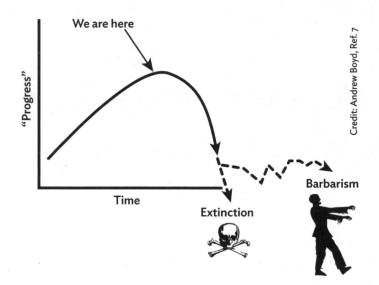

Then, what line-graphs *can* we live by?

A stepped-descent rapids, where we're just trying to get a paddle in here and there to steer clear of the worst?:

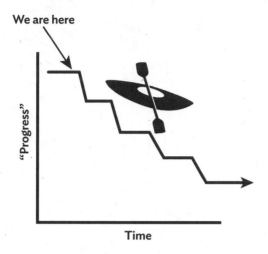

Or, do we return to the notion (popular in ancient times) of History as an endless series of rise-and-fall cycles:

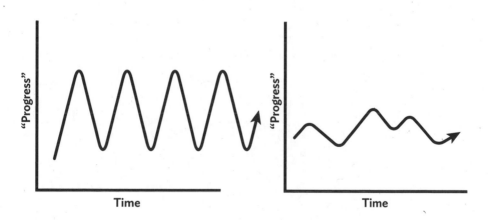

Or is the human story best understood as a mythic journey that Returns us home, again and again, for Eternity:

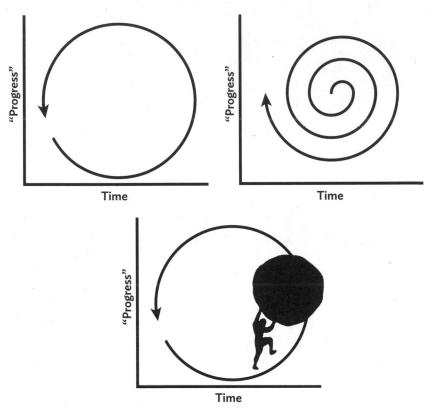

Or is life (and History) more like this?:

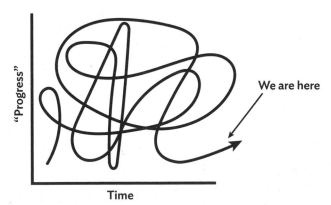

Some days (and maybe some centuries) it certainly feels that way.

At this point, all bets are off, so feel free to draw your own, and consider the story it tells.

Let go of your iPhone.

We expect our iPhones to always go up in number.[3] That's just how the world works, right? I currently have an iPhone 5. When it craps out, I'll trade up to an iPhone 6 or 7 or 8—and get more storage, features, and pixels. But consider this: what if that's not how the world works anymore? What if—metaphorically speaking—those iPhone numbers are going to start going down? What if "progress" is an artifact of a very specific set of temporary historical conditions—particularly the abundance of cheap fossil-fuel energy, and the ability to treat the Earth as a production "externality"—which are now hitting hard limits? Faced with a choice between gracefully powering down our civilization, or suffering a violent decline, how do we proceed? Well, besides sharing the pain equally (see Climate Justice; principle of), it's all about expectation management.

Imagine if in the future, when our iPhone 5 craps out—again, metaphorically speaking—our only option is to go to an iPhone 4 or earlier model. What would you do? Well, you'd take the iPhone 4 and find a way to be happy, and it really shouldn't be that hard. After all it's like having all the computing power that existed in the entire world in the 1960s in a sleek black glass rectangle in *your* pocket. If cars had experienced the same tech advances as computers, you'd now be able to drive from New York to Los Angeles in three seconds on a thimbleful of gas. Well, you and your "new" iPhone 4 can still do that.

But I know what you're thinking: the numbers are heading towards zero, and, eventually, you're going to lose or break your last iPhone 1 and there'll be nothing to "upgrade" to. Then what? Maybe you'll go outside and play. Or stay inside and meditate. Or learn to paint. And, who knows, it might be the best thing that ever happened to you.

Now, not all of us are going to be able to pull this off. Most of us are not going to be happy that our iPhone upgrades are going backwards. But, here's the thing: you don't have to be happy about it. All you have to do is find a way to not be so unhappy about it that you start voting iPhone deniers into office, or burn your neighbor's house down looking for an extra iPhone 6, or blame the Jews for the iPhone problem, or round up all those iPhone-stealing Mexicans in camps. What does victory look like in an era of civilizational unravelling? Easy: Finding within us just enough humanity to not do any of those things because of an iPhone.

Give pessimism a chance.

Pessimism has a bad rap. We think of pessimists as sour Debbie Downers, lazy, prone to depression, even un-American. Pessimism is linked in the popular mind with a "glass is half empty" outlook, and is assumed to be primarily a psychological disposition, and an unhealthy one at that. In actuality, however, pessimism has a long and provocative philosophical tradition, spanning everyone from Heraclitus to Camus, with much to offer our contemporary moment.

In the 60s we "gave peace a chance." In the 21st century we might want to "give pessimism a chance."

The German philosopher Arthur Schopenhauer (1788–1860) is arguably the most notable proponent of extreme philosophical pessimism. We live, thought Schopenhauer, in "the worst of all possible worlds," a world constantly on the brink of destruction. Schopenhauer believed that "The worst is yet to come." He was clearly ahead of his time.

For Schopenhauer, happiness is an illusion, and life is essentially suffering with no meaning or purpose. His goal was a bearable life. The key to making life bearable: extremely low expectations.

How might Schopenhauer counsel us if he were alive today? He might remind us that what ultimately plunges us into sadness and anger isn't disappointment, but hope. He might therefore counsel us to break our hearts now, before life (or climate catastrophe) does it for us at a moment of its own choosing. Remember, life is suffering, he might tell us. If you don't believe that yet, you soon will. Why not get ahead of the curve?

Not only did Schopenhauer believe we were living in the worst of all possible worlds, he believed that existence itself was a mistake and it'd be preferable if humans did not exist at all. However, since we're already here, the next best thing we can do is strive (primarily through ascetic denial, but also artistic contemplation) for a state of being in which the world becomes nothing.

This extreme pessimism is summed up with a contemporary flourish by one of Schopenhauer's modern-day avatars, Detective Rustin Cohle, Matthew McConaughey's character on *True Detective*, Season 1, Episode 1:

> I believe human consciousness is a tragic misstep in human evolution. We became too self aware. Nature created an aspect

of nature separate from itself. We are creatures that should not exist by natural law. We are things that labor under the illusion of having a self. The secretion of sensory experience and feeling, programmed with total assurance that we are each somebody, when in fact everybody is nobody. I think the honorable thing for our species to do is deny our programming: stop reproducing. Walk hand in hand into extinction. One last midnight, brothers and sisters, opting out of a raw deal.

And the amazing thing is, if you want to take Detective Cohle by the hand and walk into his one last midnight, you can: by joining the Voluntary Human Extinction Movement, an environmental movement calling for "all people to abstain from reproduction to cause the gradual voluntary extinction of humankind." Yes, that's a thing. And in spite of being founded by a fellow who goes by the pseudonym Les U. Knight, it's a very real and serious thing, whose logic, as one observer put it, is "as absurd as it's unassailable."[4]

Logical or not, with human population still rising exponentially, and scheduled to top ten billion by 2050, the movement is not catching on. So, what do you do in the meantime? Schopenhauer has a few practical suggestions for making life bearable:

- Live in the present, and try to make it as painless as possible.
- Set limits—on anger, desires, wealth, and power. Limitations lead to something like happiness.
- Accept misfortunes; only obsess about them if we're responsible.

Which all sounds pretty level headed and doable. Except that the last bit of advice—don't dwell on misfortune unless we're responsible—is a bit tricky to navigate these days given the epic misfortune (climate change, eco-catastrophe, mass extinction, etc.) our species is currently visiting upon the planet and upon ourselves. If you think the denialists and Lords of Carbon and our corrupt politicians are the only ones responsible, well then, live in the present and set your sights on "something like happiness." However, if you think you, too, at least to some degree, are also responsible, well, time to face the music: not only are you living in the worst of all possible worlds, with an even worst-er world yet to come, but, according to extreme philosophical pessimism, you're morally obligated to think about it. A lot.

What would Marcus Aurelius do?

In contemporary parlance, a stoic is a person who shows strength in misfortune. "We've got to be stoic about it," someone might say when a family member gets cancer, or when a breadwinner loses their job. It's a spur to put on a brave face, keep it together, tough it out. Stoicism has the aura of squareness, of stolidness, of some upright someone (often a male someone) with clenched jaw bearing something with courage and grace.

In contemporary culture, "he's being so stoic about it," carries a whiff of suspicion. We wonder if there's a bit of pathology involved, a joyless fixation on this thing that must be borne. We wonder whether they haven't been quite creative enough to avoid it, or flexible enough to finesse it, or had enough therapy to get past it. But given the darkness on our collective horizon, we would do well to set these prejudices aside and appreciate stoicism in its more classical, philosophic sense.

Philosophical stoicism understands free will as the "voluntary accommodation to what is in any case inevitable."[5] Ethically, it holds us to the triple virtues of objective judgement, unselfish action, and willing acceptance of all external events. Teleologically, it suggests we view History as a series of cosmic cycles, each ending in a periodic conflagration (*ekpyrosis*). Metaphysically, it asks us to accept the idea, to borrow a phrase from Alexander Pope, that "whatever is, is right"—especially misfortune and death. And given the unfortunate circumstances of our historical moment (can you say "*ekpyrosis*"?)—and the likelihood of a major uptick in misfortune and death in the years ahead—stoicism deserves our renewed attention, if not a grand revival.

Marcus Aurelius is arguably history's preeminent stoic. Emperor of Rome from 161 to 180 CE, he was considered one of the "good emperors" and an exemplar of Plato's ideal of the philosopher-king. While navigating court intrigue, battling German barbarians, preventing civil war, and enduring personal tragedies, he kept counsel with himself via a series of writings never meant for publication. They have come down to us as the "Meditations"—a fancy title added later to this grab bag of miscellaneous reflections, spiritual exercises, and notes to self ("the philosophic equivalent of 'Phone Dr. Re: appt. Tuesday?'" as one commentator put it[6]) which is nonetheless considered one of the most concise and affecting summations of stoic philosophy.

Among its pages are many snippets of wisdom strikingly relevant for us and our climate predicament. Cherry-picking through the *Meditations* turns up the following advice:

- Hear unwelcome truths.
- Write off your hopes.
- Be optimistic in adversity.
- Always be self-reliant.
- Remember that the wrongdoer only does wrong because he thinks it is right.
- Be kind.
- Do not mistrust the future.
- Accept death in a cheerful spirit.
- Take up your post like a soldier who needs no oath or witness.
- Do your best, trusting that all is *for* the best.

Threaded throughout is the core commandment *amor fati*, love your fate. "What is unique to the good man?" Marcus Aurelius asks. And answers: "To welcome with affection what is sent by fate." With *affection*, he says. Affection? In our era of approaching catastrophe, an extremely daunting task. How do we go about it? One answer might be found in the philosophy of *Hózhó*, the (stoic-ish) life-way of the Dine' people. In *Sacred Clowns*, the first in a series of Tony Hillerman novels set in the four-corners region of America's Southwest, fictional Navajo detective Jim Chee explains it like so:

> This business of Hózhó...I'll use an example. Terrible drought, crops dead, sheep dying. Spring dried out. No water. The Hopi, or the Christian, maybe the Moslem, they pray for rain. The Navajo has the proper ceremony done to restore himself to harmony with the drought. You see what I mean. The system is designed to recognize what's beyond human power to change, and then to change the human's attitude to be content with the inevitable.

Hózhó, writes resilience.org blogger Alan Wartes,[7] "encompasses the Navajo ideal of living in harmony with all that is, of being in right relationship with the world." Hózhó "advises that adjusting ourselves to reality is a much easier (less stressful) and more balanced way to live than trying to bully the world back in line with our program." Hózhó does not counsel capitulation or surrender; it counsels creative, resilient, harmonious adaptation.

Jim Chee and Marcus Aurelius are asking us to do more than just pull ourselves together here. More than just get through the bad times and survive. They are asking us to face our grand misfortunes with grace, equanimity, and courage. Where others command us to "keep hope alive!" Jim Chee and Marcus Aurelius suggest instead that we "set hope aside!" and find a way to love whatever life throws at us no matter how terrible, climate catastrophe included.

Homo notsosapien.

Even the most foundational stories can be changed.
— Rebecca Solnit

As the apex predator in the food chain, and the only species currently able to write down words, we got to name all the species, including our own.

We are *Homo sapiens*, the sole surviving (nonextinct) member of the genus *Homo*. Or so one member of our species, Carolus Linnaeus, the father of modern biological classification, named us in 1758. In Latin, "homo sapien" means "wise man."

Naming ourselves—an act of nepotism rife with all the self-dealing and virtue-pandering you might expect—is a perilous art, and it's possible we've gotten it dead wrong. Fortunately, our official name has not prevented various observers of the scene from coining unofficial ones.

Aristotle thought of us as *Homo politicus*, political man. Our essential quality, what most fundamentally distinguished us from our brother and sister animals, was our ability to form complex societies. For Marx, we were *Homo faber*, tool-making man: man as producer, as creator. For cultural historian Johan Huizinga, it was *Homo ludens*, game-playing man. Whether love, war, poker, or theoretical physics, we're the species that loves to play, whose very existence is the "game of life." For novelist T. H. White, it was *Homo ferox*, ferocious man, the species that all other species are afraid of, including our own.

Each of these stories elevates one aspect of human existence as definitional, and in so doing, provides an essential insight into who we are. Each perspective is also a product of its time: for Aristotle, the political flourishing and chaos of the Greek city-states; for Marx, the extraordinary explosion of human productivity during the Industrial Revolution; for T. H. White, the cruel fox hunts and even crueler world wars of 20th century Britain.

What of our time? In the teeth of a self-inflicted mass extinction event, what does our time teach us about the essential nature of humanity?

Maybe it is more accurate to think of ourselves as *Homo notsosapiens*? Unwise man. We might be clever, boundlessly clever, but you'd be hard-pressed to call any species that managed to work themselves into the self-defeating predicament we've worked ourselves into "wise".

What name would you give our species? What "story of us" would you tell?

Maybe *Homo myopicus*? Short-sighted man. Man whose actions and wants have consequences far beyond our ability or willingness to see? Or *Homo malafide*? Bad-faith man. Man who refuses to act on what he knows to be true. Or possibly *Homo perdita*? Lost man. Man who has cut himself off from his own surrounds, from his brother and sister species and the rhythms of the cosmos, even from his own nature. This man is a stranger to himself and his world.

Or *Homo deus*? Man with the power of gods—to split atoms and undo hundreds of millions of years of geologic time in a cosmic second—who now, to paraphrase Stewart Brand, had better get good at it. Or *Homo ubercomplicaticus*. Overly complicated man. The creature who can (and everything else being equal, will) make things more complicated than it can handle. Or even, *Homo postsapiens*. The creature that—either by despoiling its own habitat, or inventing its own AI-enhanced superior[8]—is determined to extinguish itself.

There's certainly many worse things you could say: *Homo culus*. Asshole man. *Homo somebodyelsesproblemicus*. *Homo wereallyfuckedthisupicus*.

And to a degree, it's all true: We are irresponsible destructive assholes. We are the lost, myopic, overly complicated creature, who, wielding god-like powers yet unable to operate in good faith, is destroying itself and the world.

Can we imagine ourselves differently? Can we put some *sapiens* back into *Homo sapiens*, and earn those stripes? Instead of *Homo culus*, asshole man, why not *Homo supercoolus*? Man blessed with wonder. The creature who looks around at the world and finds it so supercool that we treat it with care and respect? Instead of *Homo somebodyelsesproblemicus*, can we become *Homo ibrokeitsoillfixiticus*? Man who takes responsibility for his own mess, who tries to heal the world he has broken. Instead of *Homo wereallyfuckedthisupicus*, can we evolve into *Homo letsunfuckthisupicus-andfast!*?

In the end, all we have is our name.

Vent your contradictions!

When two aspects of a situation are both true, yet at odds with each other, you've got a contradiction. Our climate predicament is full of them, and throughout this book we've seen many:

- Our climate news is "impossible"—AND—It must be told and heard.
- We're all in this together—AND—We're *not* all in this together.
- The Green New Deal must happen ASAP—AND—A just transition of livelihoods and infrastructure cannot be rushed.

That last one made a brief appearance earlier in the book (see page 158). This big tension, which many organizers fighting for a Green New Deal have encountered, was crystallized into a clear, two-sided contradiction by a project called Vent Diagrams, which adapts the Venn Diagram form to name and navigate the many contradictions we're facing, climate and otherwise.

Here are some of my favorites:[9]

Credit: Rachel Schragis and Elana Eisen-Markowitz

Co-creators Rachel Schragis and Elana "E.M." Eisen-Markowitz playfully refer to these visual stratagems as *vent* diagrams because they capture an ultimately unresolvable tension. "Making vent diagrams," they say, "helps us recognize and reckon with contradictions and keep imagining and acting from the intersections and overlaps. Venting is an...outlet for our anger, frustration, despair...an emotional release of stale binary thinking in order to open up a trickle of fresh ideas and air."[10] They deliberately don't label the overlapping middle section in order to "imagine what's actually in the overlap every time we see and feel the vent."[11]

If you have an unresolvable tension you want to get off your chest (and at the end of the world, who doesn't), you might want to give this approach a try. There are many more examples of vents, and tips on making vents, both alone and in community with others, on the Vent Diagrams website (ventdiagrams.com).

As the co-creators remind us, "none of these contradictions are going to be solved—AND—we cannot allow them to keep us from moving forward."

Let the venting begin!

Imagine your utopia.

Do you ever feel like the sheer reality of *what is* prevents you from imagining *what could be*? Welcome to the "tyranny of the possible," where the weight of the past and our worst assumptions about the future prevent us from pursuing—or even imagining—or even feeling like we deserve—a better world.

So, channeling the utopian impulse of visionary fiction (shout out to adrienne maree brown here, see page 253), let's break that trance![12] Instead of doomsplaining or wallowing in the "reflexive impotence" of Capitalist Realism (see page 278) let's tell a different story. Let's imagine a radically better world. *But, wai*—No, you wait! Enough about how everything is already too fucked up to undo. Enough about how nothing good is possible anymore.

Just for a moment, forget about how all the carbon that's already in the atmosphere has already crushed our beautiful dreams. Just forget the fact that we really needed to start building that bridge to the future 30 years ago. Forget how the evil nexus of fossil fuel lobbyists, gerrymandering, voter suppression, and a Senate full of conservative old white guys is going to block all meaningful progress at the federal level. Forget also about exactly how we're going to pay for all the nice things we need. Just set aside all that for a moment. Just forget it. Instead, let's imagine what's possible, especially if it's impossible, because, then—and only then—might it ever actually become possible.

But that's not—Shhh! Remember: utopia is not a blueprint, it's not a step-by-step plan, it's not even a strategy; it's an act of imagination and an act of will. So let's give our imagination some breathing room. What world do you *really* want? What is your utopia?

In my utopia, art is free and artists are paid. (*But wait, how*—Duuude, just google "Universal Basic Income" or ask Canada's National Film Board, they're already halfway there.) In my utopia, traffic fines are prorated according to the net wealth of the driver. (Finland[13] is actually already doing this: In 2004 the 27-year-old heir to a Finnish sausage fortune was fined $204,000 for driving the equivalent of 50 miles per hour in a 25 mph zone.) In my utopia, there are no slum lords, but if there were, and they broke the law, they'd be sentenced to live in their own bug-infested, terminally unrepaired hovels. (In 1985, a Los Angeles municipal judge did just that![14]) In my utopia, rivers and ecosystems have legal personhood. (And Ecuador, New Zealand, and, yes, Toledo, Ohio, are already starting to make this happen.[15]) In my utopia, the Green New Deal transforms America into a world leader in clean green tech, making West Virginia and Detroit and everywhere else great again for real. In my utopia, instead of trying to colonize Mars, we use that same bold visionary techno-inventiveness to save the one planet we already have, by like, oh, I don't know, building, say, an expanse of solar arrays in the Sahara so massive they'll not only provide clean green energy to the whole continent of Africa, but in the shade of their panels, help the desert bloom again. (Yep, it turns out some folks have a plan[16] to do just that. Climate catastrophe vs. Wakanda: game on!)

What's your utopia?

Hold a group meeting in
the halfway house of your soul.

Most of us tell ourselves multiple different stories each day.

— Naomi Klein[17]

In this time of unravelling we feel anxious and unmoored. There's something uncanny, eerie, about it all that we can't quite put our finger on. We nervously sample perspectives, try on different outlooks, hoping something fits. This schizophrenia makes sense. The depth of our civilizational crisis has dropped us into unknown territory; we don't know quite what to think or feel, or by what scheme to act.

Chuck Collins, who we met in the "Sartre is my whitewater rafting guide" chapter, describes a conversation that often happens inside his head: two voices bargaining over whether the human species deserves to survive, or whether Gaia would be better off without us. So far, he's decided he likes people too much to abandon us to our worst natures. Being considerably more neurotic than Chuck, my head is less a two-way debate than a halfway house anyone can crash in.

I might start the day drafting a boosterish "We can still #ActInTime" email. Maybe later that morning, after reading a particularly pessimistic report on the likelihood of frozen ocean-floor methane being released into the atmosphere, I might slide into a "But, but…it's not over till it's over, right?" bargaining with the future. Then, say, if a midday errand takes me past the hundreds of chainsawed tree stumps in my local park, and one too many just-built luxury condos and glittering midtown Manhattan boutiques full of unnecessary things, while the decaying subway breaks down a couple times (plus, I'm still thinking about that ocean-floor methane), I might snap into darker and more bitter thoughts: *Wash us away! Let's start again from first principles. Nothing else will do.* Finally, towards the end of the day, refinding my equilibrium, I might be telling myself: "I've just got to do everything I can…I've just got to do everything I can." And I'll believe it. And follow my own advice. And get back to work.

Whatever the sequence, all these attitudes have a call upon my heart; they all conspire to set my moral compass. Each feels real and true enough when it's the one driving; each impacts the decisions I make, where I direct my life force, how I show up in the world.

"You must be an extremely complex person," someone once said to psychologist M. R. Davies,[18] who jauntily replied, "No, actually, I'm four or five different people, and they're all quite simple." Or, as a friend of mine put it: "A person is a crowded place." Indeed.

So, rather than try to press-gang our psyche into line,[19] let's pull together a group meeting in the halfway house of our soul, tack one of those chore wheels up on the wall, crack open a few beers, and let everyone say their piece.

All who show up are us, but let's not expect us to agree with each other. Each voice is a split-self trying to reckon (or refusing to reckon) with our existential crisis in their own particular way. Some will likely whisper, others shout; some have been waiting many years to speak, others may have crash landed in our soul just yesterday. All have something to say, and all have a role to play.

Fist step: call the meeting.

Second step: try to get all your selfs to fill out that chore-wheel.

Are you a YES or a NO kind of person?

Start where you are. Use what you have. Do what you can.

— Arthur Ashe

Are you a Yes kind of person? Or are you a No kind of person? Whichever you are, the climate justice movement needs you. Because to create a just and livable world, we have to both stop some bad stuff and build some good stuff. Using this handy flowchart you can get started now:

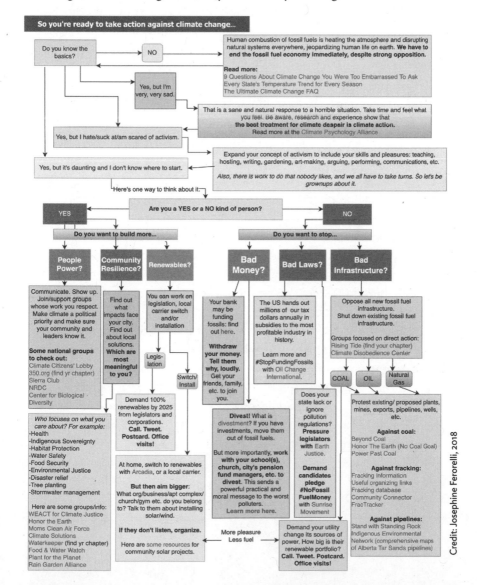

Credit: Josephine Ferorelli, 2018

Let the eyes of the future bore uncomfortably into your skull.

*The eyes of the future are looking back at us and they
are praying for us to see beyond our own time.*

— Terry Tempest Williams

Research[20] shows that how we think of our "legacy" can be one of the most effective motivators of "helping behavior." A series of studies conducted at Columbia University over a two-year period indicates that "being asked to consciously think about the way they would be remembered by future generations caused people to act in more environmentally friendly ways."

Unfortunately, most of us don't have researchers in lab coats asking us these helpful behavior-modifying questions. Nor do we have anyone yet to be born future centuries from now conveniently on hand to look us in the eye and hold us to account. Instead, we need to exercise our own moral imagination and do that for ourselves. But how? As with most things of this kind, via an unpleasant and emotionally overwrought creative visualization exercise. Luckily, you can do this in the privacy of your own home.

Take a moment to quiet your mind. Breathe. Close your eyes. Damn! Open your eyes. Find the goddam thing making that buzzing noise. Turn it off. Since you're up, hunt down all the devices you own that might possibly buzz in the next 30 minutes, and turn them off. If you have a roommate or a partner, turn theirs off, too.

OK, sit back down. Close your eyes. Breathe. Look into the future.

Bringing along what you know of science fiction, of climate projections, of human possibility...go one, two, three, five, as far as seven generations into the future. Notice your surroundings. There are real people there. Let them look at you. Feel the eyes of the future looking back at you. Feel the prayers and pleadings of those eyes boring hard into your skull. What are they trying to tell you?[21]

Should I bring kids into such a world?

*You've thrown the worst fear / That could ever be hurled /
Fear to bring children / Into the world...*

— Bob Dylan, "Masters of War"

The decision to have a kid used to be basically a question of: Am I with the right partner? Can I afford this right now? Do I have it in me to be a good mom or dad? How many diapers am I willing to change so I don't have to die alone? Now, thanks to climate change, on top of these questions, we have even weightier ones: Is it right to bring a child into such a world? Will she be able to live out her full life given the catastrophes that await? Am I being selfish to add yet another outsized American carbon footprint onto an already overburdened ecosystem?

My mom once said to me, "We had you boys when Kennedy was President. Things were still hopeful. If I'd known how it was all gonna go, we probably wouldn't have." I was a bit shocked. *Mom, aren't you supposed to—once you've had me, at least—not be able to imagine your life without me?* I was pretty glad to have been had, of course, and glad my mom didn't realize how terrible the world was going to get until it was too late. (Victory! I exist!) But I could see her point. And she was only worrying about nuclear war—something that might not even happen. And now here we are five decades later and nuclear war still hasn't happened, but we know some kind of climate catastrophe will.

I've never hankered for kids. That instinct just never kicked in. I can never decide whether this makes me lucky or unlucky. But maybe I'm doing myself—and them—a favor. Maybe I'm protecting all of us from a future that is "only dark and darkening further"?

For me, there's a year—2050—that floats out there in the future. A membrane between the bearable and the unbearable. In 2050 I'll be 86. Both my parents died in their 86th year. This year becomes a strange boundary, marking a certain limit to my emotional exposure. When I hear how the second half of the 21st century is likely to be go, I think: *At least I won't be around for* that. Am I proud of this feeling? No. It feels like a failure of my soul. And yet, it gives me a certain comfort, a certain distance from the worst of it. On the other side of 2050, nothing can happen to me.

These emotional calculations are, of course, kind of insane. And this attempt to limit my emotional exposure falls apart very quickly. I have many dear friends 20 years or more younger than me. That gets us to 2070. I'm an honorary uncle to two lovely young humans, now 19 and 18 years old. Which brings us to 2085. And then there's my affection for humanity in general, which gets us to, well, forever. I am along for the whole ride.

Having kids used to be an optimistic move. A vote of confidence in the future. A way to sail our hopes and dreams forward. I don't know how much my pessimism about the future held me back from having kids, but maybe being childless allows me see more clearly; I can more "objectively" weigh out my hopes and hopelessnesses and see where I land. (Which is helpful, I suppose, when writing a book about the end of the world.)

But what about people who—civilizational collapse-trajectories aside— really want to have kids, or, whoops!, just got pregnant without quite planning or meaning to?

A few months after she'd learned she was pregnant with her first (and to date, only) child, I spoke with Meg McIntyre, then in her mid-30s. Her dilemmas were fierce and riven with paradox:

> Having a child is the most hopeful thing you can do, but climate chaos has scrambled that calculation, made it all weird. Now there's hope for us, but not for them. My life in the next 40 years will be sort of OK, but what about hers? But if I really thought the world was going to be ruined in my daughter's lifetime, then why the fuck would I be having a kid? If I actually feel as defeated as I think I do, then I wouldn't be having a kid. So there's got to be some kind of hope. And I don't want my daughter to think I'm a defeated nihilist. That doesn't feel like good behavior to model. I'm choosing to see my decision to have a kid as a hopeful maneuver, by me, on myself. I'm under more obligation now to seek out that hope, manufacture it, cultivate it. But climate change really challenges that, because we're really close to the point where there's no hope anymore.

Meg saw her decision to have a kid as "a hopeful maneuver" by her, on herself. Paul Kingsnorth, father of two young children, has a simpler explanation: "We're animals," he says. "I love my wife and I wanted to have kids with her." After a little more reflection he added: "If people were all acting rationally, we probably wouldn't have children. We wouldn't fly in planes. We wouldn't have any problems." His conclusion: "It's another one of those paradoxes you hold." [22]

In *This Changes Everything*, Naomi Klein movingly documents her own struggles to conceive a child alongside the severe impacts that our reliance on fossil fuels is having on the planet's reproductive systems:

> In species after species, climate change is creating pressures that are depriving life-forms of their most essential survival tool: the ability to create new life and carry on their genetic lines. Instead, the spark of life is being extinguished, snuffed out in its earliest, most fragile days: in the egg, in the embryo, in the nest, in the den. [23]

Leatherback sea turtles, to take just one heartbreaking example, have "survived so much," says Klein. They've been around for 150 million years, making them the longest-surviving marine animals on earth. They've even survived the asteroid that likely wiped out the dinosaurs, but, says Klein, "it's not clear that they're going to be able to survive even incremental climate change." These sea turtles bury their eggs in the sand, but even with just 1°C rise, "the eggs are not hatching; they're cooking in the sand." [24] In light of this gradual collapse of the planet's reproductive systems, Klein proposes "the right to regenerate" as a new fundamental right. [25]

She describes climate change as "intergenerational theft." "Kids," she says, "are growing up in a mass extinction, robbed of the cacophonous company of being surrounded by so many fast-disappearing life forms.... What a lonely world we are creating." [26] Naomi herself held off having kids until she was a "pretty late." Part of what held her back was not being able to imagine anything other than a dystopian future. Seeing signs of hope in the world—and particularly in the social movements around her—helped her decide to become a parent. [27] However, she cautions against the idea that people with kids care about the future more: "Some of the most caring people that I know don't have kids," she says. [28]

In a 2012 interview with *The Phoenix*, speaking of the "carbon-industrial complex" and "the mess that had been made of this world" and the role it played in holding her back for so long from having a kid, she said: "I'd rather fight like hell than to give these evil motherfuckers the power to extinguish the desire to create life. We don't all have to do it. But if we want to do it, if we want to be part of this amazing process that we share in common with all living things, I'm certainly not going to give these guys the power to take that away from me."[29]

Mass extinctions, carbon footprints, "hopeful maneuvers" on ourselves. Trying to decide whether to have a kid at the end of this world is its own very particular kind of crazy. Luckily, there's an organization that can help us puzzle our way through it: Conceivable Future (http://con ceivablefuture.org), a women-led network "telling the stories of climate change's impact on our reproductive lives" in order to "build moral power for climate action."

Conceivable Future has collected almost a hundred testimonies, mostly from women in their 20s and 30s. While co-founder Josephine Ferorelli is constantly surprised at how subjective each person's process is, she identifies two main groupings: those worried about the future safety of their child, and those feeling guilty about how their kid is going to contribute to the problem. Many are concerned about both. Josephine herself, childless and in her mid-30s, had never dreamed about a wedding or marriage, never pictured a family or child or any particular outcome, and believes this was largely due to the climate crisis and her awareness of how rotten a future might be in store. "If I don't have a kid," she reasoned, it feels like "I'm somehow not accountable to the future in the same way. I won't have to feel as guilty about my inability to forestall the worst in the future."

Many concerned about their child's future carbon footprint consider adoption, including international adoption. But things get complicated fast. On average, the carbon footprint of a kid growing up in Somalia is 150 times less than a kid growing up in the USA. In Vietnam, it's six times less.[30] Haunted by this, one woman very consciously opted for a domestic adoption. That way, she reasoned, the child's carbon footprint would stay the same; she wouldn't be taking a low-carbon baby and giving it a high-carbon upbringing.

Another woman, keen enough to have a child that she was using fertility treatments to help the process along, was nonetheless struggling[31] over the terrible future she feared she was sentencing her child to. In the end, she concluded: "If we have a baby, I will do so knowing, though maybe not really believing, that a life can be any length."

The stakes are higher for a lot of people of color in the US. Typically, you're already up against a lot of external forces that are not supportive. An African-American woman—mid-30s and leaning towards "No"— came to one of the organization's Chicago gatherings. Her dilemma: *Is it better for the environment if I don't have a kid? Yes. Is it better for the Black community if I don't have a kid? No.* With high rates of incarceration and death by gun violence in the Black community, and the historical trauma of slavery and forced sterilization never far away, the pressures were acute in both directions. For her it was a real balancing act.

Meanwhile, in the UK, led by singer Blythe Pepino, hundreds of women (and men) who've chosen not to have children because of looming climate breakdown are taking their difficult personal decisions and making them strategically political by declaring a "#BirthStrike."[32] It's "a very hopeful act," says Pepino. "We're not just making this decision, hiding it and giving it up." Instead, she sees it as a way to channel her grief "into something more active and regenerative and hopeful".

Many of the women going on #BirthStrike or testifying with Conceivable Future are choosing not to have kids because of a looming sense of ecological catastrophe. But what if you think the apocalypse is coming and you *already* have kids? In our meeting, Tim DeChristopher described more than a few parents he knows who take the attitude, *I have kids, so I can't even let myself think about this.* He shared the story of two of his close colleagues, both fathers, who were going through a period of climate-induced soul searching and despair. Their wives insisted that they go see a psychiatrist. Their families pressured them into taking antidepressants. According to Tim:

> There's this sense that "our child will catch this." At this formative age, it will somehow instill itself into our child. Maybe there's legitimacy to that. For me, that's one of the real wild cards: When do people reach the age of maturity? At what age are they ready to grapple with the complexity of our situation, the open hopelessness of it? I'm very opposed to the climate movement treating everybody like children, like we're only capable of simplistic, either/or, black-and-white thinking. However, I'm not necessarily opposed to treating children like children.

One father[33] I spoke with took a more pragmatic approach: "I have kids, and this is the world they live in, so I'm going to raise them: (1) amidst social movements so they can experience for themselves how people-power can change things for the better; (2) with hard skills for adapting and surviving; and (3) to always remember that poisoned rivers and mass extinctions are not normal."

When I spoke to Gopal about why he had kids, he said, "First, not having children is a terrible strategy for the survival of the species." And second, that climate justice is a multi-generational fight, and his whole family is in it together:

> My younger kid is a serious action junkie. I can't keep him away. And when he's at the action, it's a signifier that I think it's right action. If I did not think it was right action, I would not bring my children. I believe so much in what I'm doing that my children are here. If you decide to throw tear gas at my children, you're picking an entirely different fight with me. That is not acceptable, and I will come after you. The platform upon which I stand has a generational logic, it's a logic of transition that is about preparing ourselves—and our children, and our families—for what's coming by what we do today.

Gopal's whole family is like a multi-generational nonviolent action squad, fighting together for a livable future. The parents in Tim's story—in an attempt to psychically quarantine off hopelessness—are telling their spouses: "We have kids, we can't afford to even think about any of this." And yet it was hope—specifically signs of hope in the world around her—that helped Naomi Klein decide to become a parent. Meanwhile, Josephine opens herself to the possibility of having children by letting go of any expectation that they will survive the future she's involuntarily signing them up for.

It seems there's as many paths through this dilemma as there are people confronting it. Realizing she was about to bring a kid into a future she had very little faith in, Meg decided to treat her pregnancy as a "hopeful maneuver on herself." But you don't have to get pregnant to do that. All of us, childless me included, are anxiously eyeballing that same future, entwined in our shared human story, trying every day to pull a hopeful maneuver on ourselves.

ANOTHER END OF THE WORLD IS POSSIBLE

Catastrophe is not a matter of fate.
It's a matter of choice.

— George Monbiot

Hope and hopelessness, both.

When I try to pull a hopeful maneuver on myself, it often requires one of those chaotic group meetings in the halfway house of my soul (see page 321).

Some of my selves need a reminder. Others need to be coaxed out of their rooms. But eventually a quorum of us circle up in the kitchen, along with a motley crew of role models. We make some popcorn and talk it all out.

Activist-me calls the meeting to order. Going through the agenda, he tries his level best to keep everyone focused on "fixing" the mess we've made. Stoic me trudges up from his basement room, pours shots of whiskey all round, and suggests we "welcome with affection what is sent by fate." Jaded nihilist me just sits there, certain of the one burnt-to-the-ground future he can already smell. Utopian me sketches out dreams of what could be (or at least could have been) and holds them up for all to see. While extreme philosophical pessimist me anxiously hums the *Ode to Joy*, self-soothing his way through this worst-of-all-possible house meetings.

Am I depressed? Ye-e-s, so depressed. I have every reason to be. The facts of our situation are overwhelmingly depressing. Am I angry? Hell, yes! Fighting mad angry! Due to greed and short-sightedness, we are visiting a horrific ecocide upon ourselves and all of life. Do I grieve? Yes. Am I disgusted? Yes. Do I swear a lot? Yes. Do I wax philosophical? Yes, how could I not at such a cusp-moment in human history?

If the meeting goes well, my philosophizing balances me. My grief grounds me. My anger empowers me. My depression (in a marginally self-harming way) protects me. And my discipline refocuses me.

Not all of me's fill out the chore-wheel, but many of me do.

Like the 12-year-old girl in Disney's *Inside-Out*, I need the gaggle of all my homunculi in order to feel the full truth of myself and the world.

Likewise, to find my way through the end of the world these last eight years, I've needed the wisdom of all the people I've met along this journey. If it's indeed a quest I've been on, these remarkable hopers and doomers have been my spirit guides.

And what sorrowful yet fierce counsel they have given!

A doomer scientist stepped me through the near-term extinction of humanity over Skype. And yet Guy McPherson, even in the face of certain extinction, counsels honesty, kindness, and solidarity.

An ex-con divinity student suggested I cultivate "death-eyes" to face the terrible truth of our predicament. Yet Tim DeChristopher, knowing it's "too late" to save the world, is still fighting hard for the new world he imagines rising upon its ashes.

An eco-Buddhist workshop leader told me and all assembled to give up hope. And yet even as Meg Wheatley suggests we "give in," she encourages us to "not give up."

A grassroots strategist admitted we were in for Collapse. Yet, in every Shock and Slide of that Collapse, Gopal Dayaneni sees an opportunity for organized people to wrest back control of our future from an ecocidal system.

A beloved wisdom teacher and systems theorist was unwilling to say whether we were hospice workers at the end of the world or midwives of a new one. And yet Joanna Macy gamely suggests we prepare for both.

A collapse psychologist who gives us humans "two, no more than three generations, tops," proposed I look at the whole story of our species' demise through tragic eyes. Even so, Jamey Hecht, ever the therapist, tries to help collapse-aware people find a way to "be happy at the end of the world."

A visionary community organizer described her Detroit home as already "post-apocalyptic." Yet, adrienne maree brown, even though she's convinced our civilization is in for a hard fall, helps people across the country build up their trust and skills so that our communities can "fall together, not apart."

A celebrated Indigenous botanist shared her "moral terror" of the moment we finally realize that the plants we've been manhandling as objects are actually "holy people"—but by then it's too late to repair the world. And yet here is Robin Wall Kimmerer, grief-stricken to her bones by the extinction crisis, making a stand at the confluence of two great rivers: *what do I love too much to lose?* and *what am I going to do about it?*

I'd begun this quest a hopeless-hopeful schizophrenic. And now—eight meetings, thousands of miles, and two flowcharts later—here I am at the end, still anguished by how much we're likely to lose but still striving to lose as little of it as possible. In the end, I am an optimist and pessimist both.

Many homes across Canada have something called a "tickle trunk." A chest full of dress-up items: hats and capes and boas and fake mustaches and the like. These days, if you open my tickle trunk (it's the existential kind) and dig down past the hoodie of despair and the hopium-tinted glasses and the mustache of doom, you'll find a cape of "tragic optimism" and a pair of not-at-all-rosy "can-do pessimism" goggles.

Whatever tint is the opposite of rosy, that's what color the goggles of can-do pessimism are. They let me see—in plain brutal terms how fucked we are. They also focus me on where I can help. When I put them on I become stubborn and purposeful; my eyes narrow in on the prize. Our task, Antonio Gramsci said, "is to live without illusions without becoming disillusioned." These goggles help me do that.

And when I'm brokenhearted, but still want to attempt large things, I reach for my cape of tragic optimism. I put it on and feel, well, brokenhearted—but now in a more soulful and empowered way. It will all end badly, I tell myself, but not quite as badly as it would if I did not act. So let me do something. While expecting nothing. While still longing for everything. And I call upon that grand longing—for everything wrong to be set right and everything broken to be set whole—to help me act in as large and openhearted a way as I can, even in the face of inevitable failure. Putting on my cape, I feel a commitment to truth, beauty, solidarity, and, yes, strategic thinking.

Poet Gary Snyder was once asked,[1] "Why bother to save the planet?" He replied with a grin: "Because it's a matter of character and a matter of style!" That spirit—and on a good day, that grin—flow through me when I put on these goggles and cape and step out into the world. Tragic optimism and can-do pessimism are costumes in my existential tickle trunk, tools in my strategic-philosophical toolbox. I pull one or the other out depending on the task before me, or the particular crisis of faith that's got me by the throat.

To some folks, these approaches will come across as another variation on climate defeatism. But they are neither defeatist nor unique to our climate predicament. Writing about a different time (the 19th and 20th centuries), and a different fight (class struggles in Brazil), the novelist Joao Ubaldo Ribeiro, walks a similar path:

> We don't have enough weapons to overcome oppression and never will, although it's our duty to fight whenever our survival and our honor have to be defended [...] This is a fight that will go on across centuries, because our enemies are very strong. [...] The bullwhip still prevails, poverty increases, nothing has changed. [...] But we are making this revolution of small and great battles, some bloody, some muffled, some secret, and this is what I do...[2]

So let us go forth—be-goggled, en-caped, clothed in whatever power-garments each of us can pull from our existential tickle trunks—and make all the difference we can.

AEOTWIP!

Une autre fin du monde est possible.

— Graffito, France, 2017

"There Is No Alternative." That's what the fathers[3] of global capitalism have wanted us to believe ever since the fall of the Soviet Union in 1991. It's neoliberalism or nothing, as far as the eye can see. The world as it is, we're told, is the only world that can be. This shining world of sound fiscal policy (enforced austerity), shrewd privatization (gutted public services), wide-open deregulation (ecological ruin), and global market efficiencies (where everything sacred—from our water to our democracy to our precious cultural treasures—is for sale to the highest bidder) is not only inevitable, but right and good.

Beginning, however, with the Zapatista revolt in 1994, a counter-cry went up: "Another World Is Possible!" Another world that puts the Earth and its people before corporate profits; a world where local communities and small countries have "the right to make their own mistakes"; a world that cultivates the Commons as a treasury for all. This other, better world "is not only possible," sang Arundhati Roy, "she is on her way. On a quiet day, I can hear her breathing." And it's pretty much been AWIP v. TINA ever since.

What about me? On some of those quiet days—via books, journalism, and the prankish arts—I've tried to help that other world to breathe. On some of the louder days—in Seattle in '99, say, or Occupy Wall Street in '11—I've been one of the ones in the streets shouting AWIP! AWIP! AWIP! Until the last few years, that is, when I began the slow, heartbreaking process of reckoning with our climate future, which slid me into a deep state of NWIP. No World Is Possible.

By day it was still AWIP! AWIP! But by night, I'd read the Dark Mountain blog and skulk around eco-collapse user groups. My AWIP!/NWIP schizophrenia couldn't last, and nor did it actually make much sense, because one way or the other we're going to have a world, even if it's a badly collapsed world. So where did that leave me? The AWIP v. TINA battle lines were still there, only now—whether either side acknowledges it or not—we're actually fighting over *how* the world is going to end. And I am in this fight 100%, shouting the decidedly more difficult to get behind (and pronounce) acronym, AEOTWIP! Another End Of The World Is Possible!

Why fight over this, you might ask? Because *how* the world ends matters! In fact, it might be the most important fight we humans will ever fight. Is it going to end in an ugly, dog-eat-dog Terminator Capitalism on a badly collapsed world? Or some kind of wiser, kinder, shared simplicity on an only-partly-collapsed world?

As Gopal Dayaneni and adrienne maree brown both made clear, every battle must be fought (and won) twice. Once over the problem (alerting society to it, framing what's at stake, convincing a critical mass of actors that serious action must be taken); and again over the solution (what's the plan?, what specific remedies?, who decides? who benefits?, how do we share the burdens?). We're close to winning the first battle and we're currently in the thick of the second. But even though we must absolutely win both, some of us have been fighting so long and hard just to get the problem of climate change taken seriously, that we forget how crucial this second battle over solutions is.

To illustrate, let me share the following transcript of a conversation between you and an unnamed middle-of-the-road politician that was recently teleported to me from the very near future:

> You: Climate change is a serious problem!
> Politician: You're so right.
> You: I am?
> Politician: Yes, it's a very serious problem.
> You: Right, well, um…we need solutions now!
> Politician: I agree.
> You: You do?
> Politician: Yes, and they need to be bold, radical, immediate, no-nonsense solutions!
> You: I, uh, couldn't agree more.
> Politician: Ha, look, now you're agreeing with me. (Politician smiles.)
> You: But—
> Politician: To get to zero-carbon immediately and protect our way of life, we need market incentives for wind and solar; a business-smart carbon tax; corporate subsidies for a massive build-out of nuclear; designated sacrifice zones; synthetic bees to replace our lost keystone pollinators; forced conscrip-

tion of prison labor to fight forest fires; a militarized, climate
refugee-proof border; AI-enhanced protein farms; and given
the latest overshoot projections, we really shouldn't take
forced sterilization of the homeless and illegals off the table,
because if—

You: But, look—

Politician: No, you look! We're talking about our survival here,
there's simply no alternative.

There's only one way the world can end, this guy is saying, and that's his
way. It's the same old TINA!, just dressed up for the end of the world.
But just because Capitalism would sooner end the world than end itself,[4]
that doesn't mean we have to play along. And just because Capitalism
turns us into economic cannibals and *then* tells us that human nature is
economic cannibalism, doesn't mean we have to believe it. In fact, we'd
better unlearn it quick, because at the end of the world economic cannibalism is going to get real ugly real fast. Luckily, there are so many other
ways the world can end. Or only almost end. Or not quite even end.
And we've looked at many throughout this book, from Drawdown to
Degrowth to Deep Adaptation; from Gopal's Just Transition to Joanna's
Great Turning to Robin's Honorable Harvest.

The good news: many, many other ends of the world are possible.
The bad news: it's still the end of the world. Because even if the most
ambitious and optimistic of the approaches above succeed, we still won't
"save" the world, in the way we usually mean it—the glaciers will still
melt away, the Great Barrier Reef will bleach and die, there will be untold
human suffering and chaos. It's sobering to reckon with, but our goal
here is not so much to "save" the world as to just barely salvage it.

Shit, you're probably thinking (at least that's what I always think at
this point in the conversation), not only do we have to reinvent our civilization and imagine our way past a capitalism that's done all in its power
to thwart us from imagining anything beyond it, but we also have to
make up for all those wasted decades not solving the most complex and
dangerous problem humanity has ever faced—in order to "just barely
salvage" the world?!

Yes. No one said it would be easy—or even possible—but someone[5]
said: When the going gets tough, the tough get going. And someone else
(Hunter S. Thomson[6]) said: When the going gets weird, the weird turn
pro. (Which might explain how I got a book deal.) And, though no one

ever said this, I'm saying it now: When the going is full of terrifying trade-offs and unsolvable predicaments, even the toughest and weirdest among us need goofy, easy-to-remember names for the paradoxes we're facing. Which is why this book is full of them.

Because, we can't fix a predicament. And we're all in this together—not! And it's gonna take everyone to change everything, um, yesterday. And despair might indeed be our only hope. And somehow we're going to need to do the impossible, because what's merely possible is gonna get us all killed.

And then there's always—especially in the face of such daunting circumstances—a third battle: The battle over why bother at all? This battle happens in our hearts, in our guts, in our souls, in our consciences. It happens when we ask ourselves the question: "What do I love too much to lose?" Or: "What kind of ancestor do I want to be?" Or: "How can I lose hope when other people can't afford to?" Or, simply: "What good can I still do?"

Each of us must ask and answer these difficult questions for ourselves. As we do, it can help to have role models, both actual and archetypal. Earlier in these pages, we met Congolese park ranger Rodrigue Katembo, who endures death threats, poachers, illegal oil developers, and lengthy separations from his family in order to protect the elephants and gorillas in the national park he loves. Many of us have been inspired by Greta Thunberg, the autistic teenager whose "disability" is actually a gift of plain sight.[7] Simply to have a life later this century, she realizes she must go on strike against her government's failure to act.

Greta chooses the way of the Rebel. And millions of young people are following her on this path, rising up in a global rebellion against their own extinction. But your path might be that of the Healer, striving to "undo the damage we've done," or the Artist, whose heightened sensitivities are like "antennae for the species." It hurts to feel the future so deeply, but keeping that channel open is your way to serve.

Some paths force a choice: Do we follow the Prepper (going it alone in our rugged individualist bunker); or the Good Neighbor (like adrienne, helping our community become more resilient, believing "we can get through this together"). Do we take orders from the Capitalist, who sees in the apocalypse a gravy train of high-dollar contracts, whether for sea walls or gated communities, he doesn't much care; or do we conspire with the Visionary Engineer, who's trying to build humanity a technological bridge to the 22nd century?

The way ahead may be narrow, but many paths lead through it. Whether you're hopeless or hopeful, punk rock or more square, a "joiner" or not so much, there's a path for you if you're willing to seek it out. Just consider the different sensibilities of, say, the Sunrise Movement and Extinction Rebellion. The former keeps a fierce, eyes-on-the-prize hope, channeling Apollo-god-of-reason (and recruiting all the A-students) to craft an inside-outside strategy to steer the Green New Deal through a broken system to victory. The looser Extinction Rebellion runs riot through the streets with Dionysus, keening their grief, demanding that the necessary-but-impossible be done now!, telling the truth no matter the psychic cost, and turning their hopelessness into a by-any-means-necessary force to disrupt business-as-usual, all the while dancing to whatever beautiful now we still have left.

Different folks might be drawn more to one than the other, but to have a chance at a livable future, humanity needs both Apollo and Dionysus on our team. To find our way, humanity also needs the Trickster, who's made such trouble across these pages, and whose shape-shifting fondness for paradox and the in-between can help us to straddle the worlds (one dying, the other struggling to be born) across which we must transition.

This transition will be treacherous. There's no avoiding that now. But, as the hopers and doomers we've met make so clear, we can still choose a path with heart. Because the end of the world is not the end of the world, full stop (Kingsnorth); and we can still turn new normals—even new abnormals—into new beautifuls (Greer); and if we must fall, we can at least fall as if we were holding a child on our chest (brown).

Is there is going to be suffering? Yes, but we can try to distribute that suffering more equitably (Dayaneni). Have we broken Nature? Almost, yes, but it's not too late to "re-story" who we are and learn to be good partners again (Kimmerer). Are we in for a catastrophe? Yes. Likely a whole string of catastrophes in the "Long Emergency" that lies before us. But there are worse catastrophes and better ones; ones we can recover from and ones we can't. So let's bring to this fight all the courage, kindness, and wise planning we can muster, so we get the least terrible and most compassionate catastrophe still available. The world is going to end one way or another—let's make sure it's another.

NOW IS WHEN
YOU ARE NEEDED MOST

Even a wounded world is feeding us.
Even a wounded world holds us,
giving us moments of wonder and joy.
I choose joy over despair.
Not because I have my head in the sand,
but because joy is what the Earth gives me daily
and I must return the gift.

— Robin Wall Kimmerer

Skagit County, Washington, Summer 2018

John Michael Greer suggests we try to turn new normals into new beautifuls.[1] Hard advice to follow, but I try. And I'm trying especially hard at one of my last ports of call: my friend Lois's farm in western Washington, where I've come to gather my thoughts and try to finish off this manuscript.

Here along the Skagit River, in the foothills of the Cascades, the early morning mist mingles with air smoky from forest fires on the far side of the range. The West this summer (and, it seems now, every summer), is on fire. Our new anemic winters no longer have snap-freezes cold enough to kill off pine bark beetles,[2] and so they spread northwards, a literal plague, infesting millions of coniferous trees across the West. This is bad enough, but then our climate-changed summer comes—hotter and drier than before—and lightning (or a careless match) hits the dry and dying forests and...fwooom!

And now much of the North Cascades, and swaths of the broader West are on fire, and, even here, 50 miles from the nearest blaze, smoke (including particles small enough to be drawn into our lungs and through our bloodstream and deposited in vulnerable organs) hangs in the air.

The setting sun some nights is a blood-red orange: the orb of our possible salvation, refracted—eerie, and, yes, strangely beautiful—through the shrouds of our self-destruction.

In the day I write, taking breaks to help Lois can green beans or pickle cucumbers or pick quart after quart of blueberries. At the end of every afternoon, after a long day of trying to wrangle the manuscript, I swim naked in the fast cold currents of the Skagit. Climbing back up the bank, I stand bare and momentarily triumphant amidst the rough grasses and dandelions, as I let the warm air dry me off. There's a bend in the river in each direction, and I read and write as the rushing water chatters past several fallen trunks, lashed and held close to the bank by steel cables. Knowing they would fall to the Skagit's raging autumn floods, and hoping to slow the erosion of her land, Lois had lassoed them in years past while they were still standing trees.

For the last many seasons now, you see, the river has not been itself. In springs and autumns past, the river would gradually swell with storms or snowmelt, but now, with our new rising temperatures, it becomes suddenly and violently engorged with rain that used to fall as snow. Every year now, the river eats away up to half an acre of Lois' farmland and floods most of the rest. By the time she's 80, in spite of her best efforts, the Skagit might be at her doorstep, and she'll have to leave this place she wished dearly to be her last home.

This ground under me is disappearing. This same time next year, it may be gone. And it's here, by the edge of this hungry river, upon this disappearing earth, under this fire-tinged air, that I try to feel humanity's moment, going and coming.

The sun is setting, bathing the white and blue stones on the opposite bank in the "butter light" of photographers' dreams. Across the river, the shadow line slowly climbs Sauk Mountain's forested slopes, until the last of the light vanishes from the pock-marked glaciers far to the east. These glaciers, too, just like the ground underneath my feet, will soon be gone forever.

During the last few years, as I've tried to sift through all that I'd learned, I've found myself at many a place like Lois's Skagit Valley farm: extraordinary, and under extraordinary pressures. If someone asked what I was up to, I'd sometimes tell them, half in sorrow, half in jest, that I'd been going from beautiful place to beautiful place writing about terrible things.

Here I am, at one such troubled and beautiful place. I began it at another: the edge of Manhattan, haunted by ghosts from the future and the question "how should we live, knowing that catastrophe is coming?"

Along the way, there were many such places. At the Mesa Refuge, head buried in the desolate doomscape of *The Road*, I'd look up from those pages and out across Tomales Bay, and simply be grateful that the sun still shined, and the villagers had no plans to eat me.

In Joshua Tree, we'd rented a little cabin with a backyard of scrub, dotted with its namesake trees, each transfigured into a grave, multi-limbed crucifix pointing not sure where. One 105°F afternoon, writing on a weather-beaten desk beached in our desert backyard, a grey fox loped by. A Trickster in our midst.

In Gloversville, New York, a town that used to boast 500 tanneries that literally supplied half the world's gloves, but was now a rust-belt shadow of its former industrial self, I was trying to write on the back porch of a place my girlfriend and I—possibly very foolishly—had rented sight unseen. A yard crew came by unannounced, noise-blocking head-phones on, diesel-fueled leafblowers strapped to their backs, blowing off of my newly rented porch the one or two leaves that had accumulated since whenever the last time they'd been through. Diesel wafted into my nostrils; the blasting sounds of compressed air rattled my coffee and key-board. How was this helping anyone?

At Lacawac, Pennsylvania, a protected nature sanctuary and sci-ence monitoring station, young ecologists took constant readings of the untouched glacial lake and its surrounding air and forests. (No swim-ming was permitted in the lake to prevent even the tiniest bit of sun-screen from spoiling the pristine water.) Their instruments inescapably registered how during the handful of years I've been on this journey, the parts per million of atmospheric carbon has gone from a dangerous 397[3] to an even more dangerous 407 (and now in 2022 it's pushing a yet more dangerous 419). Though their instruments didn't show it, during that same time, my mother died, we marched in New York, we got the treaty signed in Paris, Trump was elected, we lost the battle of Standing Rock, Greta sailed across an ocean, the Green New Deal arose, then sputtered, and the battle continues.

Over those same years, I've met with many remarkable hopers and doomers. Where do they stand now?

I sent a follow-up email to Guy McPherson: *Are you sure there are no lifeboats on our Titanic?* He wrote back: "Nada. Nilch. Zero." He, for one, remains sure how this story ends.

Tim DeChristopher and the ten others arrested blocking the Enbridge pipeline in West Roxbury argued their case on the grounds that new fossil-fuel infrastructure posed an existential threat to humanity, and it was thus *necessary* to take direct action to stop it. The presiding judge, swayed by their closing statements, found all not guilty, an unexpected victory that established the "necessity defense"[4] as a national precedent for other such actions. Tim is now a famer on Wabanaki territory in Maine, while continuing his involvement with the Climate Disobedience Center.

Meg Wheatley continues training people in the Shambhala warrior tradition, and writing on how to move beyond hope and fear, including the short essay, "Freeing Ourselves from the Addiction of Hope."[5]

Gopal Dayaneni continues his work "at the intersection of ecology, economy and extractivism," and is still living in his Oakland intentional community of nine adults and eight children (though the chickens are gone, and a dog has arrived). He's now shifted onto the board of Movement Generation, is teaching part-time at San Francisco State University's Race and Resistance Studies program, and is still very active in climate justice and other movement work.

Joanna Macy stays mum on whether she thinks we're ultimately hospice workers or midwives. Now over 90, she has retired, but her Work That Reconnects carries on in the hands of a next generation of trainers and grief counselors. She recently co-translated a new edition of Rilke's *Letters to a Young Poet*.

Jamey Hecht has moved from LA to Brooklyn. He's still seeing patients, quoting Aristotle ("Happiness is not amusement; it is good activity"), writing poetry about RFK's assassination, and blogging about everything from regenerative agriculture to how Cuba survived its oil crunch.

After our meeting, adrienne maree brown set aside some of her many hats to try to take a restful sabbatical. On her return, she noted[6] that "the world ended at least twice while i was away." She is now writer-in-residence at the Emergent Strategy Ideation Institute, has launched the podcast, *Octavia's Parables*, and brought out two new books, including: *We Will Not Cancel Us* and *Grievers*, the first novella in a trilogy on the Black Dawn imprint.

Seven years after its publication, *Braiding Sweetgrass* made it to the *New York Times* bestseller list. To this wider audience, Robin Kimmerer continues her soulful advocacy on behalf of plants.

A dozen honking geese make a wide bank over the Skagit. A kingfisher darts into the river in search of dinner. Two turkey vultures, circling the smoke-thickened thermals, survey the beauty and destruction below.

At night, Lois asks me whether I'd seen the *Times Magazine* piece about how we could have stopped climate change 30 years ago?[7] I tell her yes, and with a sigh, ask her whether she'd seen the article in *The Guardian* about how scientists are starting to see signs of a "domino effect"[8] of compounding and accelerating climate change?

Here we go, she says.

Shoot me now, I say.

No, she says. *This is when we need you most.*

That night, in the pitch black of the new moon, we could hear apples drop from the trees. Our fate thudding to the grass.

And yet, the next day we knock on doors for Initiative 1631,[9] a statewide ballot initiative that would put a carbon fee on big polluters, and use the revenue to invest in clean energy and protections for workers in fossil-fuel industries. Big Oil and Coal are putting in over $30 million to defeat it.[10] But we have an army of neighbors across the state. Game on.

There are signs all around us that what we've most dreaded is actually beginning to happen. But this battle is far from over. And this land (even with its smoke-tinged air and pock-marked glaciers and eroding riverbanks and eerie blood-orange sun) is still beautiful. And now is when we are all needed most.

PASSING THE TORCH

What time is it on the clock of the world?

— Grace Lee Boggs

New York City, Fall 2019

It's hard to write a book about the end of the world. It's also hard to finish writing a book about the end of the world. Especially when you're also trying your damnedest to stop the world from ending. Case in point:

One afternoon in September 2019—an afternoon I had earmarked as one of the many afternoons I would indeed finish writing the book (having not finished it, nor even come close to finishing it, at Lois's farm)—an email arrived in my inbox: "Greta wants a clock." Greta? *That* Greta?

Yes, that Greta. She needed someone to build her a clock that would show the time we had left to act before we blew through our extremely limited carbon budget with catastrophic consequences for the ecosphere and human civilization.

She wanted to hold up this clock during her speech at the UN General Assembly…in nine days!

The email had come to me and my friend Gan Golan, a frequent collaborator on many climate-arts projects.

(Q: Why were we—neither of us clockmakers—suddenly being asked to build a climate countdown clock for Greta Thunberg? A: Because a year earlier, our somewhat insane idea to set up climate countdown clocks in city centers all over the world (see page 6) had been totally ignored by the same folks who were now asking us to build one in nine days.

Are you kidding me?

But this clock was for Greta.

And nobody does anything unless they're under a deadline. Whether that's building a weird clock. Or saving the world.

So, I stuffed the manuscript back in its drawer, and we put out the bat signal to our circle of creatives, and by the next evening we'd pulled together a crackerjack team of coders, makers, designers, artists, and campaigners to not just make a clock for Greta but, buoyed by the attention her speech would inevitably garner, launch the full www.Climate Clock.world effort we'd originally had in mind.

Our premise: To meet our climate deadline, everyone across the planet needed to "synchronize their watches" around our remaining time window for action. This deadline was, arguably, "the most important number in the world." But wait, what number was that exactly? What number should go on this clock?

It's a conundrum familiar to anyone who's reckoned with our climate predicament. A moment when facts collide with the human need for story, science struggles to fit its calipers around the most complex (and consequential) problem humanity has ever faced, and once again hope and hopelessness do their slippery, awkward dance.

The previous year, at the 2018 meeting of the Intergovernmental Panel on Climate Change (IPCC), the world's largest and most authoritative gathering of climate scientists had announced:[1] "Global net human-caused emissions of carbon dioxide (CO_2) would need to fall by about 45 percent from 2010 levels by 2030, reaching 'net zero' around 2050." Which is science-talk for "ACT NOW, IDIOTS!" Or as the BBC summarized it[2] the day of the announcement: "Going past 1.5°C is dicing with the planet's liveability." Our "1.5°C temperature 'guard rail' could be exceeded in just 12 years, in 2030."

Understandably, the media[3] picked up on the notion that there were "12 years left to save the world." And it was this number that lodged, with a mixture of alarm and hope, in the public's mind. It was also the number we'd originally planned to put on the clock. The science is complex, the uncertainties treacherous, but humans need a goal to focus on.

However, this goal, in spite of capturing the public's attention, being championed by AOC,[4] and even getting written into Bernie Sanders' Green New Deal proposal, was just one milestone among many, cherry-picked out of the 2018 IPCC report by the media, and neatly refocused by activists into a "we must cut our emissions in half by 2030 to save the world" deadline.

And Greta would have none of it. Over the course of a few follow-up emails, it was clear she had a far stricter set of parameters in mind. Whether she's leading school strikes or scolding Prime Ministers, "listen to the scientists," has always been her message, and one of the scientists she was listening to here was Dr. Ottmar Edenhofer, advisor to Angela Merkel and the world's foremost authority on carbon budgets. He and his Berlin-based Mercator Research Institute on Global Commons and Climate Change (MCC) had calculated the remaining carbon we could burn and still have a 67% chance of remaining under 1.5°C. His calculations were stricter than the consensus IPCC report, which was based on just a 50% chance. His carbon budget only left us eight years and three months.[5]

Suddenly, we were three years closer to the end of the world!

But those were the numbers Greta insisted on and that's the clock we built for her. We delivered it to her the night before her big speech. "Let's do this!," came back her smiling video message. You don't see her smile all that often. I thought: maybe it's too late to save the world, but at least I did something to make this fierce, extraordinary girl smile.

The next morning, we were all poised at our laptops ready to hit launch on the just-built website and flood the zone with #ClimateClock and #ActInTime hashtags as soon as Greta, clock by her side, gave her speech to the world.

UN security, however, refused to let the clock through.

Oh, come on!

Why would security care about wires peeking out from the back of a mysterious block of bright phosphorescent LEDs counting down? *Oh right*. And in the end, I guess it actually is a kind of bomb: a slow-motion carbon time-bomb we are dropping on ourselves and all of Nature.

In the end, her six-minute speech, heralded as "the Gettysburg Address of Climate," had all the intensity and moral clarity our existential crisis deserves, and in no way suffered for lack of a clock, though we were, of course, disappointed.

The ironies here almost speak for themselves. A symbolic "bomb" meant to alert the world to the very real carbon bombs we keep detonating is mistaken as a terrorist threat. The insistent mantra of "listen to the scientists" suddenly puts us three years closer to the end of the world. And maybe the biggest irony of all: at a gathering of the world's most powerful people, the only adult in the room was 16 years old.

"How dare you?!" said Greta, directly into their faces (and in a sense, all our faces, because even us adults who've been trying to fix things have failed her and her generation).

In her speech, she laid down a gauntlet of generational responsibility:[6]

> This is all wrong. I shouldn't be up here. I should be back in school on the other side of the ocean. Yet you all come to us young people for hope. How dare you! You have stolen my dreams and my childhood with your empty words.[7]

Pivoting to the crucial difference between a 50% risk of exceeding 1.5°C warming and a 67% risk of doing so that she had insisted be reflected in the clock's figures, she told her assembled elders, "Fifty percent may be acceptable to you. But…a 50% risk is simply not acceptable to us—we who have to live with the consequences."

"Those numbers," she continued, "rely on my generation sucking hundreds of billions of tons of your CO_2 out of the air with technologies that barely exist." She closed out her speech:

> You are still not mature enough to tell it like it is. You are failing us. But the young people are starting to understand your betrayal. The eyes of all future generations are upon you. And if you choose to fail us, I say: We will never forgive you.[8]

Her ferocity makes sense. Greta and her generation are angry that us old people have so badly fucked up a future that mostly they will have to live in (and maybe even worse: that we continue to fuck it up, excuse after excuse, COP26 being only the latest example (see page 6). Young people have way more skin in this game. Their anger at their elders is more than understandable.

I am 59 years old. I'll most likely be dead by 2050, and won't have to live through the cascading climate impacts that will play out across the second half of this century. I'm simply not in the same existential position as 18-year-old Greta Thunberg or 19-year-old Xiye Bastida or 18-year-old Jerome Foster II, or any of the other extraordinary young climate activists I've had the honor of working with.

Jerome Foster went from climate striking outside the White House for 58 weeks to being the youngest member of the White House Environmental Justice Advisory Council. He believes, "The youth of this world…

Credit: Jake Ratner for 350.org, April 2021

Climate Striker Jerome Foster stands with a Climate Clock

are being used and neglected by those in power who choose profits over life."[9] "Young people," he says, realize "the system doesn't work for us and it's not going to work for our future, so it must change."[10] And from the streets to the corridors of power to a glittering array of climate initiatives he has founded during his brief 18 years on this planet, he's doing his level best to change it.

Mexican-American climate striker Xiye Bastida, in the lead essay in the 2020 best-selling anthology *All We Can Save*, says: "To me and a lot of other young people, it feels like we are rooted in awareness while the adults around us live in obliviousness."[11] "Human civilization," she says, "needs to mature."[12] But she's not trying to get any "OK, boomer" digs in. In fact, she acknowledges that "we need to work intergenerationally" and that her Indigenous ancestors and elders (she is a descendent of the Otomi-Toltec people) have the "principles that humanity needs in these critical times."

Yes, the elders are needed; we veteran activists have some hard-won experience to offer; the eight remarkable Hopers and Doomers I met with on my journey have some essential wisdom and strategies to share; but it makes sense that young people like Greta and Jerome and Xiye are increasingly leading the fight to unfuck a world they will have to live in.

As we prepared the clock for Greta in the manner she'd requested, we haggled over whether there were 12 years to "save the world" or just eight (now less than seven), but the sorry truth of our predicament is that whatever number goes on the clock, we've already blown way past our ideal deadline. We can "curate" the science to tell ourselves a story that feels more hopeful, but the reality of our situation is that we're already living on borrowed time. We don't actually have a "carbon budget" in the sense that there's a certain amount of carbon we can "safely" emit. Rather, we're engaging in—and have been for decades—out-of-control carbon deficit spending, and it's wrecking the world.

And now, in 2022, it's that much worse. The daily headlines read like biblical curses. Record-breaking, fatal heat waves in Oregon, Germany, France; droughts in California and Madagascar; rainfalls so concentrated that cities can no longer absorb them, promising a new normal of catastrophic floods. Meanwhile, the clock is clicking down towards six years but it might actually be running behind reality. Feedback loops, melting permafrost, carbon bubbling up from the depths of the oceans, 70°F temperatures in the Arctic, the Amazon threatening to flip from rainforest to savannah, heralded by a chorus of scientists saying, "This is all happening much faster than predicted."[13]

The emergency is now upon us. The rest of the world is quickly catching up with Greta, Jerome, and Xiye, and with the eight hopers and doomers I met on my journey. Even as we mourn all we're losing, we need to take their deep lessons to heart and figure out how to survive the future. Together, we must craft a better catastrophe.

Appendix: Stuff You Can (Still) Do

Wherever you fall on the EndTimes Enneagram®, there's no shortage of stuff you can (still) do that matters:

Warrior
Sunrise Movement—sunrisemovement.org
Climate Justice Alliance—climatejusticealliance.org
Movement Generation—movementgeneration.org
NDN Collective—ndncollective.org

Engineer
Project Drawdown—drawdown.org
Climate Engineering—ucsusa.org/resources/what-climate-engineering

Healer
Climate Awakening—climateawakening.org
Dr. Daniel Foor and Ancestral Healing—ancestralmedicine.org
The Climate Ribbon—theclimateribbon.org
Make Beauty in Wounded Places—radicaljoy.org

Prepper
The Prepared—theprepared.com
The Provident Prepper—theprovidentprepper.org

Good Neighbor
Transition Town Network—transitionnetwork.org
Resilience—resilience.org

Policy Ninja
Richard Heinberg—richardheinberg.com/bookshelf
Beautiful Solutions—beautifulsolutions.info

Storyteller
Dark Mountain—dark-mountain.net
Yale Program on Climate Communication—climatecommunication.yale.edu
Center for Story-Based Strategy—storybasedstrategy.org

Rebel

Extinction Rebellion—rebellion.earth
Greta and Fridays for Future—fridaysforfuture.org

Artist

Artists Unite for a Green New Deal—usdac.us/gnd
Artists & Climate Change—artistsandclimatechange.com

Philosopher

John Michael Greer—ecosophia.net
Timothy Morton—tinyurl.com/TimothyMortonLongRead

Seer

Greenfaith—greenfaith.org
Interfaith Power & Light—interfaithpowerandlight.org
Starhawk—starhawk.org

Elder

Third Act—thirdact.org
Wisdom of the Elders—tinyurl.com/WisdomOfTheElders[1]
Turtle Lodge—turtlelodge.org
Original Instructions—innertraditions.com/books/original-instructions
Raging Grannies—raginggrannies.org

Timekeeper

Climate Action Tracker—climateactiontracker.org
Climate Clock—climateclock.world

Trickster

Crazy Wisdom—tinyurl.com/CrazyWisdomSummary[2]
Beautiful Trouble—beautifultrouble.org

Figure References

1. The Hubble Heritage Team. *GPN-2000-000933*. Photograph. Washington, DC, June 3, 1999. NASA.

2. Azote. Planetary Boundaries Pollutants 2022 Update. Stockholm Resilience. University of Stockholm, 2022. https://www.stockholmresilience.org/research /planetary-boundaries.html.

   Enescot. Projected Carbon Dioxide Emissions and Atmospheric Concentrations over the 21st Century for Reference and Mitigation Scenarios. Wikimedia Commons. Wikipedia, July 19, 2013. https://commons.wikimedia .org/wiki/File:Projected_carbon_dioxide_emissions_and_atmospheric_con centrations_over_the_21st_century_for_reference_and_mitigation_scenarios .png.

   Hanno. Global Temperature 1ka. Wikimedia Commons. Wikipedia, November 9, 2005. https://commons.wikimedia.org/wiki/File:Global_temp erature_1ka.png.

3. Amakuha. *Crossbones Vector (PSF)—Single Color*. Wikimedia Commons. Wikipedia, August 9, 2009. https://commons.wikimedia.org/wiki/File: Crossbones_vector_(PSF)_-_Single_color.svg.

   Eddo. *Sun Icon*. Wikimedia Commons. Wikipedia, January 6, 2011. https://commons.wikimedia.org/wiki/File:Sun_icon.svg.

   Pixabay. *Zombie-156055*. Wikimedia Commons. Wikipedia, June 11, 2016. https://commons.wikimedia.org/wiki/File:Zombie-156055.svg.

4. Chart adapted from: Frase, Peter. *Four Futures*. London, UK: Verso, 2015.

5. Chart was originally posted by a reddit user to a subreddit that has since been shut down. We made all attempts to appropriately credit the person who made this chart. It too is adapted from: Frase, Peter. *Four Futures*. London, UK: Verso, 2015.

6. Davis, Donald. Planetoid Crashing Into Primordial Earth. Wikimedia Commons. Wikipedia, March 27, 1991. https://commons.wikimedia.org/wiki /File:Planetoid_crashing_into_primordial_Earth.jpg.

   Devcore. *Icon Atomic 256x256*. Wikimedia Commons. Wikipedia, May 11, 2013. https://commons.wikimedia.org/wiki/File:Icon_Atomic_256x256.png

7. Amakuha, *Crossbones Vector*. https://commons.wikimedia.org/wiki/File :Crossbones_vector_(PSF)_-_Single_color.svg.

   Pixabay, *Zombie-156055*. https://commons.wikimedia.org/wiki/File: Zombie-156055.svg.

8. Malaspina, Gabriele. *Kayak—The Noun Project*. Wikimedia Commons. Wikipedia, December 18, 2017. https://commons.wikimedia.org/wiki/File: Kayak_-_The_Noun_Project.svg.

Flowcharts Bibliography

"Al Gore Likens Global Warming to Nazi Threat," *Los Angeles Times*, July 8, 2009. https://latimesblogs.latimes.com/washington/2009/07/al-gore-likens-global-warming-to-nazi-threat.html.

"Antonio Gramsci." In Oxford Essential Quotations, edited by Ratcliffe, Susan.: Oxford University Press, https://www.oxfordreference.com/view/10.1093/acref/9780191843730.001.0001/q-oro-ed5-00018416.

Bloch, Nadine, and Rae Abileah. "Activism during the Coronavirus Pandemic," *The Commons*, February 21, 2022. https://commonslibrary.org/beautiful-troubles-guide-to-activism-during-the-coronavirus-pandemic/.

Brod, Max. *Franz Kafka: A Biography*, translated by Richard Winston and G. Humphreys Roberts. New York, NY: Schocken Books, 1960. 75.

Christian Wahl, Daniel. "Midwives of the Regeneration: On the Fertile Edges of the More Beautiful World," *Medium*. Age of Awareness, May 6, 2018. https://medium.com/age-of-awareness/midwives-of-the-regeneration-on-the-fertile-edges-of-the-more-beautiful-world-4a28a9c6496f.

Christensen, Miyase. "Slow Violence in the Anthropocene: An Interview with Rob Nixon on Communication, Media, and the Environmental Humanities," *Environmental Communication* 12, no. 1 (2017): 7–14. https://doi.org/10.1080/17524032.2017.1367178.

Fisher, Mark. *Capitalist Realism: Is There No Alternative?* Winchester, UK: Zero Books, 2010.

"From Banks And Tanks To Cooperation and Caring: A Strategic Framework for a Just Transition." Berkeley, CA: Movement Generation, 2016.

Hausdoerffer, John, Brooke Parry Hecht, Melissa K. Nelson, and Katherine Kassouf Cummings. *What Kind of Ancestor Do You Want to Be?* Chicago, IL: University of Chicago Press, 2021.

Hawken, Paul. "You Are Brilliant, and the Earth Is Hiring," *NAMTA Journal* 38, no. 1, 2013. 269–271.

Hopkins, Rob. *The Transition Companion: Making Your Community More Resilient in Uncertain Times*. Cambridge, UK: Green Books, 2013.

Jensen, Derrick. "Beyond Hope," *Orion Magazine*, July 29, 2020. https://orionmagazine.org/article/beyond-hope/#:~:text=hope%20is%20a%20longing%20for%20a%20future%20condition%20over%20which%20you%20have%20no%20agency.

Krassner, Paul. "Why I'm Optimistic about the Future," *The 3rd Page*, 2003. https://www.emptymirrorbooks.com/thirdpage/krassner2.html.

Macy, Joanna. *Active Hope: How to Face the Mess We're in without Going Crazy.* Novato, CA: New World Library, 2020.

Miller, Henry, and Thomas H. Moore. *Henry Miller on Writing.* New York, NY: New Directions Press, 1964. 96.

Mooney, Chris. "Rex Tillerson's View of Climate Change: It's Just an 'Engineering Problem,'" *The Washington Post,* October 27, 2021. https://www.washingtonpost .com/news/energy-environment/wp/2016/12/13/rex- tillersons-view-of-climate -change-its-just-an-engineering-problem/?utm_term=.67efe02d753a.

Morgan, Piers, and David Frost. "Other," Piers Morgan Interviews David Frost. New York, New York: CNN, April 26, 2011.

Nietzsche, Friedrich Wilhelm, Helen Zimmern, and J. M. Kennedy. *Human, All-Too-Human; A Book for Spirits, Pt. 1.* Edinburgh, UK: T. N. Foulis, 1909. 87.

Sainato, Michael. "Stephen Hawking, Elon Musk and Jeff Bezos Think the Earth Is Doomed," *Observer,* June 30, 2017. https://observer.com/2017/06/colonizing -mars-elon-musk-stephen-hawking-jeff-bezos/.

Schneider, F., G. Kallis, and J. Martinez-Alier. 2010. "Crisis or opportunity? Economic degrowth for social equity and ecological sustainability," *Journal of Cleaner Production* 18(6) 511–18.

Scranton, Roy. "Learning How to Die in the Anthropocene," *The New York Times,* November 10, 2013. http://opinionator.blogs.nytimes.com/2013/11/10/learning -how-to-die-in-the-anthropocene/?_r=0.

Solnit, Rebecca. "Everything's Coming Together While Everything Falls Apart," *Le Monde Diplomatique,* December 23, 2014. https://mondediplo.com/openpage /everything-s-coming-together-while-everything.

Solnit, Rebecca. *Hope in the Dark: Untold Histories, Wild Possibilities.* New York, NY: Nation Books, 2004

Thomas-Muller, Clayton. "The Rise of the Native Rights-Based Strategic Framework," *Honor The Earth,* 2016. https://www.honorearth.org/the_rise_of_the _native_rights_based_strategic_framework.

Usher, Shaun. *Letters of Note: An Eclectic Collection of Correspondence Deserving of a Wider Audience.* San Francisco, CA: Chronicle Books, 2014. 10.

Ustinov, Peter. *Dear Me.* London, UK: Arrow Books Ltd, 2006.

Václav, Havel, and Karel Hvížďala, *Disturbing the Peace: A Conversation with Karel Hvížďala.* New York, NY: Vintage Books, 1991, 181–182.

Wheatley, Margaret. "Beyond Hope & Fear." Presentation at the Beyond Hope and Fear Workshop, Boulder, Colorado, Aug 12–14, 2016.

Williams, Jeremy. "Mitigation, Adaptation and Suffering," *The Earthbound Report,* September 13, 2021. https://earthbound.report/2019/10/31/mitigation -adaptation-and-suffering/.

Zinn, Howard. *You Can't Be Neutral on a Moving Train: A Personal History of Our Times.* Boston, MA: Beacon Press, 2018.

Notes

Prologue: It's the End of the World. Now What?

1. Paul Rogat Loeb, "The Impossible Will Take a While," *The Tyee*, October 11, 2004, https://thetyee.ca/Citizentoolkit/2004/10/11/TheImpossibleWillTake Awhile/.

2. "Majority of New Renewables Undercut Cheapest Fossil Fuel on Cost," IRENA, June 22, 2021, https://www.irena.org/newsroom/pressreleases/2021 /Jun/Majority-of-New-Renewables-Undercut-Cheapest-Fossil-Fuel-on -Cost#:~:text=%E2%80%9CToday%2C%20renewables%20are%20the%20 cheapest%20source%20of%20power%2C%E2%80%9D%20said%20IRENA %E2%80%99s%20Director%2DGeneral%20Francesco%20La%20Camera.

3. Eileen Myles, "Eileen Myles on the Pointless Demolition of Manhattan's East River Park," The online edition of Artforum International Magazine, December 17, 2021, https://www.artforum.com/slant/the-pointless-demolition-of -manhattan-s-east-river-park-87431.

4. The Climate Clock in New York's Union Square was co-created by Gan Golan, Andrew Boyd, Katie Peyton Hofstadter, and Adrian Carpenter in September 2020.

5. Oliver Milman, "Joe Manchin Leads Opposition to Biden's Climate Bill, Backed by Support from Oil, Gas and Coal," *The Guardian*, October 20, 2021, https://www.theguardian.com/us-news/2021/oct/20/joe-manchin-oil-and -gas-fossil-fuels-senator.

6. Bill McKibben, "We Need to Literally Declare War on Climate Change," *The New Republic*, August 15, 2016, https://newrepublic.com/article/135684 /declare-war-climate-change-mobilize-wwii.

7. Benjamin Hulac, "Pollution from Planes and Ships Left out of Paris Agreement," *Scientific American*, December 14, 2015, https://www.scientific american.com/article/pollution-from-planes-and-ships-left-out-of-paris -agreement/.

8. Carolyn Gramling, "Earth Will Blow Past Climate Targets Even with Current Pledges to Cut Emissions," *Science News*, October 26, 2021, https://www .sciencenews.org/article/climate-earth-warming-emissions-gap-pledges.

9. Peter Walker, "COP26 'Literally The Last Chance Saloon' to Save Planet— Prince Charles," *The Guardian*, October 31, 2021, https://www.theguardian .com/environment/2021/oct/31/cop26-literally-the-last-chance-saloon-to-save -planet-prince-charles.

10. United Nations Environment Programme (2019). Emissions Gap Report 2019. UNEP, Nairobi.

11. "Cut Global Emissions by 7.6 Percent Every Year for Next Decade to Meet 1.5°C Paris Target—UN Report," UN Environment, November 26, 2019, https://www.unep.org/news-and-stories/press-release/cut-global-emissions -76-percent-every-year-next-decade-meet-15degc.

12. Ayesha Jain, "'Architects of the Crisis': Fossil Fuel Industry Is Largest Delegation at COP26," *TheQuint*, November 10, 2021, https://www.thequint .com/climate-change/at-cop26-fossil-fuel-delegates-outnumber-any-single -national-delegation.

13. Shivani Singh, Aaron Sheldrick, and Noah Browning, "'Down' and 'out'? COP26 Wording Clouds Way Ahead on Climate," Reuters, November 15, 2021, https://www.reuters.com/business/cop/business-usual-global-fossil -fuel-firms-now-after-un-climate-deal-2021-11-15/.

14. Compassionate Nihilism was first developed in an earlier book, *Daily Afflictions*, by the author. Can-Do Pessimism is new to the current work. Tragic Optimism was originated by existential psychologist Victor Frankl decades ago, most famously in his bestselling *Man's Search for Meaning*, and takes on renewed significance in the context of the climate crisis. Andrew Boyd, *Daily Afflictions: The Agony of Being Connected to Everything in the Universe*, (New York, NY: W.W. Norton, 2002). Viktor E. Frankl, *Man's Search for Meaning*, (New York, NY: Pocket Books, 1984).

Chapter 1: Impossible News

1. As yet, we have no evidence of life elsewhere in the universe, never mind sentient life; however, in his 2018 book, *Searching for Stars on an Island in Maine*, Alan Lightman has suggested that "the odds that no life exists on the billions and billions of other habitable planets would be as improbable as no fires ever starting in a billion trillion dry forests." So, let's keep our eyes out and try to survive long enough to find out. Alan P. Lightman, *Searching for Stars on an Island in Maine*. (New York, New York: Vintage Books, 2019).

2. The author would like to thank Charles Eisenstein for inspiring this wise clarification of "we," which he deployed well on page 23 of *Climate: A New Story*. Charles Eisenstein, *Climate: A New Story*, (Sydney, Australia: Read-HowYouWant, 2019). 23.

3. "Ecological Footprint," Global Footprint Network, 2022, https://www.foot printnetwork.org/our-work/ecological-footprint/.

4. "How Many Earths? How Many Countries?," Earth Overshoot Day, March 9, 2022, https://www.overshootday.org/how-many-earths-or-countries-do-we -need/.

5. This quick summary of the Climate Change data and science is drawn from several sources, in particular: Oliver Milman, Alvin Chang, Rita Liu, and Andrew Witherspoon. "The Climate Disaster Is Here—This Is What the Future Looks Like," *The Guardian*, October 4, 2021. https://www.theguardian .com/environment/ng-interactive/2021/oct/14/climate-change-happening -now-stats-graphs-maps-cop26.

6. Nathaniel Rich, "Losing Earth: The Decade We Almost Stopped Climate Change," *The New York Times*. 2018, https://www.nytimes.com/interactive/2018/08/01/magazine/climate-change-losing-earth.html.

7. "That's How Fast the Carbon Clock Is Ticking," Remaining carbon budget. Mercator Research Institute on Global Commons and Climate Change (MCC), March 16, 2022, https://www.mcc-berlin.net/en/research/co2-budget .html.

8. Jonathan Watts et al., "Half World's Fossil Fuel Assets Could Become Worthless by 2036 in Net Zero Transition," *The Guardian*, November 4, 2021, https://www.theguardian.com/environment/ng-interactive/2021/nov/04 /fossil-fuel-assets-worthless-2036-net-zero-transition.

9. "Science," Climate Clock, March 2022, https://climateclock.world/science.

10. https://climateactiontracker.org/

11. Bill McKibben, "How the World Would Look in 2050 If We Solved Climate Change," *Time,* September 12, 2019, https://time.com/5669022/climate -change-2050/#:~:text=we%20won%E2%80%99t%20be%20able%20to%20 have%20civilizations%20like%20the%20ones%20we%E2%80%99re%20 used%20to.

12. Jeff Tolleson, "How hot will Earth get by 2100?" *Nature*, 22 April 2020, https://www.nature.com/articles/d41586-020-01125-x.

13. Naomi Klein, *This Changes Everything: Capitalism vs. the Climate*, (New York, New York: Simon & Schuster, 2015).

14. https://www.theguardian.com/environment/2020/may/08/sea-levels-could -rise-more-than-a-metre-by-2100-experts-say.

15. A study published in July 2015 "written by James Hansen, NASA's former lead climate scientist, and 16 co-authors, many of whom are considered among the top in their fields, concludes that glaciers in Greenland and Antarctica will melt 10 times faster than previous consensus estimates, resulting in sea level rise of at least 10 feet in as little as 50 years." "Hansen's study does not attempt to predict the precise timing of the feedback loop, only that it is 'likely' to occur this century. The implications are mindboggling: In the study's likely scenario, New York City—and every other coastal city on the planet—may only have a few more decades of habitability left." Eric Holthaus, "Earth's Most Famous Climate Scientist Issues Bombshell Sea Level Warning," *Slate*, July 20, 2015, https://slate.com/news-and-politics/2015/07/sea-level-study -james-hansen-issues-dire-climate-warning.html.

16. "The Beat of Courage; the Shape of Hope: Creative Resistance in the Face of Catastrophe," Labofii, May 14, 2014, https://labofii.wordpress.com /2014/05/09/the-beat-of-courage-the-shape-of-hope-creative-resistance-in -the-face-of-catastrophe/.

17. Amy Westervelt, "Big Oil's 'Wokewashing' Is the New Climate Science Denialism," *The Guardian*, September 9, 2021, https://www.theguardian .com/environment/2021/sep/09/big-oil-delay-tactics-new-climate-science -denial.

18. Paul Kingsnorth, *Confessions of a Recovering Environmentalist and Other Essays*, (Minneapolis, MN: Graywolf Press, 2017). 143.

19. Kingsnorth, *Confessions*, 143.

20. John Michael Greer served as Grand Archdruid of the Ancient Order of Druids in America (AODA) from 2003–2015. He's the author of many books, and blogs regularly at ecosophia.net. He lays out the core distinction between problems and predicaments in the appropriately named mini-essay, "Problems and Predicaments" in an August 31, 2006 posting to The Archdruid Report. John Michael Greer, "The Archdruid Report—Archive," The Archdruid Report: Problems and Predicaments, Ancient Order of Druids of America, August 31, 2006, https://thearchdruidreport-archive.200605.xyz /2006/08/problems-and-predicaments.html.

21. Greer, "Problems and Predicaments."

22. Greer, "Problems and Predicaments."

Interview: Dr. Guy McPherson

1. Guy McPherson's primary public platform is his blog Nature Bats Last. The evidence and arguments behind his claim that we are in the midst of abrupt climate change that will result in the near-term human extinction (NTHE) is summed up (and regularly updated) in a "monster climate-change essay" at the top of his site. Guy McPherson, "Climate-Change Summary," Nature Bats Last (Guy McPherson, August 2, 2016), https://guymcpherson.com/climate -chaos/climate-change-summary-and-update/.

2. Atmospheric scientist Michael Tobis and science educator Scott K. Johnson are two of the more notable examples of scientists who strongly dispute Guy McPherson's scientific conclusions: Scott K. Johnson, "How Guy McPherson Gets It Wrong," Fractal Planet, October 15, 2015, https://fractalplanet.word press.com/2014/02/17/how-guy-mcpherson-gets-it-wrong/. Michael Tobis, "McPherson's Evidence That Doom Doom Doom," Planet 3.0, March 13, 2014, http://www.planet3.org/2014/03/13/mcphersons-evidence-that-doom -doom-doom/index.html.

3. Albert Bates has been charting the views of notable climate thinkers along an axis from Ecotopia to Collapse at resilience.org. The "deep end" is literally where McPherson shows up. There have been seven iterations of his original diagram since its publication in 2014. Albert Bates. "Charting Collapseniks." Resilience. Post Carbon Institute, January 15, 2014. https://www.resilience .org/stories/2014-01-15/charting-collapseniks/. The diagrams have since been removed in the archived version of the original article but may be accessed below: https://damnthematrix.files.wordpress.com/2014/02/collapsenik.jpg.

4. All interviews in the book are heavily edited for length, clarity and style.

5. Dr. McPherson lays out the evidence and arguments for his claim of Near Term Human Extinction here: https://guymcpherson.com/extinction_fore told_extinction_ignored.

6. Changing Precipitation Regimes and Terrestrial Ecosystems: A North American Perspective, Edited by Jake F. Weltzin and Guy R. McPherson, University of Arizona Press, 2003.

7. T. J. Garrett, "No way out? the double-bind in seeking global prosperity alongside mitigated climate change," Earth System Dynamics 3, no. 1 (May 2012): pp. 1–17, https://doi.org/10.5194/esd-3-1-2012.

8. The author interviewed Guy McPherson via Skype on August 2, 2016.

Interview: Tim DeChristopher

1. "Posing as a Bidder, Utah Student Disrupts Government Auction of 150,000 Acres of Wilderness for Oil & Gas Drilling," *Democracy Now!*, December 22, 2008, https://www.democracynow.org/2008/12/22/posing_as_a_bidder _utah_student.
2. Power Shift Network, "Tim DeChristopher | Power Shift 2011 Keynote" Youtube Video, 00:08:27, April 17, 2011, https://m.youtube.com/watch?v=81 EZUkYzrxU.
3. The author interviewed Tim DeChristopher in New York City on July 25, 2016.
4. Terry Tempest Williams. "What Love Looks Like," *Orion Magazine*, December 2011, https://orionmagazine.org/article/what-love-looks-like.
5. My research did not turn up any publication where Dr. Root officially substantiated her (terrifying) claim that we are likely on track for 6°C warming, a projection that—according to 2022 IPCC climate modelling—is still an outlier, and near-worst-case scenario.
6. These attacks, as a 2012 Union of Concerned Scientists commissioned study has shown, have had a quantifiable impact on the way scientists report climate data. Francesca Grifo et al., "A Climate of Corporate Control: How Corporations Have Influenced the U.S. Dialogue on Climate Science and Policy," The Union of Concerned Scientists, May 2012, https://www.ucsusa .org/sites/default/files/2019-09/a-climate-of-corporate-control-report.pdf.
7. Michael E. Mann. "I'm a Scientist Who Has Gotten Death Threats. I Fear What May Happen Under Trump," *The Washington Post*, December 16, 2016, https://www.washingtonpost.com/opinions/this-is-what-the-coming-attack -on-climate-science-could-look-like/2016/12/16/e015cc24-bd8c-11e6-94ac-3d 324840106c_story.html?noredirect=on&utm_term=.58d1e1b5315d.
8. John H. Richardson, "When the End of Human Civilization Is Your Day Job," *Esquire*, July 20, 2018, https://www.esquire.com/news-politics/a36228/ballad -of-the-sad-climatologists-0815/.
9. Richardson, "Day Job."
10. Robinson Meyer. "Are We as Doomed as That *New York Magazine* Article Says?" *The Atlantic*, August 17, 2017, https://www.theatlantic.com/science /archive/2017/07/is-the-earth-really-that-doomed/533112/.
11. Meyer, "Doomed."
12. Nathan Thanki, "Fuck Your Apocalypse," *Medium*, published in The World At 1°C, July 11, 2017, https://worldat1c.org/fuck-your-apocalypse-c82696 b533d9.
13. "The Yale Program on Climate Change Communication," Yale Program on Climate Change Communication, February 15, 2022, https://climatecommuni cation.yale.edu/.
14. Geoff Blackwell and Ruth Anna Hobday, *200 Women: Who Will Change the Way You See the World*, (San Francisco, CA: Chronicle Books, 2019). 304.

Chapter 2: The Five Stages of Climate Grief

1. Per Espen Stoknes, *What We Think about When We Try Not to Think about Global Warming: Toward a New Psychology of Climate Action*, (White River Junction, VT: Chelsea Green Publishing, 2015).

2. Daniel Smith, "It's the End of the World as We Know It…and He Feels Fine," *The New York Times*, April 17, 2014, https://www.nytimes.com/2014/04/20/magazine/its-the-end-of-the-world-as-we-know-it-and-he-feels-fine.html.

3. Jill Hammer, "The Climate Ribbon As a Tree of Life." The Climate Ribbon, November 3, 2015, http://www.theclimateribbon.org/stories/2015/11/4/the-climate-ribbon-as-a-tree-of-life.

4. At our meeting, Guy said "we're already taking down with us a couple hundred species every day to extinction." That is the consensus view of the very highest estimate of the number of species lost daily. It's a hard number to pinpoint as the actual total number of species on Earth is unknown and estimates vary widely, from as few as two million to as many as 100 million species co-existing with us on Earth. What we *do* know is that "we're not only losing species at a much faster rate than we'd expect, we're losing them tens to thousands of times faster than the rare mass extinction events in Earth's history." For a deep dive into the numbers behind Earth's past five great extinctions, as well as our current man-made sixth, see the Extinctions/Biodiversity page of Our World in Data: Hannah Ritchie and Max Roser, "Extinctions," Our World in Data, April 15, 2021, https://ourworldindata.org/extinctions#are-we-heading-for-a-sixth-mass-extinction.

5. Norminton admits that, "this conceit [that we must walk our five stages of grief in reverse] may be nothing more than a provocation. Yet which of us, awake in the early hours, has not had similar misgivings?" Gregory Norminton, "Climate Change and the Five Stages of Grief," *Transition Free Press*, 2014, sec. Talkback, pp. 14–14, https://issuu.com/transitionfreepress/docs/tfp_2014-05_issuu_a01.

6. Mary Carson, "What Shell Knew about Climate Change in 1991—Video Explainer," *The Guardian*, February 28, 2017, https://www.theguardian.com/global/video/2017/feb/28/what-shell-knew-about-climate-change-in-1991-video-explainer.

7. Kathy Mulvey and Seth Shulman, "The Climate Deception Dossiers," The Union of Concerned Scientists, July 2015, 8. https://www.ucsusa.org/sites/default/files/attach/2015/07/The-Climate-Deception-Dossiers.pdf.

8. Much of the cataloguing of contradictory denialist rationales is drawn from the Global Warming rationalwiki page: "Global Warming," RationalWiki (RationalWiki, 2022), https://rationalwiki.org/wiki/Global_warming.

9. https://www.amazon.com/Merchants-Doubt-Handful-Scientists-Obscured/dp/1608193942.

10. Emma Green. "Half of Americans Think Climate Change Is a Sign of the Apocalypse," *The Atlantic*, November 22, 2014, http://www.theatlantic.com/politics/archive/2014/11/half-of-americans-think-climate-change-is-a-sign-of-the-apocalypse/383029/.

11. Donald Worster, *Dust Bowl: The Southern Plains in the 1930s* (New York, NY: Oxford University Press, 2012).

12. Hannah Hickey, "Limiting Warming to 2 C Requires Emissions Reductions 80% Above Paris Agreement Targets," UW News, February 9, 2021. https://www.washington.edu/news/2021/02/09/limiting-warming-to-2-c-requires-emissions-reductions-80-above-paris-agreement-targets/.

13. David L. Chandler, "Explaining the Plummeting Cost of Solar Power," MIT News, November 20, 2018, https://news.mit.edu/2018/explaining-dropping -solar-cost-1120.

14. "Renewable Electricity Use in Scotland: Eir Release," Scottish Government, January 12, 2022, https://www.gov.scot/publications/foi-202100253288/.

15. https://cleanenergynews.ihsmarkit.com/research-analysis/germany-launches -fullscale-renewable-power-transition-in-easte.html.

16. Bjorn Carey, "Stanford Engineers Develop State-by-State Plan to Convert U.S. to 100% Clean, Renewable Energy by 2050," Stanford News Release (Stanford University, June 8, 2015), https://news.stanford.edu/pr/2015/pr-50states -renewable-energy-060815.html.

17. Whit Gibbons, "The legend of the boiling frog is just a legend." Ecoviews 2007 (2002).

18. *Getting to Yes: Negotiating Agreement Without Giving In* is a best-selling 1981 nonfiction book by Roger Fisher and William Ury. Subsequent editions in 1991 and 2011 added Bruce Patton as co-author. All of the authors were members of the Harvard Negotiation Project. Robert Fisher, William Ury, and Bruce Patton, *Getting to Yes: Negotiating Agreement without Giving In*, (Harmondsworth: Penguin Books, 1986).

19. For more unexpectedly positive twists on depression, climate and otherwise, see: Joanna Macy. "Working through environmental despair," *Ecopsychology*: Restoring the earth, healing the mind 2, 1995. 40–259. Margaret Klein Salamon, "Opinion: Facing the Climate Emergency: Grieving the Future You Thought You Had," *Common Dreams*, June 23, 2019, https://www.common dreams.org/views/2019/06/23/facing-climate-emergency-grieving-future-you -thought-you-had.

20. Walter Benjamin, *Illuminations*, ed. Hannah Arendt (New York, NY: Harcourt, 1968).

21. Michael Winship, "Naomi Klein: Climate Change 'Not Just about Things Getting Hotter... It's about Things Getting Meaner,'" BillMoyers.com, February 3, 2016, https://billmoyers.com/story/naomi-klein-climate-change -not-just-about-things-getting-hotter-its-about-things-getting-meaner.

22. Damian Carrington, "From Congo Child Soldier to Award-Winning Wildlife Ranger—a Life in Danger," *The Guardian*, April 25, 2017, https:// www.theguardian.com/environment/2017/apr/24/from-congo-child-soldier -to-award-winning-wildlife-ranger-a-life-in-danger?CMP=Share_iOSApp _Other.

23. K.R. Anthony et al., "Ocean Acidification Causes Bleaching and Productivity Loss in Coral Reef Builders," *Proceedings of the National Academy of Sciences* 105, no. 45 (November 2008): pp. 17442–17446, https://doi.org/10.1073/pnas .0804478105. Will Steffen et al., "Trajectories of the Earth System in the Anthropocene," *Proceedings of the National Academy of Sciences* 115, no. 33 (June 2018): pp. 8252–8259, https://doi.org/10.1073/pnas.1810141115.

24. Institute for Economics and Peace, Global Peace Index 2021: Measuring Peace in a Complex World (Sydney: IEP, 2021).

25. Ben Smee, "Great Barrier Reef: 30% of Coral Died in 'Catastrophic' 2016 Heatwave," *The Guardian*, April 18, 2018, https://www.theguardian.com

/environment/2018/apr/19/great-barrier-reef-30-of-coral-died-in-catastrophic
-2016-heatwave.

26. Woody Guthrie's words deserve to be quoted in full here: "I hate a song that makes you think you are not any good. I hate a song that makes you think that you are just born to lose. Bound to lose. No good to nobody. No good for nothing. Because you are too old or too young or too fat or too slim. Too ugly or too this or too that. Songs that run you down or poke fun at you on account of your bad luck or hard traveling. I'm out to fight those songs to my very last breath of air and my last drop of blood. I am out to sing songs that will prove to you that this is your world and that if it has hit you pretty hard and knocked you for a dozen loops, no matter what color, what size you are, how you are built, I am out to sing the songs that make you take pride in yourself and in your work. And the songs that I sing are made up for the most part by all sorts of folks just about like you.

"I could hire out to the other side, the big money side, and get several dollars every week just to quit singing my own songs and to sing the kind that knock you down farther and the ones that poke fun at you even more and the ones that make you think you've not got any sense at all. But I decided a long time ago that I'd starve to death before I'd sing any such songs as that. The radio waves and your movies and your jukeboxes and your songbooks are already loaded down and running over with such no good songs as that anyhow."

"Woody Guthrie's Biography," the official Woody Guthrie website. Woody Guthrie Publications, Inc., n.d. https://www.woodyguthrie.org/biography /woodysez.htm.

27. Whether Mitchell's insight is itself gallows humor or actually meta-gallows humor, it remains a constant source of comfort to this author.

Interview: Dr. Margaret "Meg" Wheatley

1. The author participated in Meg Wheatley's Beyond Hope & Fear Workshop in Boulder, CO, Aug 12–14, 2016.
2. Several posts on the site NewAgeFraud.org, throw the provenance of the "There is a river…" prophecy from "Hopi Elders" into question. Here is the most relevant passage: "The Hopi made a statement a few years ago, that all this only pertained to the Hopi and had absolutely nothing to do with anyone else. This one attributed to 'The Elders,' is some say a Hopi Prayer. But it is in English. It started showing up around 1999 on the internet, a favorite of nuagers. The last line "We are the ones we been waiting for" has been attributed to so many people, it's hard to say. Some have attributed it to Twylan Nitsch, then picked up by nuagers. Some say the first was late poet June Jordan's 'Poem for South African Women.' That was about 1980. But since the Hopi have a very long oral history it is probably theirs. Now everyone is using it. It's a good example of exploitation." http://www.newage fraud.org/smf/index.php?topic=1890.0.
3. Chogyam Trungpa, author of the early 70s classic *Cutting Through Spiritual Materialism*, was celebrated in the 1970s and 1980s for making Tibetan

Buddhist teachings more accessible to Western audiences. He was also controversial for some of his more radical "crazy wisdom" practices which bordered on abuse. More concerningly, his appointed successor, his son Sakyong Mipham Rinpoche, who has run the Boulder Shambhala Center for years, has been accused of "financial mismanagement, verbal abuse, sexual abuse and physical assault." The author was unaware of these allegations at the time of his visit. "Opinion: Naropa Needs to Face the Skeleton in the Closet," *Boulder Beat*, January 3, 2021, https://boulderbeat.news/2020/06/12 /-opinion-naropa-needs-to-face-the-skeleton-in-the-closet/.

4. Buddhism's Second Noble Truth concerns the origins of suffering. In that context, the origin (Pali: samudaya) of suffering (Pali: dukkha) is commonly explained as craving (Pali: tanha) conditioned by ignorance (Pali: avijja). Donald S. Lopez and Brian Duignan, "Four Noble Truths," Encyclopædia Britannica (Encyclopædia Britannica, Inc., March 14, 2017), https://www .britannica.com/topic/Four-Noble-Truths.

Chapter 3: Existential Crisis Scenario Planning

1. "The Manifesto," Dark Mountain, June 12, 2018. https://dark-mountain.net /about/manifesto/.
2. David Loy, *Ecodharma Buddhist Teachings for the Ecological Crisis*, (Somerville, MA: Wisdom Publications, 2019). 137.
3. The news is dreary, yes, but the leap from massive changes to near-term human extinction is not one often made by climate scientists: Phoebe Weston, "Top Scientists Warn of 'Ghastly Future of Mass Extinction' and Climate Disruption," *The Guardian*, January 13, 2021, https://www.theguardian .com/environment/2021/jan/13/top-scientists-warn-of-ghastly-future-of-mass -extinction-and-climate-disruption-aoe.
4. Roy Scranton, "Learning How to Die in the Anthropocene," *The New York Times*, November 10, 2013, http://opinionator.blogs.nytimes.com/2013/11/10 /learning-how-to-die-in-the-anthropocene/?_r=0.
5. Tsunetomo Yamamoto and Barry D. Steben, *The Art of the Samurai: Yamamoto Tsunetomo's Hagakure* (London, UK: Duncan Baird Publishers, 2008).
6. Scranton, "Learning How to Die."
7. Derrick Jensen, "Beyond Hope," *Orion Magazine*, July 29, 2020, https://orion magazine.org/article/beyond-hope.
8. Eisenstein, *A New Story*, 74.
9. Scranton, "Learning How to Die."
10. Richard Heinberg, *Power Down: Options and Actions for a Post-Carbon World* (Gabriola Island, BC: New Society Publishers, 2005).
11. Richard Heinberg, "#204: Timing and the Post Carbon Manifesto," April 2009, https://richardheinberg.com/204-timing-and-the-post-carbon-manifesto.
12. David Fleming, Shaun Chamberlin, and Jonathon Porritt, *Lean Logic: A Dictionary for the Future and How to Survive It* (Hartford, VT: Chelsea Green Publishing , 2016). xix.
13. Fleming, *Lean Logic*, xviii.

14. Ibid.

15. Fleming, *Lean Logic*, xix.

16. Ibid.

17. Roger Atwood, "Organic or Starve: Can Cuba's New Farming Model Provide Food Security?," *The Guardian*, October 28, 2017, https://www.theguardian.com/environment/2017/oct/28/organic-or-starve-can-cubas-new-farming-model-provide-food-security. Carmen Diana Deere, "Reforming Cuban Agriculture," *Development and Change* 28, no. 4 (1997): pp. 649–669, https://doi.org/10.1111/1467-7660.00059.

18. John Michael Greer, *The Long Descent: A User's Guide to the End of the Industrial Age* (Danville, Illinois: Founders House, 2019).

19. Chris Mooney, "Rex Tillerson's View of Climate Change: It's Just an 'Engineering Problem'", *The Washington Post,* October 27, 2021, https://www.washingtonpost.com/news/energy-environment/wp/2016/12/13/rex-tillersons-view-of-climate-change-its-just-an-engineering-problem/?utm_term=.67efe02d753a.

20. Timothy Morton. *Hyperobjects: Philosophy and Ecology After the End of the World* (Minneapolis, MN: University of Minnesota Press, 2013).

21. "Climate Change 'Biggest Threat Modern Humans Have Ever Faced', World-Renowned Naturalist Tells Security Council, Calls for Greater Global Cooperation | Meetings Coverage and Press Releases," United Nations, February 23, 2021, https://www.un.org/press/en/2021/sc14445.doc.htm.

22. Horst W. Rittel and Melvin M. Webber, "Dilemmas in a General Theory of Planning," *Policy Sciences* 4, no. 2 (1973): pp. 155–169, https://doi.org/10.1007/bf01405730.

23. Jeffrey E. Conklin, *Dialogue Mapping: Building Shared Understanding of Wicked Problems* (Chichester: J. Wiley, 2006).

24. Michael Parrish, "New Buzz in Automobiles: The Industry as a Whole Is Beginning to Plug into Electric Cars," *Los Angeles Times*, December 2, 1991, https://www.latimes.com/archives/la-xpm-1991-12-02-fi-580-story.html.

25. "Gasoline Vehicle Phaseout Advances Around the World," Coltura, 2021, https://www.coltura.org/world-gasoline-phaseouts.

26. K. Levin, B. Cashore, Steven Bernstein, and G. Auld. "IOPscience," *IOP Conference Series: Earth and Environmental Science* (IOP Publishing, February 1, 2009), http://iopscience.iop.org/article/10.1088/1755-1307/6/50/502002/meta.

27. E. F. Schumacher, *A Guide for the Perplexed* (New York, NY: Harper Colophon, 1978).

28. The phrase "climate change" has long been mired in labelling warfare. As highlighted in the movie *Vice*, the Republican pollster Frank Luntz encouraged the George W. Bush administration to use the phrase "climate change" rather than "global warming." Matthew Cantor, "Could 'Climate Delayer' Become the Political Epithet of Our Times?," *The Guardian*, March 1, 2019, https://www.theguardian.com/environment/2019/mar/01/could-climate-delayer-become-the-olitical-epithet-of-our-times. Yale researchers recount a secret memo in which Luntz pointed out that a focus group participant

felt "climate change 'sounds like you're going from Pittsburgh to Fort Lauderdale,'" whereas "global warming has catastrophic connotations." Oliver Burkeman, "Memo Exposes Bush's New Green Strategy," *The Guardian*, March 4, 2003, https://www.theguardian.com/environment/2003/mar/04/usnews.climatechange.

29. "Why Global Warming Can Mean Harsher Winter Weather," *Scientific American*, February 25, 2009, https://www.scientificamerican.com/article/earthtalks-global-warming-harsher-winter/.

30. Jon Erdman, "New England Record Snow Tracker: Boston Breaks All Time Seasonal Snow Record in 2014–2015," The Weather Channel, June 25, 2015, https://weather.com/news/news/new-england-boston-record-snow-tracker.

31. The 2018 focus group which evaluated "Climate Chaos" as a catch-all term was quite informal, comprising myself, a bottle of wine, and my partner at the time, Katie Peyton Hofstadter. In fact, if one were to really be a stickler about it, one might call it a "dinner conversation" and not a focus group at all.

32. For a short glossary of the changes *The Guardian* made to their style guide, for use by their journalists and editors when writing about the environment, see both: Sophie Zeldin-O'Neill, "'It's a Crisis, Not a Change': The Six Guardian Language Changes on Climate Matters," *The Guardian*, October 16, 2019, https://www.theguardian.com/environment/2019/oct/16/guardian-language-changes-climate-environment. Damian Carrington, "Why the Guardian Is Changing the Language It Uses about the Environment," *The Guardian*, May 17, 2019, https://www.theguardian.com/environment/2019/may/17/why-the-guardian-is-changing-the-language-it-uses-about-the-environment.

33. Among the those who pioneered the term "Climate Emergency," few stand out more than the founder of The Climate Mobilization, Margaret Klein Salamon. Margaret Klein Salamon, "Introducing the Emergency Climate Movement," The Climate Psychologist, September 28, 2015, http://theclimatepsychologist.com/introducing-the-emergency-climate-movement/.

34. An emergency, by definition, ends. When does the climate emergency end? For more on the problems with this term, it's worth reading Dougald Hine's piece, Emergency Democracy. Dougald Hine, "Notes from Underground #4: Emergency Democracy," Bella Caledonia, December 5, 2019, https://bellacaledonia.org.uk/2019/12/05/notes-from-underground-4-emergency-democracy/.

35. Jordain Carney, "Schumer Calls for Biden to Declare Climate Emergency," The Hill, January 26, 2021, https://thehill.com/homenews/senate/535811-schumer-suggests-biden-should-declare-climate-emergency.

36. Klein, *This Changes Everything*, 1.

37. Gillian King, "Boxing a Glacier," March 29, 2012, http://thisnessofathat.blogspot.com/2012/03/boxing-glacier.html.

38. Gillian King, "Shell Pisses in the Swimming Pool," September 18, 2012, http://thisnessofathat.blogspot.com/2012/09/shell-pisses-in-swimming-pool.html.

39. Gillian King, "How Much Air in the Tank?," July 27, 2012, http://thisnessofathat.blogspot.com/2012/07/how-much-air-in-tank.html.

40. A paraphrase of the classic (and historically brutal) military phrase, "Kill them all; let God sort them out."

41. If there was a DEA equivalent to police America's oil addiction, it might look something like this: The Oil Enforcement Agency (OEA): https://www .youtube.com/watch?v=5DelJwNXv8Q. Larry Bogad. "Radical Simulacrum, Regulation By Prank: The Oil Enforcement Agency," *Contemporary Theatre Review* 17, no. 2 (2007): 261–264.

42. Michael E. Mann, *The New Climate War: The Fight to Take Back Our Planet* (New York, NY: PublicAffairs, 2021).

43. Adam Frank, "We Need a New Mythic-Scale Story to Tell about Climate Change," Literary Hub, https://lithub.com/we-need-a-new-mythic scale -story-to-tell-about-climate-change/.

44. Olivia Foster Rhoades, "Blue-Green Planet: It's a Cyanobacterial World, and We Just Live on It," *Science in the News* (Harvard University, January 27, 2020), https://sitn.hms.harvard.edu/flash/2019/blue-green-planet-its -a-cyanobacterial-world-and-we-just-live-on-it/#:~:text=Around%203%20 billion%20years%20ago,about%2020%25%20oxygen%20gas.

45. Klein, *This Changes Everything*, 21.

46. Klein, *This Changes Everything*, 5.

47. Gillian King, "After Pearl Harbour," June 8, 2012, http://thisnessofathat.blog spot.com/2012/06/after-pearl-harbour.html.

48. https://teachinghistory.org/history-content/ask-a-historian/24088.

49. King, "Pearl Harbour."

50. Bill McKibben, "We Need to Literally Declare War on Climate Change," *The New Republic*, August 15, 2016, https://newrepublic.com/article/135684 /declare-war-climate-change-mobilize-wwii.

51. Mckibben, "War on Climate Change." This section deserves to be quoted at length: "Unlike Adolf Hitler, the last force to pose a planet-wide threat to civilization, our enemy today is neither sentient nor evil. But before the outbreak of World War II, the world's leaders committed precisely the same mistake we are making today—they tried first to ignore their foe, and then to appease him.... But Hitler was playing by his own set of rules, which meant he had contempt for the political 'realism' of other leaders. (Indeed, it meant their realism wasn't.) Carbon and methane, by contrast, offer not contempt but complete indifference: They couldn't care less about our insatiable desires as consumers, or the sunk cost of our fossil fuel infrastructure, or the geo-strategic location of the petro-states, or any of the host of excuses that have so far constrained our response to global warming. The world came back from signing the climate accord in Paris last December exactly as Chamberlain returned from Munich: hopeful, even exhilarated, that a major threat had finally been tackled. [...But] This is, simply put, as wrong as Chamberlain's "peace in our time." Even if every nation in the world complies with the Paris Agreement, the world will heat up by as much as 3.5 degrees Celsius by 2100—not the 1.5 to 2 degrees promised in the pact's preamble."

52. Charles Eisenstein, *Climate: A New Story* (Sydney, Australia: ReadHowYou- Want, 2019). 49.

53. Paul Kingsnorth, "What If It's Not a War?," Dark Mountain, May 23, 2018), https://dark-mountain.net/what-if-its-not-a-war/.

54. Sharon Blackie, "Finding Our Way out of the Wasteland: Reclaiming the Voices of the Wells," sharonblackie.net, August 26, 2021, https://sharonblackie.net/finding-our-way-out-of-the-wasteland-reclaiming-the-voices-of-the-wells/.

55. In this personal email communication with the author, Heinberg goes on to say: "There may be some good points in the anti-vaxxers' screeds about the benefits of natural health measures to boost immune systems, the greed of the pharmaceutical companies, and the possibility that the virus escaped from a research lab. But even taking all that into account, all the 'Fauci is just like the Nazis' rhetoric is not just wrong, it's damaging to whatever is left of our social fabric."

56. Rebecca Solnit, *A Paradise Built in Hell: The Extraordinary Communities That Arise in Disaster* (New York, NY: Penguin Books, 2010).

57. Amanda Schupak, "Is Remote Working Better for the Environment? Not Necessarily," *The Guardian*, August 2, 2021, https://www.theguardian.com/environment/2021/aug/02/is-remote-working-better-for-the-environment-not-necessarily.

58. Jack Goodman and Flora Carmichael, "Coronavirus: Bill Gates 'microchip' conspiracy theory and other vaccine claims fact-checked," BBC News, May 30, 2020, https://www.bbc.com/news/52847648.

59. Gregor Aisch, Jon Huang, and Cecilia Kang, "Dissecting the #PizzaGate Conspiracy Theories," *The New York Times* , December 10, 2016, https://www.nytimes.com/interactive/2016/12/10/business/media/pizzagate.html.

60. Michael Chabon (@michaelchabon), "Today I was reading over…falling apart," *Twitter*, September 8, 2021. https://twitter.com/michaelchabon/status/1435454409899655170.

61. "This Is What 3°C of Global Warming Looks Like." *The Economist*, November 1, 2021, https://www.economist.com/films/2021/10/30/this-is-what-3degc-of-global-warming-looks-like.

62. "U.S. Coastline to See Up to a Foot of Sea Level Rise by 2050." National Oceanic and Atmospheric Administration (NOAA, February 15, 2022), https://www.noaa.gov/news-release/us-coastline-to-see-up-to-foot-of-sea-level-rise-by-2050.

63. Josh Holder, Niko Kommenda, and Jonathan Watts. "The Three-Degree World: Cities That Will Be Drowned by Global Warming," *The Guardian*, November 3, 2017, https://www.theguardian.com/cities/ng-interactive/2017/nov/03/three-degree-world-cities-drowned-global-warming.

64. World Bank Group. "Groundswell Report." World Bank, September 13, 2021, https://www.worldbank.org/en/news/press-release/2021/09/13/climate-change-could-force-216-million-people-to-migrate-within-their-own-countries-by-2050.

65. "UN Report: Nature's Dangerous Decline 'Unprecedented'; Species Extinction Rates 'Accelerating,'" United Nations Sustainable Development, United Nations, May 6, 2019, https://www.un.org/sustainabledevelopment/blog/2019/05/nature-decline-unprecedented-report.

Interview: Gopal Dayaneni

1. The author interviewed Gopal Dayaneni in Oakland, California, on Sept. 30, 2016.

2. At the time of the interview, the Movement Generation collective was currently in the process of documenting the ideas they'd been developing over the past several years into a new booklet that would come out at the end of 2016 called, "From Banks and Tanks to Cooperation and Caring: The Just Transition Framework." Our interview—fortuitously—served Gopal as an opportunity to "workshop" those ideas out loud in real time. The booklet can be viewed and downloaded here: https://movementgeneration.org/wp-content/uploads/2016/11/JT_booklet_Eng_printspreads.pdf.

3. The notion of "Peak Everything" is a concept shared with Movement Generation by one of its co-founders, Dave Henson of Occidental Arts and Ecology Center. This notion of "Peak Everything," and ecological limits, led to MG's strategic thinking about "peak incrementalism."

4. ETC Group monitors the impact of emerging technologies and corporate strategies on biodiversity, agriculture, and human rights. https://www.etcgroup.org.

5. The notion that all economies are nested in ecosystems was core to Gopal and MG's analysis. "If your economy extracts faster than the capacity of the system to regenerate, then your economy will collapse," Gopal said. "And the only way to extract faster than the system can regenerate is to further concentrate control over work—every kind of work, including when we take energy from the sun and convert it into the power to do work. It doesn't matter if those sun rays are eight minutes old when they hit the leaves (or a solar panel), or if the work the sun's rays did is millions of years old when we pull it out of the ground as a fossil fuel, it is still just the labor of the living world. By the same token, we humans are embedded in an economy. And in that sense, there are no 'natural' disasters. A 5.0 earthquake in San Francisco will rattle dishes, but a 5.0 earthquake in Bangladesh could collapse a six-story sweatshop and kill 1,500 people. We can only ever experience anything through an economy. (Which is itself embedded in ecology.)"

6. Gopal Dayaneni, Dave Henson, Michelle Mascarenhas-Swan, Jason Negrón-Gonzales, Mateo Nube, and Carla Perez, "Three Circles Strategy Chart," Movement Generation Justice & Ecology Project, 2010. https://joshuakahnrussell.files.wordpress.com/2008/10/three-circles-strategy.pdf.

7. At the time of the interview with Gopal, Movement Generation's first principle of Just Transition was "What the hands do, the heart learns." It has since been reformulated as, "What we feed grows," to avoid using an ableist metaphor.

8. "From Banks and Tanks to Cooperation and Caring," Movement Generation, https://movementgeneration.org/wp-content/uploads/2016/11/JT_booklet_Eng_printspreads.pdf.

9. It's possible that most of "what's been done to us," we've actually "done to ourselves." Then again, maybe not. To join the debate, see the chapter, "We have met the enemy and he is us. No, them! But also us. But mostly them."

10. Karl Marx, *The Eighteenth Brumaire of Louis Bonaparte*, translated from the German by Eden Paul and Cedar Paul (New York, NY: International Publishers, 1926).

11. Over May 16–19, 2019, Post Carbon Institute and Institute for Policy Studies convened 16 individuals at the Blue Mountain Center with a diverse range of relevant expertise to explore a profoundly urgent question: How can humanity navigate the interrelated energy, economic, environmental, and equity crises of the 21st century?

12. In *Man's Search for Meaning*, existential psychologist and Nazi death camp survivor Victor Frankl tells us: "Everything can be taken from a man [sic] but one thing: the last of human freedoms—to choose one's attitude in any given set of circumstances, to choose one's own way." Viktor E. Frankl, *Man's Search for Meaning* (New York, NY: Pocket Books, 1984).

13. Michael F. Leonen, "Etiquette for Activists," *YES! Magazine*, May 20, 2004, http://www.yesmagazine.org/issues/a-conspiracy-of-hope/etiquette-for-activists.

14. Some sharp thinking about the challenges of uniting together around a Green New Deal come from a draft document that was circulated to me in March of 2019 called "Shoring Ourselves Up For Cultural Work On The Green New Deal" with the following explanation and credit: "This is an open source draft document written collaboratively by: Desiree Kane, Gan Golan, Janice Gan, Jayeesha Dutta, Josh Yoder, Kate McNeely and Rachel Schragis. Vent Diagrams designed by Janice Gan, with Desiree Kane, Jordan Mudd, Ryan Madden and Rachel Schragis. Convened and edited by Rachel Schragis. For a digital version or questions email: schragis.rachel@gmail.com." It can be found at: https://www.gndcreate.org/GND_Cultural_Work.pdf.

Chapter 4: How to Be White at the End of the World

1. For interesting discussion on the provenance of this saying see: https://quoteinvestigator.com/2016/10/24/privilege.

2. Tracy Jan, "White Families Have Nearly 10 Times the Net Worth of Black Families. and the Gap Is Growing," *The Washington Post*, November 24, 2021, https://www.washingtonpost.com/news/wonk/wp/2017/09/28/black-and-hispanic-families-are-making-more-money-but-they-still-lag-far-behind-whites/?utm_term=.0b46cc838bb7.

3. Laura Mather. "Dear White Men: Five Pieces of Advice for 91 Percent of Fortune 500 CEOS," The Huffington Post, August 4, 2016, http://www.huffingtonpost.com/laura-mather/dear-white-men-seven-piec_b_7899084.html.

4. "Full List of Stuff White People Like," Stuff White People Like, March 14, 2013, https://stuffwhitepeoplelike.com/full-list-of-stuff-white-people-like.

5. https://www.russellsage.org/publications/parents-to-children.

6. "Nabila Espanioly (Israel)," WikiPeaceWomen English, PeaceWomen Across the Globe (PWAG), accessed March 9, 2022, https://wikipeacewomen.org/wpworg/en/?page_id=2957.

7. Christian Parenti, *Tropic of Chaos: Climate Change and the New Geography of Violence* (New York, NY: Nation Books, 2012).

Chapter 5: Is There Hope?

1. Rebecca Solnit, "Everything's Coming Together While Everything Falls Apart," *Le Monde Diplomatique*, December 23, 2014, https://mondediplo.com /openpage/everything-s-coming-together-while-everything.
2. Macy, *Active Hope*, 3.
3. Stoknes, *What We Think About.*
4. Stoknes, *What We Think About*, 221.
5. Václav Havel and Karel Hvížďala, *Disturbing the Peace: A Conversation with Karel Hvížďala* (New York, NY: Vintage Books, 1991), 181–182.
6. Piers Morgan and David Frost. *Other*. Piers Morgan Interviews David Frost. (New York, New York: CNN, April 26, 2011).
7. From a personal conversation with the author on September 10, 2018 in Santa Rosa, CA.
8. Kathleen Dean Moore, Mary DeMocker, and Brooke Parry Hecht, "If Your House Is on Fire: Kathleen Dean Moore on the Moral Urgency of Climate Change, 2021," Center for Humans and Nature, February 15, 2022, https:// humansandnature.org/if-your-house-is-on-fire-kathleen-dean-moore-on-the -moral-urgency-of-climate-change-2021/#:~:text=we%20respond%20to%20a -,lack%20of%20hope,-.%20A%20person%20could.
9. Rebecca Solnit, *Hope in the Dark: Untold Histories, Wild Possibilities* (Chicago, IL: Haymarket Books, 2016).
10. Solnit, *Hope in the Dark.*
11. Solnit, *Hope in the Dark.*
12. Solnit, *Hope in the Dark.*
13. Solnit, *Hope in the Dark.*
14. Howard Zinn, *You Can't Be Neutral on a Moving Train: A Personal History of Our Times* (Boston, MA: Beacon Press, 2018).
15. James Everette Richardson, *Vectors: Aphorisms & Ten-Second Essays* (Keene, NY: Ausable Press, 2001). 105.
16. Jim Wallis, "Spirituality Quotation," Spirituality and Practice, accessed March 13, 2022, https://www.spiritualityandpractice.com/quotes/quotations /view/23942/spiritual-quotation.
17. Jane Johnson, Richard Hummerstone, and Bruce Robertson, *Little Giant Encyclopedia* (New York, NY: Sterling Innovation, an imprint of Sterling Publishing Co., Inc., 2009). 373.
18. "Antonio Gramsci," in *Oxford Essential Quotations*, edited by Susan Ratcliffe, Oxford University Press, https://www.oxfordreference.com/view/10.1093/acre f/9780191843730.001.0001/q-oro-ed5-00018416.
19. Bill McKibben, "Winning Slowly is the Same as Losing," *Rolling Stone*, December 1, 2017, https://www.rollingstone.com/politics/politics-news/bill -mckibben-winning-slowly-is-the-same-as-losing-198205.

Interview: Joanna Macy

1. The author interviewed Joanna Macy in Berkeley, CA, on September 4, 2016.
2. Joanna Macy, *Active Hope: How to Face the Mess We're in without Going Crazy* (Novato, CA: New World Library, 2020). 2.

3. Macy, *Active Hope*, 3.
4. Loy, *Ecodharma*, 139.

Interview: Dr. Jamey Hecht

1. The author interviewed Jamey Hecht in Los Angeles, CA, on September 6, 2016.
2. Peter Fimrite, "Great Pacific Garbage Patch Is Now Nearly 4 Times the Size of California." *San Francisco Chronicle*, March 26, 2018, https://www.sfchronicle.com/nation/article/Huge-garbage-sprawl-in-Pacific-ocean-is-much-12773818.php.
3. Scott Wilson, "Decoding the Redwoods," *The Washington Post*, February 7, 2018. https://www.washingtonpost.com/news/national/wp/2018/02/07/feature/as-climate-changes-threaten-californias-giant-redwoods-the-key-to-their-salvation-might-be-within-them/?noredirect=on&utm_term=.70848f7c9f07.
4. Phil Ochs, "The World Began In Eden And Ended In Los Angeles." Track 7, *Rehearsals for Retirement*. A&M Records, 1969, Vinyl.
5. "Progress Cleaning the Air and Improving People's Health," Environmental Protection Agency, March 9, 2022, https://www.epa.gov/clean-air-act-overview/progress-cleaning-air-and-improving-peoples-health.
6. Gabrielle Canon, "Los Angeles' climate future hangs in the balance as city votes for new mayor," *The Guardian*, June 2, 2022. https://www.theguardian.com/us-news/2022/jun/02/california-los-angeles-mayor-elections-climate
7. "100% Clean Energy (SB100)," focus.senate.ca.gov (California State Senate, September 7, 2017), https://focus.senate.ca.gov/sb100/faqs.
8. Jamey Hecht, "5 Reasons Why Some People Insist on Discussing Collapse, and Even Extinction," Nature Bats Last, September 26, 2014, https://guymcpherson.com/5-reasons-why-some-people-insist-on-discussing-collapse-and-even-extinction/.
9. Hecht, "5 Reasons."
10. Jamey Hecht, "Collapse Awareness and the Tragic Consciousness," Carolyn Baker, July 7, 2013, https://carolynbaker.net/2013/07/07/collapse-awareness-and-the-tragic-consciousness-by-jamey-hecht/.
11. Hecht, "Collapse Awareness."
12. Hecht, "Collapse Awareness."
13. Hecht, "Collapse Awareness."
14. Hecht, "Collapse Awareness."
15. Hecht, "Collapse Awareness."
16. Hecht, "Collapse Awareness."
17. Biochar is a promising carbon-capture technology that is "produced from biomass sources [i.e., wood chips, plant residues, manure or other agricultural waste products] for the purpose of transforming the biomass carbon into a more stable form (carbon sequestration)." Kurt Spokas, "Biochar," ARS/USDA, March 3, 2020, https://www.ars.usda.gov/midwest-area/stpaul/swmr/people/kurt-spokas/biochar/.
18. William James, *The Principles of Psychology* (New York: H. Holt, 1890).

19. From the author's interview with Jamey Hecht: "The book that I read that made me most liberated to make this claim is a strange and marvelous work by English neuroscientist Iain McGilchrist called *The Master and His Emissary.*"
20. Hecht, "5 Reasons."

Chapter 6: What Is Still worth Doing?

1. Daniel Smith, "It's the End of the World as We Know It…and He Feels Fine," *The New York Times*, April 17, 2014, https://www.nytimes.com/2014/04/20/magazine/its-the-end-of-the-world-as-we-know-it-and-he-feels-fine.html.
2. https://www.theguardian.com/world/2013/dec/20/brazil-salutes-chico-mendes 25 years-after-murder.
3. https://navdanyainternational.org/our-staff/vandana-shiva.
4. Smith, "End of The World."
5. "The Manifesto." Dark Mountain, June 12, 2018. https://dark-mountain.net/about/manifesto/.
6. "The Manifesto."
7. Paul Kingsnorth, "Dark Ecology," *Orion Magazine*, July 29, 2012, https://orionmagazine.org/article/dark-ecology/.
8. Kingsnorth, "Dark Ecology."
9. Kingsnorth, "Dark Ecology."
10. Kingsnorth, "Dark Ecology."
11. Kingsnorth, "Dark Ecology."
12. Kingsnorth, "Dark Ecology."
13. Kingsnorth, "Dark Ecology."
14. Kingsnorth, "Dark Ecology."
15. Kingsnorth, "Dark Ecology."
16. "The Decisive Decade for Climate Change," *The Economist*, https://www.economist.com/films/2021/09/27/the-decisive-decade-for-climate-change. Matt McGrath, "Biden: This will be 'decisive decade' for tackling climate change," BBC News, April 22, 2021, https://www.bbc.com/news/science-environment-56837927. Joyashree Roy et. al, *Critical Junctions on the Journey to 1.5°C: The Decisive Decade*, Climate Strategies, https://mission2020.global/wp-content/uploads/2021/05/Critical-Junctions-on-the-Journey-to-1.5C.pdf.
17. For an ever growing toolbox of just transition solutions, see: https://beautifulsolutions.info.
18. Lew Daly, *False Solutions*, NYRenews, https://static1.squarespace.com/static/58ae35fddb29d6acd5d7f35c/t/60351d79b4a58450d1f9dd8b/1614093694407/False+Solutions+Report+-+FINAL.pdf.
19. Ibid.
20. Ibid.
21. Ibid.
22. Jessica McKenzie, "Interview: Author and activist Bill McKibben on the 'timed test' of the climate crisis," *Bulletin of the Atomic Scientists*, December 16, 2021, https://thebulletin.org/2021/12/interview-author-and-activist-bill-mckibben-on-the-timed-test-of-the-climate-crisis.
23. McKibben, "Winning Slowly."

24. UN News, "IPCC report: 'Code red' for human driven global heating, warns UN chief," https://news.un.org/en/story/2021/08/1097362.

25. The generational dimension of the "what does it mean to be 'too late'" debate was taken up in a New York Times profile piece on Gen Z climate activists here: Cara Buckley, "'OK Doomer' and the Climate Advocates Who Say It's Not Too Late," *New York Times*, March 22, 2022, www.nytimes.com/2022/03/22/climate/climate-change-ok-doomer.html.

26. George Monbiot, "We need optimism—but Disneyfied climate predictions are just dangerous," *The Guardian*, May 13, 2022, www.theguardian.com/commentisfree/2022/may/13/optimism-climate-predictions-techno-polluters?CMP=Share_iOSApp_Other.

27. Ibid.

28. Umair Irfan, "This is what we need to invent to fight climate change," *Vox*, May 8, 2022, www.vox.com/23042818/climate-change-ipcc-wind-solar-battery-technology-breakthrough.

29. Neel Dhanesha, "There's a climate solution hiding in our walls," *Vox*, April 20, 2022, www.vox.com/recode/23016732/climate-change-buildings-insulation.

30. Jules Seidenburg, "'No Regrets' Options," ResearchGate, January 2012, www.researchgate.net/publication/269872641_'No_Regrets'_Options.

31. "Drawdown Solutions," Project Drawdown, September 1, 2021, www.drawdown.org/solutions.

32. Rob Hopkins, *The Transition Companion: Making Your Community More Resilient in Uncertain Times*, (Cambridge, UK: Green Books, 2013).

33. "Jackson, Miss. Mayor-Elect Chokwe Lumumba: I Plan to Build the 'Most Radical City on the Planet,'" Democracy Now!, June 26, 2017, https://www.democracynow.org/2017/6/26/jackson_miss_mayor_elect_chokwe_lumumba.

34. "Cooperation Jackson," Cooperation Jackson, accessed March 9, 2022, https://cooperationjackson.org.

35. "The Leap Manifesto," The Leap Manifesto (*This Changes Everything*, 2015), https://leapmanifesto.org/en/the-leap-manifesto.

36. Benedict Clements, Sanjeev Gupta, and Jianhong Liu, "$57 Trillion Additional Climate Debt Calls for Policy Action by G20," Center for Global Development, www.cgdev.org/blog/57-trillion-additional-climate-debt-calls-policy-action-g20.

37. Jason Hickel, "Clean Energy Won't Save Us—Only a New Economic System Can." *The Guardian*, July 15, 2016, www.theguardian.com/global-development-professionals-network/2016/jul/15/clean-energy-wont-save-us-economic-system-can.

38. Samuel Alexander, "We Need Economic 'Degrowth' to Stop a Carbon Budget Blowout," The Conversation, September 18, 2014, http://theconversation.com/we-need-economic-degrowth-to-stop-a-carbon-budget-blowout-31228. "Economic growth is incompatible with the rapid emissions reductions that are now necessary." Dr Samuel Alexander is a lecturer with the Office for Environmental Programs, University of Melbourne, teaching a course called Consumerism and the Growth Economy: Interdisciplinary Perspectives in

the Masters of Environment program. He is also a research fellow with the
Melbourne Sustainable Society Institute and co-director of the Simplicity
Institute. In 2015 he published two books of collected essays, *Prosperous
Descent: Crisis as Opportunity in an Age of Limits* and *Sufficiency Economy:
Enough, for Everyone, Forever*, both available from the Simplicity Institute
publications page. His other books include *Simple Living in History: Pioneers
of the Deep Future, Entropia: Life Beyond Industrial Civilisation*, and *Volun-
tary Simplicity: The Poetic Alternative to Consumer Culture.*

39. Samuel Alexander, "What is Degrowth? Envisioning a Prosperous Descent,"
    www.countercurrents.org/aexander031115.htm.
40. https://leanlogic.online/glossary/lean-economy
41. Nate Higgins, "The Great Simplification: Evolving Beyond Our Extraction
    Addiction," *Kosmos journal for global transformation*, https://www.kosmos
    journal.org/kj_article/great-simplification.
42. François Schneider, Giorgos Kallis, and Joan Martinez-Alier, "Crisis or
    Opportunity? Economic Degrowth for Social Equity and Ecological Sustain-
    ability. Introduction to This Special Issue," *Journal of Cleaner Production* 18,
    no. 6 (2010): pp. 511–518, https://doi.org/10.1016/j.jclepro.2010.01.014.
43. Alexander, "What is Degrowth?"
44. Samuel Alexander, "Life in a 'Degrowth' Economy, and Why You Might
    Actually Enjoy It." The Conversation, October 1, 2014, https://theconversa
    tion.com/life-in-a-degrowth-economy-and-why-you-might-actually-enjoy
    -it-32224.
45. Jem Bendell, "Deep adaptation: A map for navigating climate tragedy,
    Ambleside, UK: IFLAS Occasional Paper 2," (2018): 1–31.
46. The Deep Adaptation Forum, a network and social movement in formation,
    based on Jem Bendel's Deep Adaptation framework, can be found at: www
    .deepadaptation.info/the-deep-adaptation-forum-daf.
47. https://www.peterfiekowsky.com/book.
48. "Geoengineering: A Horrible Idea We Might Have to Do," www.youtube.com
    /watch?v=dSu5sXmsur4.
49. Fossil fuel air pollution responsible for 1 in 5 deaths worldwide," Harvard
    School of Public Health, 2021, https://www.hsph.harvard.edu/c-change
    /news/fossil-fuel-air-pollution-responsible-for-1-in-5-deaths-worldwide/.
50. As podcaster Amy Westervelt said in a January 27 2021 tweet that's been
    since deleted: "We're at a moment in time where radical change is necessary
    because those in power refused to make incremental change over the last 20
    years. That's as true on minimum wage as it is on climate." https://twunroll
    .com/article/1354513768030457857#.
51. The debate about the benefits of recycling has been long-running. In one
    episode, an October 2015 NYT Op-Ed by John Tierney. John Tierney, "The
    Reign of Recycling," *The New York Times*, October 3, 2015, https://www.ny
    times.com/2015/10/04/opinion/sunday/the-reign-of-recycling.html?_r=0. was
    strongly rebutted by Ben Adler at *Grist Magazine*. Ben Adler, "Is Recycling as
    Awful as the New York Times Claims? Not Remotely," *Grist*, October 6, 2015,
    https://grist.org/climate-energy/is-recycling-as-awful-as-the-new-york-times
    -claims-not-remotely/?utm_source=twitter&utm_medium=tweet&utm

_campaign=socialflow with a 20-year back-story provided by Boulder, CO-based zero waste advocates Eco-Cycle (http://www.ecocycle.org/recycling matters).

52. Karen McVeigh, "World Leaders Agree to Draw up 'Historic' Treaty on Plastic Waste," *The Guardian*, March 2, 2022, https://www.theguardian.com/environment/2022/mar/02/world-leaders-agree-draw-up-historic-treaty-plastic-waste.

53. Matthew Taylor, "Six Promises You Can Make to Help Reduce Carbon Emissions," *The Guardian*, March 7, 2022, https://www.theguardian.com/environment/2022/mar/07/six-promises-you-can-make-to-help-reduce-carbon-emissions.

54. Amanda Schupak, "Climate-friendly diets can make a huge difference—even if you don't go all-out vegan," *The Guardian*, June 4, 2022, https://www.theguardian.com/environment/2022/jun/04/meat-diets-climate-emissions-plant-based-vegan.

55. Taylor, "Six Promises."

56. Taylor, "Six Promises."

57. Jonathan Safran Foer, *Eating Animals* (London: Penguin Books, 2018).

58. Tania Bryer, "World Spends $423 Billion a Year to Subsidize Fossil Fuels, UN Research Says," CNBC, October 29, 2021, https://www.cnbc.com/2021/10/29/world-spends-423-billion-a-year-to-subsidize-fossil-fuels-un.html.

59. Some of the "Team Us vs. Team Them" framing adapted from material previously written by the author and others for the Beautiful Trouble Study Guide and the Beautiful Trouble Strategy Card Deck, specifically, the Debate: "The problem is inside ourselves vs. The problem is in the world around us." More information at http://beautifultrouble.org/game.

60. Personal conversation between Paul Kingsnorth and the author. Brooklyn, August 10, 2017.

61. Eisenstein, *A New Story*, 17.

62. Eisenstein, *A New Story*, 13.

63. "Developed Countries Are Responsible for 79 Percent of Historical Carbon Emissions," Center For Global Development, 2011, https://www.cgdev.org/media/who-caused-climate-change-historically.

64. According to an EU report, the average American emits 16.1 tons of carbon, while the average Ugandan, only 0.1 tons. Monica Crippa, Guizzardi Oreggioni, Diego Guizzardi, Marilena Muntean, Edwin Schaaf, Eleonora Lo Vullo, Efisio Solazzo, Fabio Monforti-Ferrario, Jos GJ Olivier, and Elisabetta Vignati, "Fossil CO_2 and GHG emissions of all world countries," Publication Office of the European Union: Luxemburg (2019).

65. Personal conversation between Paul Kingsnorth and the author. Brooklyn, August 10, 2017.

66. Personal conversation between Josephine Ferorelli and the author. Chicago, March 15, 2017.

67. Eisenstein, *A New Story*, 5.

68. Personal conversation between Patrick Reinsborough and the author. New York, April, 2019.

69. David Hasemyer, John H. Cushman, and Neela Banerjee, "CO2's Role in Global Warming Has Been on the Oil Industry's Radar since the 1960s," Inside Climate News, November 30, 2020, https://insideclimatenews.org /news/13042016/climate-change-global-warming-oil-industry-radar-1960s -exxon-api-co2-fossil-fuels/. To further explore Exxon's mendacity, and for action steps to hold them to account: https://exxonknew.org.

70. Chris McGreal, "'What we now know…they lied': how big oil companies betrayed us all," *The Guardian*, April 21, 2022, https://www.theguardian .com/tv-and-radio/2022/apr/20/what-we-now-know-they-lied-how-big-oil -companies-betrayed-us-all.

71. Damian Carrington and Matthew Taylor, "Revealed: the 'carbon bombs' set to trigger catastrophic climate breakdown," *The Guardian*, May 11, 2022, https://www.theguardian.com/environment/ng-interactive/2022/may/11 /fossil-fuel-carbon-bombs-climate-breakdown-oil-gas.

72. James Kanter and Andrew C. Revkin, "World Scientists near Consensus on Warming," *The New York Times*, January 30, 2007, https://www.nytimes.com /2007/01/30/world/30climate.html.

73. NASA, "Responding to Climate Change," https://climate.nasa.gov/solutions /adaptation-mitigation.

74. Fiona Harvey, "We cannot adapt our way out of climate crisis, warns leading scientist," *The Guardian*, June 1, 2022, https://www.theguardian.com/environ ment/2022/jun/01/we-cannot-adapt-our-way-out-of-climate-crisis-warns -leading-scientist.

Interview: adrienne maree brown

1. A portion of the background information for adrienne was sourced in her own words from this article: adrienne maree brown, "Change Is Divine: How Sci Fi Visionary Octavia Butler Influenced This Detroit Revolutionary," *YES! Magazine*, November 26, 2019, https://www.yesmagazine.org/issue /storytelling/2014/06/28/how-sci-fi-visionary-octavia-butler-influenced-this -detroit-revolutionary.

2. brown, "Change is Divine."

3. brown, "Change is Divine."

4. adrienne maree brown, "Adrienne Marie Brown—Articulate Compelling Vision," CommonBound by the New Economy Coalition (June 22–24, St. Louis, MO), 2016, https://commonbound.org/2016/adrienne-marie-brown -articulate-compelling-vision.

5. The author interviewed adrienne maree brown via phone on September 10, 2019.

6. The notion that access is something we create together was pioneered by Leah Lakshmi Piepzna-Samarasinha and participants in the Allied Media Projects, as an approach to collective care, and providing equal access to people with disabilities, but it applies just as much to how we show up for each other as things get worse. "Creating Collective Access," Creating Collective Access, June 30, 2012, https://creatingcollectiveaccess.wordpress.com/.

7. adrienne maree brown and Walidah Imarisha, *Octavia's Brood: Science Fiction Stories from Social Justice Movements* (Oakland, CA: AK Press, 2015). 279.

8. Kim Stanley Robinson, *The Ministry for the Future* (New York, NY: Orbit, 2021).

9. A term evidently coined by my on-again-off-again colleague and adrienne's teacher-friend Terry Marshall: adrienne maree brown, "St. Louis Racial Equity Summit 2021 Keynote," adrienne maree brown, August 6, 2021, https://adriennemareebrown.net/2021/08/06/st-louis-racial-equity-summit-2021-keynote/.

10. brown, "Equity Summit."

11. John Michael Greer, *Retrotopia* (Middletown, DE: Founders House Publishing, 2016).

12. Joshua Rothman, "Kim Stanley Robinson's Latest Novel Imagines Life in an Underwater New York," *The New Yorker*, April 27, 2017), https://www.newyorker.com/books/page-turner/kim-stanley-robinsons-latest-novel-imagines-life-in-an-underwater-new-york.

13. For more information on the Global Seed Vault: "Svalbard Global Seed Vault," Crop Trust, August 23, 2021, https://www.croptrust.org/our-work/svalbard-global-seed-vault/.

14. Damian Carrington, "Arctic Stronghold of World's Seeds Flooded after Permafrost Melts," *The Guardian,* May 19, 2017, https://www.theguardian.com/environment/2017/may/19/arctic-stronghold-of-worlds-seeds-flooded-after-permafrost-melts.

15. Fossil fuel air pollution responsible for 1 in 5 deaths worldwide," Harvard School of Public Health, 2021, https://www.hsph.harvard.edu/c-change/news/fossil-fuel-air-pollution-responsible-for-1-in-5-deaths-worldwide/.

16. brown and Imarisha, *Octavia's Brood*, 279.

17. Octavia E. Butler, *Parable of the Sower* (New York, NY: Grand Central Publishing, 2019).

18. Mark Fisher, *Capitalist Realism: Is There No Alternative?* (Winchester, UK: Zero Books, 2010).

19. Mark Fisher, "Exiting the Vampire Castle," openDemocracy, November 24, 2013, https://www.opendemocracy.net/en/opendemocracyuk/exiting-vampire-castle/.

20. School of Visual Arts, "David Graeber: On Bureaucratic Technologies and the Future as Dream-Time " Youtube Video, 01:26:12, May 27, 2021, https://www.youtube.com/watch?v=4Q84ar89Oxo.

21. The poet of utopia in question is Uruguayan journalist, novelist, essayist Eduardo Galeano (1940–2015). Here he is, sharing his beautiful explanation of the uses of utopia live, in his native toungue: https://www.youtube.com/watch?v=jsSqpATmAmU.

22. Eduardo Galeano, Jose Francisco Borges, and Mark Field, *Walking Words* (New York, NY, New York: W.W. Norton, 1997). 326.

23. School of Visual Arts, "David Graeber: On Bureaucratic Technologies and the Future as Dream-Time " Youtube Video, 01:26:12, May 27, 2021, https://www.youtube.com/watch?v=4Q84ar89Oxo.

24. The vibrancy of Indigenous Futurism is heralded by two recent anthologies, 2012's *Walking the Clouds: An Anthology of Indigenous Science Fiction* and 2020's *Love after the End: An Anthology of Two-Spirit and Indigiqueer*

Speculative Fiction. Grace L. Dillon, *Walking the Clouds an Anthology of Indigenous Science Fiction* (Tucson, AZ: University of Arizona Press, 2012). Joshua Whitehead, *Love After the End* (Vancouver, BC: Arsenal Pulp Press, 2020). Ursula Le Guin (though obviously not Indigenous, but certainly anti-colonial and intersectional) makes some similar moves to Indigenous Futurism in her novel, *Always Coming Home.* Ursula K. Le Guin, *Always Coming Home* (New York, New York: Harper & Row, Publishers, 1985).

25. "NDN" is a shortening and reclaiming of "Indian," sometimes used by Native Americans in the USA to refer to themselves. "NDN time" has been described as a conception of time where "the past, present and future are simultaneous." The NDN Collective (ndncollective.org) is an Indigenous-led organization dedicated to building Indigenous power. Also see following citation.

26. Daniel Heath Justice, review of *Love After the End*, by Joshua Whitehead, *Arsenal Pulp Press*, 2021. https://arsenalpulp.com/Books/L/Love-after-the -End.

27. In a footnote on page 27 of *Emergent Strategy*, adrienne maree brown says: "'Visionary fiction' is a term coined by my *Octavia's Brood* co-editor to describe the work of people who use fiction to advance justice and liberation." adrienne maree brown, *Emergent Strategy: Shaping Change, Changing Worlds* (AK Press, 2021). The book-jacket copy of *Octavia's Brood* states: "Whenever we envision a world without war, without prisons, without capitalism, we are producing visionary fiction." adrienne maree brown and Walidah Imarisha, *Octavia's Brood: Science Fiction Stories from Social Justice Movements* (Oakland, CA: AK Press, 2015).

28. Alexandra Rowland, who coined the term "hopepunk," explains its meaning thusly: In July of 2017, I coined the word "hopepunk," initially defined very simply in a Tumblr post: "The opposite of grimdark is hopepunk. Pass it on." When asked to clarify, I wrote: "The essence of grimdark is that everyone's inherently sort of a bad person and does bad things, and that's awful and disheartening and cynical. It's looking at human nature and going, 'The glass is half empty.' 'Hopepunk says, 'No, I don't accept that. Go fuck yourself: The glass is half full.' Yeah, we're all a messy mix of good and bad, flaws and virtues. We've all been mean and petty and cruel, but (and here's the important part) we've also been soft and forgiving and kind. Hopepunk says that kindness and softness doesn't equal weakness, and that in this world of brutal cynicism and nihilism, being kind is a political act. An act of rebellion." Alexandra Rowland. "One Atom of Justice, One Molecule of Mercy, and the Empire of Unsheathed Knives," The Stellar Beacon, 2019. Mirriam-Webster, sifting through various "punk"-suffixed genres and nomenclatures, notes (somewhat awkwardly) that hopepunk "finds its narrative motivation in the idea of optimism—embodied in acts of love, kindness, and respect for one another—as resistance." And speculates further that: The –punk suffix handed down from cyberpunk…hints toward the alienation and anti-authoritarian sentiment that is often manifested in the punk subculture. By attaching the suffix to subjects such as hope, kindness, and respect, those subjects—perhaps once considered very much as belonging

to the establishment—can be portrayed as instruments of change. "What Do 'Hopepunk' and 'Mannerpunk' Mean?," Merriam-Webster, accessed March 9, 2022, https://www.merriam-webster.com/words-at-play/what-is-hopepunk -mannerpunk-words-were-watching.

29. Greer, *Retrotopia*.

30. Editor of the 2018 solarpunk anthology *Glass and Gardens*, Sarena Ulibarri, describes the genre as "a mode of imagining futures in which we have made the big structural changes we need to avoid more ecological damage." And even after having "adapted to the changes we can't prevent," solarpunk narratives stay optimistic, imagining "a better future, not just for the wealthy and privileged, but for all of humanity and our whole interconnected ecosystem." Julia Goldberg, "Imagine There's No Climate Crisis," *Santa Fe Reporter*, May 29, 2019, https://www.sfreporter.com/news/theinterface/2019/05/29/imagine -theres-no-climate-crisis/.

31. Saint Andrewism, "What is Solarpunk?," Youtube Video, 00:08:30, December 16, 2020, https://www.youtube.com/watch?v=hHI61GHNGJM.

32. Personal conversation between Paul Kingsnorth and the author. Brooklyn, August 10th, 2017

33. Naomi Klein and Molly Crabapple, "A Message from the Future with Alexandria Ocasio-Cortez," The Intercept, April 17, 2019, https://theintercept .com/2019/04/17/green-new-deal-short-film-alexandria-ocasio-cortez/.

34. Klein and Crabapple, "A Message."

35. Klein and Crabapple, "A Message."

36. Klein and Crabapple, "A Message." The video was inspired by an earlier article by Kate Aronoff: Kate Aronoff, "With a Green New Deal, Here's What the World Could Look Like for the Next Generation," The Intercept, December 5, 2018, https://theintercept.com/2018/12/05/green-new-deal-proposal-impacts/.

37. "Naomi Klein Sunrise Tour Last Stop Washington DC May 13 2019." C-SPAN Video. 00:08:49. May, 13, 2019. https://www.c-span.org/video/?c4797994 /naomi-klein-sunrise-tour-stop-washington-dc-13-2019.

38. Geoff Dembicki. "Alexandria Ocasio-Cortez's Green New Deal: The Canadian Connection," The Tyee, June 3, 2019, https://thetyee.ca/News /2019/06/03/AOC-Boosting-Green-New-Deal-Canadians-Helped-Craft/.

Interview: Dr. Robin Wall Kimmerer

1. "Nature Deficit Disorder" is a term coined by environmental journalist Richard Louv to describe the human costs of separation from Nature. He discusses the problem here: Richard Louv, "What Is Nature Deficit Disorder." Weblog. Richardlouv.com, October 15, 2015. http://richardlouv.com/blog /what-is-nature-deficit-disorder/. And possible solutions here: Richard Louv. "No More 'Nature-Deficit Disorder,'" *Psychology Today*, January 28, 2009. https://www.psychologytoday.com/us/blog/people-in-nature/200901/no -more-nature-deficit-disorder.

2. Robin Wall Kimmerer, *Braiding Sweetgrass: Indigenous Wisdom, Scientific Knowledge and the Teachings of Plants* (Minneapolis, MN: Milkweed Editions, 2015). 3.

3. Kimmerer, *Braiding Sweetgrass*, 7.

4. The author interviewed Robin Wall Kimmerer at the Blue Mountain Center, NY, on September 19, 2018.

5. The Carlisle Indian School, known officially as the The Carlisle Indian Industrial School, was the flagship Indian boarding school in the US from 1879 through 1918. It was basically a re-education camp, where the primary commandant, General Richard Henry Pratt, attempted to "Kill the Indian: Save the Man" by any means necessary. Roxanne Dunbar-Ortiz, *An Indigenous Peoples' History of the United States* (Boston, MA: Beacon Press, 2015). 151.

6. Beyond the chapter referred to in *Braiding Sweetgrass*, Dr. Kimmerer has written and spoken on the subject of language and animacy elsewhere. Notably: Robin Wall Kimmerer, "Speaking of Nature," *Orion Magazine*, July 28, 2020, https://orionmagazine.org/article/speaking-of-nature/. And: Robin Wall Kimmerer, "Nature Needs a New Pronoun: To Stop the Age of Extinction, Let's Start by Ditching 'It,'" *YES! Magazine*, November 26, 2019, https:// www.yesmagazine.org/issue/together-earth/2015/03/30/alternative-grammar -a-new-language-of-kinship.

7. "It depends on whose numbers you look at and how they classify it and all that…" Dr. Kimmerer added, though the National Parks Service reports the same: "Tallgrass Prairie National Preserve," National Parks Service (U.S. Department of the Interior, 2022), https://www.nps.gov/tapr/index.htm.

8. Kennedy Warne, "This River in New Zealand Is a Legal Person. How Will It Use Its Voice?," *National Geographic*, April 24, 2019, https://www.national geographic.com/culture/2019/04/maori-river-in-new-zealand-is-a-legal -person/.

9. The findings of evolutionary and cognitive biologist Dr. Monica Gagliano are truly fascinating: Marta Zaraska, "Can Plants Hear?," *Scientific American*, May 17, 2017, https://www.scientificamerican.com/article/can-plants-hear/. As is her other research which may be found on her website, monicagagliano .com.

Chapter 7: Experiments on the Verge

1. Most mass-produced ribbons are one form or another of plastic. Instead of contributing to the waste stream, you can recycle old ribbons from gift wrapping, or rip an old sheet or pieces of fabric into ribbon-like strips.

2. Robin Wall Kimmerer and Kathleen Dean Moore, "Women and the Land," presentation, Geography of Hope Conference, Point Reyes, CA. July, 2015.

3. Speaking of iPhones, Paul Kingsnorth wondered "What politician is going to try to sell people a future where they can't update their iPhones ever?" in this haunting article from the August 2015 issue of *Esquire*: John H. Richardson. "When the End of Human Civilization Is Your Day Job." *Esquire*, January 22, 2021. https://www.esquire.com/news-politics/a36228/ballad-of-the-sad -climatologists-0815/.

4. "The Voluntary Human Extinction Movement," The Dabbler, September 18, 2013, http://thedabbler.co.uk/2013/09/the-voluntary-human-extinction -movement.

5. Marcus Aurelius, "The Meditations," in *The Meditations*, trans. Gregory Hays (New York, New York: Modern Library, 2003), xx.

6. Aurelius, *Meditations*, xl.

7. Alan Wartes' insightful article *Desperately Seeking Hozho*, posted on resilience .org in February 2011, helpfully unpacks the Dine' life-way of Hózhó, and its relevance for our times. Alan Wartes. "Desperately Seeking 'Hozho.'" Resilience, February 14, 2011. https://www.resilience.org/stories/2011-02-14 /desperately-seeking-hozho/.

8. Celina Ribeiro, "Beyond our 'ape-brained meat sacks': can transhumanism save our species?" *The Guardian*, June 3, 2022, https://www.theguardian .com/books/2022/jun/04/beyond-our-ape-brained-meat-sacks-can-trans humanism-save-our-species.

9. Rachel Schragis and Elana Eisen-Markowitz, "8 Big Vents About Climate Change." Vent Diagrams. Accessed March 3, 2022. https://www.ventdiagrams .com/archive#/8-big-vents-about-climate-change/.

10. Rachel Schragis and Elana Eisen-Markowitz, "What Is This about?," Vent Diagrams, 2017, https://www.ventdiagrams.com/vision and values.

11. Schragis and Eisen-Markowitz, "What is This About?"

12. Michael Uetricht, "The Beginning of the End of Capitalist Realism." Jacobin, January 30, 2019. https://www.jacobinmag.com/2019/01/capitalist-realism -mark-fisher-k-punk-depression.

13. Andrew Boyd, "Don't Wait for the Revolution-Live It," *YES! Magazine*, February 10, 2020, https://www.yesmagazine.org/issue/love-apocalypse/2013/07 /06/don-t-wait-for-the-revolution-live-it-andrew-boyd.

14. "Headliners; Landlord's Lodgings," *New York Times*, June 23, 1985, Late edition, sec. 4, pp. 9.

15. Patrick Barkham, "Should Rivers Have the Same Rights as People?," *The Guardian*, July 25, 2021, https://www.theguardian.com/environment/2021 /jul/25/rivers-around-the-world-rivers-are-gaining-the-same-legal-rights-as -people.

16. It turns out that, yes, large-scale wind and solar farms in the Sahara desert would increase rain and vegetation: Lee Tune. "Large-Scale Wind and Solar Farms in the Sahara Would..." UMD Right Now, The University of Maryland, September 6, 2018, https://umdrightnow.umd.edu/news/large -scale-wind-and-solar-farms-sahara-would-increase-rain-and-vegetation.

17. Oliver Milman, "Naomi Klein on Climate Change: 'I Thought It Best to Write about My Own Raw Terror,'" *The Guardian* August 26, 2015, https:// www.theguardian.com/books/2015/aug/27/naomi-klein-on-climate-change-i -thought-it-best-to-write-about-my-own-raw-terror.

18. Personal notes from lecture by John Gray, (March 2015).

19. Regarding "convictions" and having the courage of them, consider: (1) how William Butler Yeats describes a world in crisis in his great poem, *The Second Coming*: "Things fall apart; the centre cannot hold [...] The best lack all conviction, while the worst are full of passionate intensity." (2) Nietzsche's take: "very popular error: having the courage of one's convictions—? Rather it is a matter of having the courage for an attack on one's convictions." Friedrich

Nietzsche, *Unpublished Fragments: (Spring 1885–Spring 1886)*, trans. Adrian Del Caro (Stanford, CA: Stanford University Press, 2020).

20. Ezra Markowitz and Lisa Zaval, "Here's the Secret to Making People Care about Climate Change," *The Washington Post*, October 6, 2021, https://www.washingtonpost.com/posteverything/wp/2016/01/04/psychologists-have-discovered-the-secret-to-making-people-care-about-climate-change/?utm_term=.41323765b502.

21. I've sent myself on this journey to the future more than once. The first time, the denizens of the future looked back at me with forgiveness and compassion: "You did what you could, it just wasn't enough. I forgive you. May God and Nature forgive you, too." Another time, they shook their heads in disbelief. Trudging across the mud-cracked tundra, looking for a few scraps to feed their half-starving children, they curse my cowardice, my selfishness, my mind-boggling lack of wisdom and restraint: "You knew what was coming. And you knew that you alone had the chance to take a different path. We no longer have that chance—and now humanity never will. You could have set things right. It really wouldn't have taken all that much. A modicum of sacrifice, a touch of courage and imagination. When the river ran red two counties away, you really thought yours wouldn't be next? But you couldn't look at it squarely, you carried on with business-as-usual, you stayed stuck in the old squabbles. You knew what you needed to do, you just couldn't get your shit together to do it. If I could...." The voice trails off. The sentence is unfinishable. Whatever they wish they "could" do, they can't. These are the two voices I hear when I stage this meeting in my heart. The first feels like an extraordinary act of kindness. I feel grateful. I also feel slightly let off the hook. The second is crushing. Of course they curse me. Of course they blame me. They have perspective; they are wise to me in a way I'm not wise to myself. I did the exercise a third time. This time I see a woman, spade in hand, half kneeling beside a patchwork garden under an abandoned freeway. To her left is a wheelbarrow. Flitting from patch to patch is a small robot, in the shape of an oversized dragonfly, testing each patch for soil toxicity. "It's a life," she says, raising her eyes to meet mine. Then, holding my gaze, she nods ever so slightly. Across the generations, I think she's trying to say, "Thank you."

22. Personal conversation between Paul Kingsnorth and the author. Brooklyn, August 10th, 2017.

23. Klein, *This Changes Everything*.

24. Indre Viskontas. "Naomi Klein on How Fossil Fuels Threaten Our Ability to Have Healthy Children." *Grist*, September 29, 2014. https://grist.org/climate-energy/naomi-klein-on-how-fossil-fuels-threaten-our-ability-to-have-healthy-children/.

25. Iain Munro. "The birth of an ecological revolution: A commentary on Naomi Klein's climate change manifesto." *Ephemera* 17, no. 1 (2017): 229.

26. Naomi Klein. "Climate Change at the Great Barrier Reef Is Intergenerational Theft. That's Why My Son's in This Story | Naomi Klein." *The Guardian*, November 7, 2016, https://www.theguardian.com/environment/2016/nov/07

/climate-change-is-intergenerational-theft-thats-why-my-son-is-part-of-this
-story.

27. John Tarleton. "Interview: Naomi Klein Breaks a Taboo." The Indypendent,
September 12, 2014. https://indypendent.org/2014/09/interview-naomi-klein
-breaks-a-taboo/.

28. Tarleton, "Taboo."

29. Wen Stephenson, "Interview: Naomi Klein on Motherhood, Climate
Justice, and the Failures of the Environmental Movement," The Boston
Phoenix, December 14, 2012, https://blog.thephoenix.com/BLOGS/phlog
/archive/2012/12/14/interview-naomi-klein-on-motherhood-climate-justice
-the-failures-of-the-environmental-movement.aspx.

30. Monica Crippa, Guizzardi Oreggioni, Diego Guizzardi, Marilena Muntean,
Edwin Schaaf, Eleonora Lo Vullo, Efisio Solazzo, Fabio Monforti-Ferrario,
Jos GJ Olivier, and Elisabetta Vignati. "Fossil CO_2 and GHG emissions of
all world countries," Publication Office of the European Union: Luxemburg
(2019).

31. This woman, Kate Schapira, is no stranger to angsting about our ecological
future. She details her personal struggle here: Kate Schapira, "The Alchemists
at Home: On Pregnancy and Responsibility," The Toast, February 1, 2016,
http://the-toast.net/2016/02/01/the-alchemists-at-home-on-pregnancy-and
-responsibility/. And offers "climate anxiety counseling" here: https://climate
anxietycounseling.wordpress.com/about.

32. Elle Hunt, "Birthstrikers: Meet the Women Who Refuse to Have Children
until Climate Change Ends," The Guardian, March 12, 2019, https://www
.theguardian.com/lifeandstyle/2019/mar/12/birthstrikers-meet-the-women
-who-refuse-to-have-children-until-climate-change-ends. To hear more from
BirthStrikers in their own words and deeds: https://www.birthstrikeforfuture
.com/ and https://birthstrike.tumblr.com/.

33. Brett Fleishman, Head of Global Finance Campaigns at 350.org—from a
personal conversation 2016, and from (now taken down) blog post on http://
freerangeeconomics.com.

Chapter 8: Another End of the World Is Possible

1. Per Espen Stoknes, What We Think about When We Try Not to Think about
Global Warming: Toward a New Psychology of Climate Action (White River
Junction, VT: Chelsea Green Publishing, 2015). 219.

2. João Ubaldo Ribeiro, An Invincible Memory (London, UK: Faber and Faber,
1989). 455.

3. Not just the "fathers" of global capitalism but, crucially, one of its founding
mothers, Margaret Thatcher, for whom "There Is No Alternative" (TINA)
was a favorite personal slogan, as well as an oft-used political slogan of her
1980s-era Tory Party.

4. Mark Fisher in Capitalist Realism: Is There No Alternative?, encapsulates
capitalist realism in a quote that he attributes to both Fredric Jameson and
Slavoj Žižek: "It is easier to imagine an end to the world than an end to
capitalism."

5. According to Wikipedia, the origin of the phrase "When the going gets tough, the tough get going" has been attributed both to Joseph P. Kennedy (1888–1969), father of U.S. President John F. Kennedy, and sometimes to Norwegian-born American football player and coach Knute Rockne (1888–1931). https://en.m.wikipedia.org/wiki/When_the_going_gets_tough ,_the_tough_get_going.

6. Hunter S. Thompson, *The Great Shark Hunt: Strange Tales from a Strange Time: Gonzo Papers, Vol. 1* (New York, NY: Simon & Schuster, 2003). p. 49.

7. De Elizabeth, "Greta Thunberg Recalls Being Relieved by Her Autism Diagnosis," *Teen Vogue*, September 26, 2021, https://www.teenvogue.com/story /greta thunberg autism-diagnosis-climate-activism.

Epilogue: Now Is When You Are Needed Most

1. Nadine Bloch and Rae Abileah, "Activism during the Coronavirus Pandemic," The Commons, February 21, 2022, https://commonslibrary.org/beautiful -troubles-guide-to-activism-during-the-coronavirus-pandemic/.

2. Jim Robbins, "With Climate Change, Tree Die-Offs May Spread in the West," *The New York Times*, December 11, 2017, https://www.nytimes.com /2017/12/11/science/trees-climate-die-offs-west.html. But there's some promise to halt or slow its spread: Richard Schiffman, "How Science Can Help to Halt the Western Bark Beetle Plague," Yale E360 (Yale University, January 4, 2016), https://e360.yale.edu/features/how_science_can_help_to_halt_the_western _bark_beetle_plague.

3. According to the NOAA, carbon ppm in 2014 was 397. Caitlyn Kennedy, "2014 State of the Climate: Carbon Dioxide," (NOAA, July 15, 2015), https:// www.climate.gov/news-features/understanding-climate/2014-state-climate -carbon-dioxide#:~:text=Using%20measurements%20taken%20worldwide %2C%20scientists,over%20the%202013%20global%20average. According to the NOAA, carbon ppm in 2022 is closer to 418. "Global Monitoring Laboratory—Carbon Cycle Greenhouse Gases," Global Monitoring Laboratory (NOAA, March 25, 2022), https://gml.noaa.gov/ccgg/trends/monthly.html.

4. "The Climate Necessity Defense: A Legal Tool for Climate Activists," Climate Disobedience Center, 2017, https://www.climatedisobedience.org/necessity defense.

5. Margaret Wheatley, "Freeing Ourselves from Addiction to Hope," 2021, https://margaretwheatley.com/wp-content/uploads/2021/10/Freeing -ourselves-from-addiction-to-hope.pdf.

6. adrienne maree brown, "Returning from Away: WOW," June 8, 2020, http:// adriennemareebrown.net/2020/06/08/returning-2/.

7. Nathaniel Rich, "Losing Earth: The Decade We Almost Stopped Climate Change," *The New York Times*, August 1, 2018, https://www.nytimes.com /interactive/2018/08/01/magazine/climate-change-losing-earth.html. "…in the decade that ran from 1979 to 1989, we had an excellent opportunity to solve the climate crisis. The world's major powers came within several signatures of endorsing a binding, global framework to reduce carbon emissions—far closer than we've come since. During those years, the conditions for success

could not have been more favorable. The obstacles we blame for our current inaction had yet to emerge. Almost nothing stood in our way—nothing except ourselves."

8. Jonathan Watts, "Domino-Effect of Climate Events Could Move Earth into a 'Hothouse' State," *The Guardian,* August 7, 2018, https://www.theguardian .com/environment/2018/aug/06/domino-effect-of-climate-events-could -push-earth-into-a-hothouse-state.

9. Initiative 1631, according to *The Nation* magazine "would put a carbon fee on big polluters, using the revenue to invest in clean energy and protections for workers in fossil-fuel industries." Sasha Abramsky, "This Washington State Ballot Measure Fights for Both Jobs and Climate Justice," *The Nation,* July 20, 2018, https://www.thenation.com/article/archive/green-new-deal-evergreen -state/.

10. David Roberts, "Washington Votes No on a Carbon Tax—Again," Vox, September 28, 2018, https://www.vox.com/energy-and-environment/2018 /9/28/17899804/washington-1631-results-carbon-fee-green-new-deal.

Epi-Epilogue: Passing the Torch

1. Here is the official "Summary for Policymakers of IPCC Special Report on Global Warming of 1.5°C approved by governments" report from the 2018 IPCC gathering (The United Nations, October 8, 2018): https://www.ipcc .ch/2018/10/08/summary-for-policymakers-of-ipcc-special-report-on-global -warming-of-1-5c-approved-by-governments/.

2. Matt McGrath, "Final Call to Save the World from 'Climate Catastrophe,'" BBC News, October 8, 2018, https://www.bbc.com/news/science-environ ment-45775309. A choice quote: "Scientists might want to write in capital let- ters, 'ACT NOW, IDIOTS,' but they need to say that with facts and numbers," said Kaisa Kosonen, of Greenpeace, who was an observer at the negotiations. "And they have."

3. Jonathan Watts, "We Have 12 Years to Limit Climate Change Catastrophe, Warns UN," *The Guardian*, October 8, 2018, https://www.theguardian.com /environment/2018/oct/08/global-warming-must-not-exceed-15c-warns -landmark-un-report. Chris Mooney and Brady Dennis, "The World Has Just over a Decade to Get Climate Change under Control, U.N. Scientists Say," *The Washington Post,* July 15, 2021, https://www.washingtonpost.com/energy -environment/2018/10/08/world-has-only-years-get-climate-change-under -control-un-scientists-say/. Katherine J. Wu, "The World Was Just Issued 12-Year Ultimatum on Climate Change," Smithsonian Institution, October 8, 2018, https://www.smithsonianmag.com/smart-news/world-was-just-issued -12-year-ultimatum-climate-change-180970489/.

4. When AOC repeated the IPCC report's findings in a Green New Deal- focused speech a few months later, right-wing US media had a fit. This Think Progress article succinctly clarifies many points: Joe Romm, "Scientists Say Ocasio-Cortez's Dire Climate Warning Is Spot On," ThinkProgress, January 31, 2019, https://thinkprogress.org/scientists-defend-ocasio-cortez-12-year -ipcc-science-climate-warning-9daee90fae7b/.

5. "That's How Fast the Carbon Clock Is Ticking," Remaining carbon budget—Mercator Research Institute on Global Commons and Climate Change (MCC), March 16, 2022, https://www.mcc-berlin.net/en/research/co2-budget.html.

6. Greta makes it plain "that the children are the wise ones and the adults are the unaware kids," observes French philosopher Bruno Latour. "She has a way of reversing the generational order." Bruno Latour, "Issues with Engendering." From www. bruno-latour. fr/sites/default/files/167 (2019). 5.

7. "Transcript: Greta Thunberg's Speech at the U.N. Climate Action Summit," NPR, September 23, 2019, https://www.npr.org/2019/09/23/763452863/tran script greta thunbergs speech at the u n climate action summit.

8. "Transcript: Greta Thunberg's Speech."

9. "Jerome Foster II: The next Generation Leading the Way," League of Conservation Voters, December 20, 2019, https://www.lcv.org/article/jerome-foster-ii-next-generation-leading-way/.

10. "Teen Climate Activist on the Urgency of Acting on Climate Change," Peril and Promise (New York, New York: MetroFocus, PBS Thirteen, November 18, 2021).

11. Ayana Elizabeth Johnson and Katharine K. Wilkinson, *All We Can Save: Truth, Courage, and Solutions for the Climate Crisis* (New York, NY: One World, 2021). 5.

12. Johnson and Wilkinson, *All We Can Save*, 4.

13. Jeff Tollefson, "Climate change is hitting the planet faster than scientists originally thought," *Nature*, February 28, 2022, https://www.nature.com/articles/d41586-022-00585-7.

Appendix: Stuff You Can (Still) Do

1. David Suzuki and Peter Knudtson, *Wisdom of the Elders: Sacred Native Stories of Nature*, Bantam Books, 1993, https://www.amazon.com/Wisdom-Elders-Sacred-Native-Stories/dp/0553372637.

2. The Crazy Wisdom approach is neatly summarized by Wes "Scoop" Nisker in a blog post at *Inquiring Mind*: Wes Nisker, "Essential Crazy Wisdom," *Inquiring Mind*, Spring 2002, www.inquiringmind.com/article/1802_50_nisker_essential-crazy-wisdom.

Index

Page numbers in *italics* indicate figures.

About the Author

ANDREW BOYD is a writer, humorist, activist, and CEO (Chief Existential Officer) of the Climate Clock, a global campaign that blends art, science, and grassroots organizing to get the world to #ActInTime. He also co-created the grief-storytelling ritual the Climate Ribbon and led the 2000s-era satirical campaign Billionaires for Bush. Andrew's previous books include *Beautiful Trouble: A Toolbox for Revolution*, *Daily Afflictions: The Agony of Being Connected to Everything in the Universe* and  *Life's Little Deconstruction Book: Self-Help for the Post-Hip*. His lifelong ambition, cribbed from Milan Kundera, is "to unite the utmost seriousness of question with the utmost lightness of form." Andrew lives in New York City.

ABOUT NEW SOCIETY PUBLISHERS

New Society Publishers is an activist, solutions-oriented publisher focused on publishing books to build a more just and sustainable future. Our books offer tips, tools, and insights from leading experts in a wide range of areas.

We're proud to hold to the highest environmental and social standards of any publisher in North America. When you buy New Society books, you are part of the solution!

At New Society Publishers, we care deeply about *what* we publish—but also about *how* we do business.

- This book is printed on 100% **post-consumer recycled paper**, processed chlorine-free, with low-VOC vegetable-based inks (since 2002).

- Our corporate structure is an innovative employee shareholder agreement, so we're one-third employee-owned (since 2015)

- We've created a Statement of Ethics (2021). The intent of this Statement is to act as a framework to guide our actions and facilitate feedback for continuous improvement of our work

- We're carbon-neutral (since 2006)

- We're certified as a B Corporation (since 2016)

- We're Signatories to the UN's Sustainable Development Goals (SDG) Publishers Compact (2020–2030, the Decade of Action)

To download our full catalog, sign up for our quarterly newsletter, and to learn more about New Society Publishers, please visit newsociety.com

ENVIRONMENTAL BENEFITS STATEMENT

New Society Publishers saved the following resources by printing the pages of this book on chlorine free paper made with 100% post-consumer waste.

TREES	WATER	ENERGY	SOLID WASTE	GREENHOUSE GASES
68	5,400	28	230	29,400
FULLY GROWN	GALLONS	MILLION BTUs	POUNDS	POUNDS

Environmental impact estimates were made using the Environmental Paper Network Paper Calculator 4.0. For more information visit www.papercalculator.org

Certified B Corporation

new society
PUBLISHERS
www.newsociety.com

MIX
Paper from responsible sources
FSC® C016245

SDG PUBLISHERS COMPACT